Introduction to Logic

THE JONES AND BARTLETT SERIES IN PHILOSOPHY
Robert Ginsberg, *General Editor*

Ayer, Sir Alfred J., *The Origins of Pragmatism: Studies in the Philosophy of Charles Sanders Peirce and William James*

Ayer, Sir Alfred J., *Metaphysics and Common Sense*, 1994 reprint with introduction by Thomas Magnell, Drew University

Baum, Robert J., University of Florida, Gainesville, *Philosophy and Mathematics*

Ginsberg, Robert, The Pennsylvania State University, Delaware County Campus, *Welcome to Philosophy! A Handbook for Students*

Hanson, Norwood Russell, *Perception and Discovery: An Introduction to Scientific Inquiry*

Jason, Gary, San Diego State University, *Introduction to Logic*

Pauling, Linus, and Daisaku Ikeda, *A Lifelong Quest for Peace: A Dialogue*, Translated and Edited by Richard L. Gage

Pojman, Louis P., The University of Mississippi, *Environmental Ethics: Readings in Theory and Application*

Pojman, Louis P., The University of Mississippi, *Life and Death: Grappling with the Moral Dilemmas of Our Time*

Pojman, Louis P., The University of Mississippi, *Life and Death: A Reader in Moral Problems*

Veatch, Robert M., The Kennedy Institute of Ethics, Georgetown University, *Cross Cultural Perspectives in Medical Ethics: Readings*

Veatch, Robert, M., The Kennedy Institute of Ethics, Georgetown University, *Medical Ethics*

Forthcoming

Beckwith, Francis, University of Nevada Las Vegas, and Louis P. Pojman, The University of Mississippi, *The Abortion Controversy: A Reader*

Gorr, Michael, Illinois State University, and Sterling Harwood, San Jose State University, *Crime and Punishment: Philosophical Readings*

Heil, John, Davidson College, *First-Order Logic: A Concise Introduction*

Hirschbein, Ron, and Rich Demaree, California State University, Chico, *Science and Human Values*

Rolston III, Holmes, Colorado State University, *Biology, Ethics, and the Origin of Life*

Introduction to Logic

Gary Jason
San Diego State University

Jones and Bartlett Publishers
Boston   London

Editorial, Sales, and Customer Service Offices
Jones and Bartlett Publishers
One Exeter Plaza
Boston, MA 02116
1-800-832-0034
617-859-3900

Jones and Bartlett Publishers International
P.O. Box 1498
London W6 7RS
England

Copyright © 1994 by Jones and Bartlett Publishers, Inc.

All rights reserved. No part of the material protected by this copyright notice may be reproduced or utilized in any form, electronic or mechanical, including photocopying, recording, or by any information storage and retrieval system, without written permission from the copyright owner.

Library of Congress Cataloging-in-Publication Data

Jason, Gary James, 1949–
    Introduction to logic / Gary Jason.
       p   cm.
    Includes bibliographical references and index.
    ISBN 0-86720-950-X
    1. Logic.   I. Title.
BC108.J37   1994
160—dc20   93-40059
                                                           CIP

Acquisitions Editors: Arthur C. and Nancy E. Bartlett
Production Editor: Mary Cervantes
Manufacturing Buyer: Dana Cerrito
Design: Patricia Torelli
Editorial Production Service: Colophon
Typesetting: Northeastern Graphic Services, Inc.
Cover Design: Design Ad Cetera, Inc.
Printing and Binding: Rand McNally Book and Media Services
Cover Printing: New England Book Components, Inc.

Cover Photograph: Orion Services/FPG International

Printed in the United States of America
98 97 96 95 94 10 9 8 7 6 5 4 3 2 1

Brief Contents

Preface   *xiii*

1	Logic and Related Studies	1
2	Identifying Arguments and Dialogues	15
3	The Nature of Logical Rules	65
4	Fallacies of No Evidence	87
5	Fallacies of Little Evidence	127
6	Fallacies of Language	153
7	An Extended Example	187
8	Analogies	237
9	The Method of Truth Tables	251
10	Natural Deduction	307
11	Logic Graphs and Set Theory	351
12	The Logic of Properties	401
13	The Logic of Relations	449
Appendix	Mill's Methods	481

Glossary   *501*

Answers to Selected Exercises   *509*

Index   *553*

Contents

Preface xiii

1 Logic and Related Studies   1

    1.1   Logic and the Cognitive Sciences   1
    1.2   Logic and Puzzles   4
    1.3   Logic and Rhetoric   11
    1.4   Logic, Mathematics, and Metalogic   13

2 Identifying Arguments and Dialogues   15

    2.1   Arguments, Modalities, and Dialogues   15
    2.2   Identifying Statements   17
    2.3   The Concept of an Argument   22
    2.4   Identifying Single Arguments   23
    2.5   Interwoven Arguments   35
    2.6   Telling Arguments from Other Things   43
    2.7   The Uses of Argument   52

| | 2.8 | Arguments vs. Dialogues | 56 |
| | 2.9 | Identifying Dialogues | 60 |

3 The Nature of Logical Rules — 65

- 3.1 Conversational Orientation 65
- 3.2 Two Types of Evidential Relations 69
- 3.3 Is There a Third Evidential Relation? 75
- 3.4 Modalities 78
- 3.5 Rules of Dialogue 82
- 3.6 Refutation by Counterexample and Logical Analogy 83

4 Fallacies of No Evidence — 87

- 4.1 The Concept of a Fallacy 87
- 4.2 Fallacies of Complete Evasion 88
- 4.3 Emotional Appeals 92
- 4.4 Ignoring the Issue 111
- 4.5 Loaded Question and Begging the Question 116
- 4.6 Summary 118

5 Fallacies of Little Evidence — 127

- 5.1 Reasoning with Strings Attached 127
- 5.2 Testimony as Evidence 128
- 5.3 Reasoning by Dilemma 132
- 5.4 Reasoning by Analogy 135
- 5.5 Generalization and Particularization 139
- 5.6 Inference to Cause 143
- 5.7 The Fallacy of Special Pleading 151

6 Fallacies of Language — 153

- 6.1 Pitfalls of Language 153
- 6.2 Vagueness and Slippery Slope 154

- 6.3 Ladenness and Loaded Language   155
- 6.4 Understatement and the Fallacy of Hedging   162
- 6.5 Ambiguity and Fallacies of Ambiguity   167
- 6.6 Types of Definition   179
- 6.7 Fallacies of Technical Language   184

7 An Extended Example   187

- 7.1 A Review of the Fallacies   187
- 7.2 An Extended Example   203

8 Analogies   237

- 8.1 The Uses of Analogy   237
- 8.2 Descriptive and Definitional Analogies   241
- 8.3 Analogical Arguments   245
- 8.4 The Heuristic Use of Analogies   248
- 8.5 Models in Science   250

9 The Method of Truth Tables   251

- 9.1 What Is a Theory of Deduction?   251
- 9.2 Object Language and Metalanguage   252
- 9.3 Intended vs. Unintended Interpretation   253
- 9.4 The First Three Connectives   253
- 9.5 Calculations in Propositional Logic   258
- 9.6 An Iterative Method of Translation   261
- 9.7 The Conditional   272
- 9.8 Translating Conditionals   274
- 9.9 The Biconditional   283
- 9.10 The Assessment of Arguments   286
- 9.11 Tautologies and the Nature of Validity   293
- 9.12 Validity Explained   298
- 9.13 Variants of Propositional Logic   300

10 Natural Deduction — 307

- 10.1 Drawbacks of the Method of Truth Tables 307
- 10.2 The Notion of an Inference Rule 309
- 10.3 The First Four Inference Rules 314
- 10.4 Direct Proofs Using the Four Inference Rules 320
- 10.5 The Next Five Inference Rules 324
- 10.6 Proofs Using the Nine Inference Rules 328
- 10.7 The Rule of Conditional Proof 332
- 10.8 Reductio 336
- 10.9 Proofs Using the Complete Set of Rules 338
- 10.10 Other Inference Rules 339
- 10.11 Proving Invalidity 344

11 Logic Graphs and Set Theory — 351

- 11.1 Overview 351
- 11.2 The Concept of a Set 352
- 11.3 Basic Set Operations 353
- 11.4 Translating from English to Set Theory Notation 357
- 11.5 Two-Set Carroll Diagrams 361
- 11.6 Three-Set Carroll Diagrams 368
- 11.7 Statements in Set Theory 375
- 11.8 Categorical Statements 378
- 11.9 The Square of Opposition 382
- 11.10 Translating Natural Language into Categorical Statements 385
- 11.11 Using Carroll Diagrams to Assess Arguments 387
- 11.12 Generalizing Carroll Diagrams to More Terms 394
- 11.13 Other Types of Logic Graphs 395

12 The Logic of Properties — 401

- 12.1 Referring vs. Characterizing Expressions 401
- 12.2 Particular Statements 403
- 12.3 The Two Quantifiers 407

Contents   xi

    12.4   Translation from English to Quantificational Logic   409
    12.5   Expansions   415
    12.6   Quantifier Exchange   417
    12.7   Bondage and Scope   421
    12.8   The Concept of Instantiation   423
    12.9   Two New Inference Rules   427
    12.10   The Rule of Universal Generalization   431
    12.11   The Rule of Existential Instantiation   435
    12.12   Proofs Employing All the Inference Rules   439
    12.13   Proving Invalidity   444

13   The Logic of Relations   449

    13.1   Relations and Singular Statements   449
    13.2   Multiple Quantifiers   451
    13.3   Expansions and Instantiation   457
    13.4   The Five Quantificational Logic Rules with Relations and Multiple Quantifiers   461
    13.5   Useful Theorems   467
    13.6   Proofs of Invalidity   470
    13.7   Properties of Dyadic Relations   471
    13.8   Identity and Definite Description   472
    13.9   Inference Rules for Identity   478

Appendix   Mill's Methods   481

    A.1   The Relation of Cause and Effect   481
    A.2   Mill's Methods—An Informal Account   484
    A.3   A More Formal Approach to Mill's Methods   489
    A.4   The Inverse Method of Agreement   493
    A.5   The Combined Method of Agreement   495
    A.6   The Limitations of Mill's Methods   497

Glossary   501

Answers to Selected Exercises   509

Index   553

Preface

This textbook is based on the material I have developed over many years teaching introductory logic at diverse institutions. It presents a unique approach to teaching the subject that I believe makes it better suited for introductory courses than the many textbooks currently available.

First, this text spends much more time than is customary on the skills of identifying statements, questions and their presuppositions, single and multiple arguments, and dialogues as they occur in ordinary language. The focus on ordinary language continues throughout the text in both the chapters on informal fallacies as well as those on more formal topics.

Second, the text devotes much more attention to informal fallacies than is ordinarily given. Fallacies are discussed in great detail and are illustrated with numerous examples taken from actual sources. In Chapter 7, the review chapter, a presidential political debate of considerable historical interest is presented and analyzed.

Third, the symbolic logic is presented in a very clear and measured fashion, again with an eye to ordinary language applications. Natural deduction is presented with great emphasis on the heuristics of proof construction, using a core set of 11 simple rules.

Fourth, set theory is used to introduce traditional logic (syllogistic), and Carroll diagrams are used in place of Venn diagrams. I believe that this allows the student to develop a more intuitive understanding of syllogisms. Carroll diagrams also are very similar to the diagrams used by engineers and computer scientists in circuit design.

Finally, the inductive topics of analogy and Mill's methods are covered. Mill's methods are introduced using truth tables, allowing for a very quick and simple presentation based upon earlier material. Analogies are discussed in a standard way, but additional topics, such as the analogy problems students encounter in standardized tests, are covered as well.

Throughout this book, I have tried to use clear and concise prose, leavened with humor wherever possible.

One note regarding the way this book handles *mentioning* (as opposed to *using*) terms: As is common, if a word, phrase, or sentence is being mentioned, we either put it in single quotes or set it off on a separate line. But if the context makes it clear that we are mentioning the term—say, a symbol from a formal language—then we will omit the quote marks. For example, we will write

& expresses conjunction.

instead of

"&" expresses conjunction.

The material in this book has been reviewed and classroom tested, in whole or in part, over many years by a number of people. In particular, let me express my thanks to a number of reviewers and fellow teachers who have provided valuable advice along the way: Bob Ginsberg, Pennsylvania State University, Delaware County campus; Steve Giambrone, University of Southwestern Louisiana; Tom Carroll, Jr., Saddleback College; Bonnie Paller, California State University, Northridge; William Lawhead, University of Mississippi; Gayne Anacker, Orange Coast College; Sterling Harwood, San Jose State University, Shoshanna Abel, San Francisco State University; Leonard Berkowitz; Pennsylvania State University, York campus; Willard Humphries; Richard Blackwood; Marilyn Pearsall, San Diego State University; Sandra Wawrytko, San Diego State University; and Priscilla Agnew, Saddleback College. Any errors in the text are my own responsibility.

Both an instructor's solutions manual and a student study guide have been developed for this text. The student study guide contains additional solutions to exercises, additional tips for solving problems, and other materials of great use to the student.

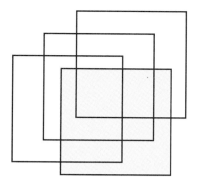

1

Logic and Related Studies

1.1 Logic and the Cognitive Sciences

The simplest questions tend to be the most difficult to answer. This is certainly true of the question "What is logic?" *Logic* is a study of reasoning—if not a 'science,' at least an academic discipline. But this definition only raises more issues. For instance, What is reasoning? Reasoning is an activity in which we commonly engage—some of us more than others. But it is a curious sort of activity, part social and part solitary. In Chapters 1 and 2, we will define 'logic' in a series of steps, each definition being more precise and adequate than the previous one.

We might focus on the social aspect of reasoning, an activity as common, say, as playing. For example, when some friends sit in a coffee shop discussing the movie they just saw, they expect each other to reason. They do not just say the movie was a bad one, they try to defend their claim. That is reasoning. Built into our language are two types of words. One type is used to signal to other people that we do not want to engage in reasoning, the other to signal that we do. Compare these sentences:

1a. I like cheese.   1b. Cheese is good for you.
2a. I didn't enjoy the book.   2b. The book was not well written.
3a. I dislike Fred.   3b. Fred is evil.

4a. I guess the business will do well.	4b. I advise you to buy stock in that business.
5a. I hope Juan wins.	5b. Juan will win.

Notice that the b-type claims set up the expectation that the person has reasons for making the claim and will provide them if asked. If someone makes a b-type claim, but refuses to argue for it and instead substitutes an a-type claim, we feel that this person has done something socially unacceptable, namely that he has not done what he implicitly promised to do.

Reasoning as a social activity can range from the very informal (in the sense of not being structured socially) to the very ritualized (even stereotypical). Chatting about relationships usually is done in an unstructured way. But many written and unwritten rules govern the reasoning involved in solving a scientific problem or settling a case at law. In any event, to say logic is a study of reasoning means it is a discipline that studies a special kind of personal and interpersonal activity: arguing for one's beliefs. Other academic disciplines study reasoning too. Cognitive psychology and neuroscience, anthropology, sociology of knowledge, linguistics, computer science (especially artificial intelligence), ethology, and history of science are among the disciplines that study reasoning.

A **cognitive science** experimentally investigates the nature of knowledge and reasoning. Is logic a cognitive science? No. The focus of any cognitive science is descriptive: its goal is to describe and analyze accurately the manner in which people (and some animals and machines) process, store, receive, transmit, and transform information. **Logic**, on the other hand, is a normative study. It lays down norms or standards of good reasoning. We do not merely describe how people reason, but prescribe how they ought to reason. Logic is similar to *ethics*, which is the study of how people ought to behave. Indeed, to the extent that reasoning is viewed as a social activity, logic can be viewed as a branch of ethics. The philosopher C.S. Peirce often called logic "the ethics of belief." In practice, description and prescription occur together. Even the most empirically oriented cognitive scientists must interpret what they see. When, say, watching the members of a tribe carry out an elaborate medical procedure using "magic," the anthropologists must assume that the activity is at least partially reasonable—and they will inevitably use their own norms in doing so.

Suppose, for illustration, that an anthropologist is studying life among the Nuer (a Sudanese people). She learns their language, and some of the things they say sound odd to her.[1] A Nuer might speak of a twin brother as "a bird." What can the anthropologist make of that? Is the Nuer simply mistaken? Is the Nuer irrational? Is the anthropologist's translation inaccurate? Is the Nuer's statement a metaphor (and what does that mean)? Is the Nuer lying to the anthropologist, or perhaps trying to joke?

Suppose the anthropologist decides that the Nuer's statement makes sense against the background of (in the context of) a belief system in which everything is infused with spirit. Still, cannot a whole belief system be irrational or unreasonable? Here again, we might mean several things. We might mean that the belief system is factually silly, as it would be unreasonable for us to deny that America exists. We might mean that the belief system is internally inconsistent, as it would be unreasonable to believe that meat is harmful to eat, but that beef, while being meat, is not harmful. Then again, we might mean that the beliefs in the system are meaningless, as it would be unreasonable to believe that the square root of two tastes delicious. Or, we might mean that the beliefs are held in a fanatical way, as it would be unreasonable to slug someone who merely disagreed with your point of view. But all of these claims that the anthropologist might make are made against the background of her own standards of logic. Thus, factual investigation of human reasoning cannot be separated completely from normative considerations.

Conversely, a widely held dictum of ethics is "ought implies can." This means that for a person to pass moral judgement fairly on another person for having failed to do something, it must have been possible for that other person to have done that thing. For example, it would be absurd for us to say of someone that he is wrong for not having ended world poverty yesterday. He could not possibly do such a thing by himself, so what sense does it make for us to say he should? A similar situation holds in logic. To lay down norms of reasoning requires determining, to some degree, how people actually do reason, for it makes no sense to say people should reason according to a specific rule if that rule is impossible to follow.

Thus, cognitive science and logic are different in the degree to which each is normative and descriptive, but they are related. We assess reasoning only after we have represented it, and, in representing reasoning, we cannot avoid applying our logical theories. In turn, incorporating empirical observation of how people reason improves logical theories. Figure 1.1 summarizes the points made in this section.

---

[1] Ernest Gellner, "Concepts and Society," in *Rationality*, edited by Bryan Wilson (New York: Harper & Row, 1970).

Figure 1.1

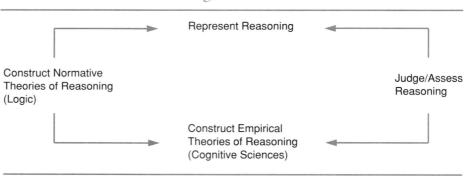

1.2 Logic and Puzzles

The conclusion of Section 1.1 is that we ought to distinguish logic, a normative study of reasoning, from the various empirical studies of reasoning, but not separate them so strongly that we overlook the interrelation between them. We now need to draw another distinction, and perhaps a good way to do so is to consider a logic puzzle—the kind found in puzzlebooks and (more ominously) on standardized exams such as the LSAT (Law School Admissions Test).

> Suppose three sisters share the same wardrobe. They have three blouses (one pink, one yellow, and one blue), and they also have three skirts (one pink, one yellow, and one blue). This morning, the sisters were in a hurry, and when they dressed, none of them ended up with skirt and blouse matching in color. Indeed, the sister who had on the pink blouse pointed this out to the one who had on the blue skirt. What color skirt is the sister in the yellow blouse wearing?

We can solve this puzzle more easily if we set up a diagram as in Figure 1.2. The capital letters refer to the blouse colors, and the lowercase letters refer to the skirt colors.[2]

We can rule out the color combinations Pp, Yy, and Bb since we are told that none of the sisters has on the same color blouse and skirt. This fact is indicated by an X in the appropriate boxes (Figure 1.3).

Next, we can rule out the combination Pb, since we are told that the sister who has on the pink blouse talked to the one who had on the blue skirt. Again, on the same diagram, this is indicated by an X (Figure 1.4).

---

[2] Martin Garder, *Aha! Insight* (New York: W.H. Freeman & Co., 1978), p. 93.

1.2 / Logic and Puzzles

Figure 1.2

	Skirt		
Blouse	p	y	b
P			
Y			
B			

Figure 1.3

	Skirt		
Blouse	p	y	b
P	x		
Y		x	
B			x

Figure 1.4

	Skirt		
Blouse	p	y	b
P	x		x
Y		x	
B			x

But this must mean that the sister wearing the pink blouse is wearing the yellow skirt. This is indicated by a / (Figure 1.5).

We can now conclude something else. Since none of the sisters are wearing the same skirt at the same time, we can rule out *By* (Figure 1.6). This is indicated by an *X*.

Figure 1.5

		Skirt		
		p	y	b
	P	x	/	x
Blouse	Y		x	
	B			x

Figure 1.6

		Skirt		
		p	y	b
	P	x	/	x
Blouse	Y		x	
	B		x	x

This leads to another conclusion. Since the blue blouse is not matched with the blue or the yellow skirt, it must be matched with the pink skirt (Figure 1.7). This is indicated by a /.

But again, since each skirt is worn by only one sister, this means we can rule out *Yp* (Figure 1.8), as indicated by an *X*.

Figure 1.7

		Skirt		
		p	y	b
	P	x	/	x
Blouse	Y		x	
	B	/	x	x

Figure 1.8

		Skirt		
		p	y	b
Blouse	P	x	/	x
	Y	x	x	
	B	/	x	x

Therefore, we are finally led to see that the sister wearing the yellow blouse is also wearing the blue skirt. Puzzle solved!

We might not have seen how to solve the puzzle until we saw the diagram device and carried it through step-by-step; but, at each step, the justification should have been intuitively obvious. We ought to distinguish logic from heuristic. **Logic**, strictly so-called, is the study of the rules governing the acceptability of reasoning, while **heuristics** is the study of how to come up with the reasoning to begin with. The word 'heuristic' comes from the Greek word 'heuriskein,' which means to find or to discover. The first-person form of that verb is 'Eureka' ("I found it!"), which is what Archimedes is supposed to have shouted when he solved a problem while taking a bath. Yet, we should not push a conceptual distinction too far, making it a complete separation in practice. We will discuss heuristics often in this book, describing the various strategies useful for the problems we encounter. Logic and heuristics will be learned together, since we learn logic by working problems, and we work problems by applying strategies.

Although the science of heuristics is still young, some of its results have become clear. George Polya has pointed out four useful strategies for approaching problems that we will use throughout this book.[3]

Strategy 1: *Working Backwards.* Often the best approach to solving a problem is to keep an eye on the outcome we need to reach. For example, specialists often design a computer program by looking at how the intended output (say, a report on the daily sales of an insurance company) should appear.

Strategy 2: *Working Forward.* Often, the best strategy to adopt is the reverse of strategy 1—to start at the initial statement of the

---

[3] See George Polya, *How to Solve It,* 2nd ed. (New York: Doubleday-Anchor, 1957).

problem and generate steps forward, checking as we go to see if any progress is made.

Strategy 3: *The Top-Down Approach.* A good idea is to break any complex difficult problem down into smaller, easier parts. This approach is commonly called 'top-down design' in computer programming. We will see repeatedly that logic is much easier to learn if we approach problems in a top-down fashion.

Strategy 4: *Looking for the Old in the New.* It is a cliché among mathematics and physics teachers that the key to learning how to solve difficult problems is to see in those problems similarities to problems successfully solved in the past. This strategy—looking for analogies—will be of especial use in our discussion of informal fallacies in Chapters 4 through 7.

EXERCISES 1.2

The problems 1–9 that follow are taken from Raymond Smullyan, What Is the Name of this Book? *(Englewood Cliffs, NJ: Prentice-Hall, 1978). Copyright 1978 Raymond Smullyan, courtesy of Prentice-Hall. (An asterisk indicates problems that are answered in the back of the book.)*

* 1. A man was looking at a portrait. Someone asked him, "Whose picture are you looking at?" He replied: "Brothers and sisters have I none, but this man's father is my father's son." ("This man's father" means, of course, the father of the man in the picture.)
Whose picture was the man looking at?

* 2. In Shakespeare's *Merchant of Venice*, Portia had three caskets—gold, silver, and lead—inside one of which was Portia's portrait. The suitor was to choose one of the caskets, and if he was lucky enough (or wise enough) to choose the one with the portrait, then he could claim Portia as his bride. On the lid of each casket was an inscription to help the suitor choose wisely. Now suppose Portia wished to choose her husband not on the basis of virtue, but simply on the basis of intelligence. She had the following inscriptions put on the caskets.

    GOLD: THE PORTRAIT IS IN THIS CASKET
    SILVER: THE PORTRAIT IS NOT IN THIS CASKET
    LEAD: THE PORTRAIT IS NOT IN THE GOLD CASKET

Portia explained to the suitor that of the three statements, at most one was true.
Which casket should the suitor choose?

1.2 / Logic and Puzzles

* 3. This is a sequel to the story of Portia's caskets. We recall that whenever Bellini fashioned a casket he always wrote a true inscription on it, and whenever Cellini fashioned a casket he always wrote a false inscription on it. Now, Bellini and Cellini had sons who were also casket makers. The sons took after their fathers; any son of Bellini wrote only true statements on those caskets he fashioned, and any son of Cellini wrote only false statements on his caskets.

Let it be understood that the Bellini and Cellini families were the only casket makers of Renaissance Italy; all caskets were made either by Bellini, Cellini, a son of Bellini or a son of Cellini.

If you should ever come across any of these caskets, they are quite valuable—especially those made by Bellini or Cellini. I once came across a casket which bore the following inscription:

THIS CASKET WAS NOT MADE BY ANY SON OF BELLINI

Who made the casket?

* 4. There is a wide variety of puzzles about an island in which certain inhabitants called "knights" always tell the truth, and others called "knaves" always lie. It is assumed that every inhabitant of the island is either a knight or a knave. Three of the inhabitants—A, B, and C—were standing together in a garden. A stranger passed by and asked A, "Are you a knight or a knave?" A answered, but rather indistinctly, so the stranger could not make out what he said. The stranger then asked B, "What did A say?" B replied, "A said that he is a knave." At this point the third man, C, said, "Don't believe B; he is lying!"

The question is, what are B and C?

* 5. We have two people, A and B, each of whom is either a knight or a knave. Suppose A makes the following statement: "If I am a knight, then so is B."

Can it be determined what A and B are?

* 6. Suppose you are visiting a forest in which every inhabitant is either a knight or a knave. (We recall that knights always tell the truth and knaves always lie.) In addition, some of the inhabitants are werewolves and have the annoying habit of sometimes turning into wolves at night and devouring people. A werewolf can be either a knight or a knave. You are interviewing three inhabitants, A, B, and C, and it is known that exactly one of them is a werewolf. They make the following statements:

    A: C is a werewolf.
    B: I am not a werewolf.
    C: At least two of us are knaves.

Our problem has two parts:

    a. Is the werewolf a knight or a knave?
    b. If you have to take one of them as a traveling companion, and it is more important that he not be a werewolf than that he not be a knave, which one would you pick?

*   7. When Alice entered the Forest of Forgetfulness, she did not forget everything; only certain things. She often forgot her name, and the one thing she was most likely to forget was the day of the week. Now, the Lion and the Unicorn were frequent visitors to the forest. These two are strange creatures. The Lion lies on Mondays, Tuesdays, and Wednesdays, and tells the truth on the other days of the week. The Unicorn, on the other hand, lies on Thursdays, Fridays, and Saturdays, but tells the truth on the other days of the week. One day Alice met the Lion and the Unicorn resting under a tree. They made the following statements:

Lion: Yesterday was one of my lying days.

Unicorn: Yesterday was one of my lying days, too.

From these two statements, Alice (who was a very bright girl) was able to deduce the day of the week. What day was it?

*   8. An enormous amount of loot has been stolen from a store. The criminal (or criminals) took the heist away in a car. Three well-known criminals—A, B, C—were brought to Scotland Yard for questioning. The following facts were ascertained:

    1. No one other than A, B, C was involved in the robbery.
    2. C never pulls a job without using A (and possibly others) as an accomplice.
    3. B does not know how to drive.

    Is A innocent or guilty?

*   9. On a certain island, half the inhabitants have been bewitched by voodoo magic and turned into zombies. The zombies of this island do not behave according to the conventional concept: they are not silent or death-like—they move about and talk in as lively a fashion as do the humans. It's just that the zombies of this island always lie and the humans of this island always tell the truth.

    So far, this sounds like another knight-knave situation in a different dress, doesn't it? But it isn't! The situation is enormously complicated by the fact that although all the natives understand English perfectly, an ancient taboo of the island forbids them ever to use non-native words in their speech. Hence whenever you ask them a yes-no question, they reply "Bal" or "Da"—one of which means "yes" and the other of which means "no."

    I once met a native of this island and asked him, "Does 'Bal' mean 'yes'?" He replied, "Bal."

    a. Is it possible to infer what "Bal" means?
    b. Is it possible to infer whether he is a human or a zombie?

*   10. Three students in a logic class were quietly talking when the professor walked into the room. He said, "In this box I've got three hats, each of which is either red or green. Close your eyes while I put them on you." They did so, and when they opened them again, he asked them to raise a hand if any saw a red hat. (Nobody could see her own hat.) All three students raised their hands. Next the professor asked if anybody could deduce the color of her own. One student raised her hand.

What color was the hat, and how did she know it?

[For a discussion of how this last problem can be generalized, see Martin Gardner, "Mathematical Games," *Scientific American* (May 1977).]

1.3 Logic and Rhetoric

Logic should not be equated with **rhetoric**, which is the study of persuasive communication. The two studies are different by definition, because what is logically solid may not succeed in persuading people, while what is persuasive to people may not be logically solid at all.

What should we make of the fact that the two studies are different? Some have ventured the opinion that logic is nearly useless for the 'practical person'—the politician, the salesperson, the lawyer, and anybody else who has to get others to act. These people (it is argued) are better off learning **sophistry**, that is, argumentation that may be illogical but which gets people to do what the arguer wants them to do. You cannot reason with people, it is added, you can only lead them like the cattle they are. Such a point of view is unduly cynical. People are not as incapable of reasoning as is made out—demagogues often lose their support in the long run. Moreover, sophistry loses its power when the audience learns logic. A public trained in logic is much less likely to be bamboozled by illogical rhetoric. Indeed, one of the benefits of studying logic is that it increases your immunity to such manipulation. In a sense, then, logic can act as a device to elevate the level of public discourse. This is especially important in a democracy.

The relationship between rhetoric and logic will come up repeatedly in our discussions. To get us thinking about the topic, consider two points.

First, effective rhetoric often employs devices that obscure the logical structure of an argument. It is effective rhetoric for a speaker to be *ironical* (that is, to deliberately say the opposite of what the speaker means) or *hyperbolic* (to deliberately exaggerate the point). Effective rhetoric also omits mention of the assumptions the speaker shares with the audience (since that will cause them to focus less upon the central point the speaker is trying to make). Both these rhetorical devices mask the logical structure of an argument. This tension between rhetorical and logical expression will require considerable attention in Chapter 2.

Second, people have an emotional side as well as a rational side—supposing, for simplicity, that we can separate the two components. Moving people to action often involves making them feel, as opposed to just believing things. But a question we need to address is whether an emotional appeal can ever constitute evidence for a belief. We will examine this issue in Chapter 4.

Figure 1.9

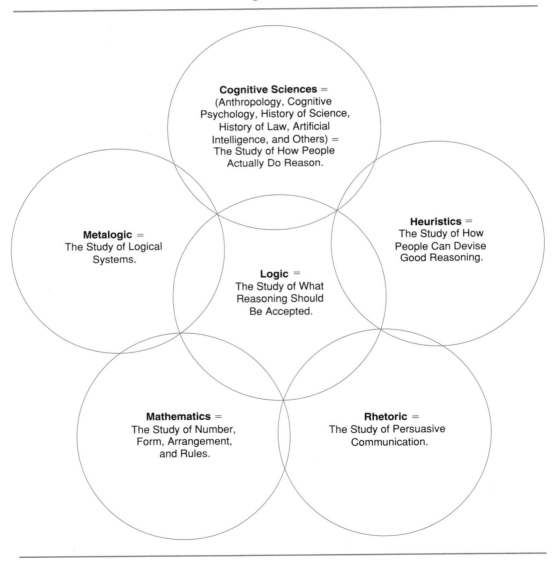

1.4 Logic, Mathematics, and Metalogic

So far, we have distinguished logic from cognitive science, heuristics, and rhetoric. Next, we must distinguish it from two other subjects with which it is easily confounded—mathematics and metalogic.

Mathematics is the study of number, form, arrangement, and rules. We saw in Section 1.2 that a diagram set up in the right way greatly aided our search for the solution to the puzzle. We will soon see that geometry, algebra, graph theory, and other branches of mathematics are of great help in analyzing reasoning. But that should not lead us to equate the two. We cannot very well do modern physics without sophisticated mathematics, but that hardly means that physics is the same as mathematics.

We must not push the distinction between mathematics and logic too far. Symbolic logic (which will be covered in Chapters 9 through 13) borders closely on mathematics—so much so that some people call symbolic logic "mathematical logic."

One other distinction should be drawn. Logic proper is the development of methods for assessing reasoning. But logic, as any other subject, can itself be an object of study. **Metalogic** is the assessment of logical systems, involving the development of methods for assessing logical methods. (*Meta* is a Greek prefix that means "after" or "over.")

Metalogic, it turns out, borders on cognitive science just as does logic. Therefore, in order to answer questions about which logic is "correct," which logics are "equivalent," which logics are "more natural," and so on, we have to take into account the background of the people who are using these logics, and the way patterns of reasoning help people deal with the world.

We can put all the definitions from this chapter in a diagram (Figure 1.9). The overlap of the circles indicates the common areas of interest of the subjects defined.

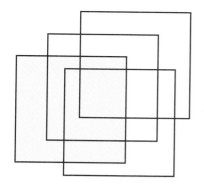

2

Identifying Arguments and Dialogues

2.1 Arguments, Modalities, and Dialogues

In Chapter 1, **logic** was defined as the study of what constitutes good reasoning. That definition helped to clarify the notion of 'logic,' by distinguishing logic from the closely related studies of cognitive science, mathematics, metalogic, and heuristics. But now we must develop a more precise characterization of the subject.

Imagine this conversation:

Adam: Why are you so upset?
Kelli: Oh, I'm mad at Joni. She stole my logic text.
Adam: Are you sure?
Kelli: I'm absolutely positive. I left my text in the classroom, and only she and Heather were in there when I left.
Adam: How do you know that Heather wasn't the one who took it?
Kelli: I looked in her book bag just after she left the classroom.

Clearly, *reasoning* is going on here. But that term can mean several things. 'Reasoning' can mean inferring, the drawing out of a conclusion from some initial facts. In the example, Kelli has inferred that Joni stole her book from the facts that only Joni and Heather were in the room, and Heather (she feels) was not the one who did it.

'Reasoning' can also mean answering a question, as when Kelli responds to Adam's queries.

Now recall (from Chapter 1) that reasoning has a social aspect. 'Reasoning' can refer to engaging in a dialogue, that is, rational exchange to jointly arrive at the truth of some matter. Think of the expression "Come, let us reason together."

Strictly speaking, we should distinguish the logic of inferences, the logic of questions, and the logic of dialogues. But since the logic of inferences is basic to the others, we will reserve the term *logic* for the study of correct inference. The normative studies of questions and dialogues will be named and described in Section 2.8. The focus now will be on inference.

Inference is the process of reasoning from starting points to some final conclusion. Since we are logicians and not cognitive scientists (psychologists, say, or brain physiologists), we are not immediately interested in the actual operations of the human brain. Instead, we are interested in the 'products' of that mental process, things called arguments. Roughly defined, an 'argument' is a claim (the one concluded) together with the reasons for it (those 'starting points'). In our example, Kelli's argument seems to be:

1. Either Joni or Heather took my book.
2. Heather didn't take it.
   _____
   ∴ Joni did take it.

The three dots stand for 'therefore' and signal the conclusion.

The 'reasons' offered for a conclusion are not necessarily good ones—indeed, they may be silly ones. The job of the logician is precisely to distinguish good reasons from bad ones.

Another way of putting the point is this: in logic, we study not the process of reasoning, but abstract representations of it. The logician represents someone's reasoning as an argument (imagining that the person began with those reasons and ended up with that claim). So far the logician is only doing some psychological guessing about what went through the arguer's mind. The logic begins when the logician assesses that argument, abstracted from the context that created it and the person who generated it. Indeed, we will sometimes create arguments "out of thin air" as examples, without pretending in the slightest that anybody would ever offer such arguments in real life.

In this chapter, we will work on developing our skill at identifying arguments, modalities, questions, and dialogues. This is a crucial skill in logic, because before we can assess arguments and dialogues, we must be able to represent fairly the reasoning we seek to judge.

2.2 Identifying Statements

We roughly defined an argument as consisting of a claim and the reasons for it. But claims and reasons are just statements to the effect that such and such is the case. What exactly is a 'statement'?

We may begin by defining 'sentence.' A **sentence** is a grammatical sequence of words. Strictly speaking, we should distinguish sentence type from sentence token. How many sentences are listed below?

I am hungry.
I am hungry.
I am hungry.
I am hungry.

In a sense, there are four—four inscriptions, we might say. But in another sense, there is only one sentence—one sentence type. When we use the term 'sentence' from now on, we will mean not sentence types, but instances or tokens of those types. Sentences fall into several categories: declarative, interrogative, exclamatory, and so on. We use sentences to do all sorts of things: convey information, request information, make jokes, christen babies, and make promises, to name just a few. Of most importance now is the use of sentences to convey information, to 'state facts.' A **statement** is the meaning of any sentence used to convey information. Some texts use the word 'proposition' in the same sense we use 'statement.'[1]

A statement can be true or false, depending on whether the information is correct or incorrect. Hereafter we will use T for "true" and F for "false." When we say that any statement is either T or F, we do not mean that the truth or falsity of any statement is known. The statement "There are 14,555,233 mountains on the moon" is T or F, but nobody may ever know which.

Since statements are the kinds of things that can be T or F, they are quite different from questions and commands. It makes no sense to answer the question "What time is it?" by "I disagree—you're absolutely wrong." Saying the question is wrong (or correct) is

---

[1] W.V. Quine has argued against the notion of 'proposition.' See his *Philosophy of Logic* (Englewood Cliffs, NJ: Prentice-Hall, 1970), pp. 1–14.

strange; questions can be loaded, stupid, based on mistaken assumptions, or misleading—but never T or F. Similarly, it makes no sense to speak of a command as being T or F; you cannot very well reply to the order "Shoot that target!" by saying "Sir, your order is false." Orders can be immoral, inappropriate, or silly, and can be disobeyed—but they are not T or F.

Harder to see is the difference between statements and sentences: we usually use a declarative sentence to make a statement, but that does not mean that the sentence is the statement. The sentence is better viewed as the tool we use to make our assertion. Here are a few points that should clarify the distinction between sentences and statements. First, different English declaratives may be used to make the same claim (statement). For example, the active and passive voice constructions "Mary loves John" and "John is loved by Mary" are used to say the same thing. But they are clearly different sentences: the first has only three words while the second has five.

Of more interest is the fact that statements often are made using nondeclarative sentences. You know of 'rhetorical questions,' which are interrogative sentences (like "Do you really believe that?") used to state claims (here, that the listener is mistaken). But there are rhetorical commands as well. "Take a walk!" likewise expresses the statement that the listener is wrong, even though a nondeclarative sentence (here, an exclamation) is used.

To further clarify the point that statements are not sentences, remember that until now we have been dealing solely with English. Sentences from different natural languages can be used to make the same claim: "John is rich" and "Juan es rico," for example. Less obvious is that conventional gestures and noises can express statements as well. A "Bronx cheer" can be used to deny (emphatically, albeit crudely) another's assertion; so can an obscene gesture.

Statements can be classified in several ways. We can classify them in terms of their purpose, that is, on the basis of the questions they answer. Such a classification would group together statements that describe what has occurred, ascribe responsibility, give an example, explain, classify, specify, and so on. Statements can also be classified by topic: politics, art, science, and so on.

Most important for us is the classification of statements on the basis of structure. We will discuss in detail the study of the structure of statements in Chapter 9; for now, we need only to distinguish direct assertions from conditional ones.

A *direct assertion* is a statement to the effect that something is so. Direct assertions can be simple:

2.2 / Identifying Statements

Kelli is happy.
Jason is afraid.
Juanita is a writer.

They can also be *compounds*, such as denials:

Kelli is not happy.
Jason is not afraid.
Juanita is not a writer.

or *conjunctions* (statements that two claims are the case):

Kelli and Kim are happy.
Jason is afraid, and so is Juanita.

or even *disjunctions* (statements that one or more of two claims are the case):

Kelli or Kim stole that car.
Either Jason is a writer or else he is neurotic.

A *conditional assertion* is a statement to the effect that something is or will be the case, given that something else is the case.

If Kim Wallace is a writer, then he should not mind writing that book.
Given that Juanita is happy, Kelli will be happy.

EXERCISES 2.2A

For each of the following sentences, reexpress the statement made by using a shorter declarative sentence. Try not to lose any important information in rephrasing, and assume that each sentence is meant sincerely, not employed as sarcasm or irony. (An asterisk indicates problems that are answered in the back of the book.)

1. Calling the Dodgers a second-rate team is a dirty lie!
* 2. Has any woman ever had a husband as nice as you?
3. In answer to your question, sir, I don't believe that we can say very precisely or even with normal precision whether or not the man died before midnight.

4. Speak softly but carry a big stick.

5. When in Rome, do as the Romans.

* 6. One considerable advantage that arises from philosophy, consists in the sovereign antidote which it affords to superstition and false religion.
> David Hume, *Dialogues Concerning Natural Religion and the Posthumous Essays,* ed. by Richard H. Popkin (Indianapolis: Hackett Publishing Company, 1980), p. 97.

7. While I certainly support your right to express critical remarks regarding my artistic ability, I find that I must respectfully disagree.

8. It is surely of tremendous benefit to the state that each person should enjoy liberty, perfectly unlimited, of expressing his or her (or—in this age of artificial intelligence—its) own sentiments on any issue of concern.

9. (From an insurance form.) In consideration of the acceptance of this note by the Company as an extension of credit to the Insured, the Insured hereby agrees that all installment payments must be received by the Company at its office as above designated on or before the due dates as above specified and any payment transmitted by mail or paid to an agent of the Company, or transmitted or paid in any other matter, shall constitute payment according to the terms of the note only if, as and when actually received by the Company at its said office within the time specified.

* 10. During the past 15 years, an increasing amount of evidence has accumulated indicating that parents play an important role in stimulating the acquisition of intellectual skills in their children. . . . Environmental stimulation of intellectual development by parents may be particularly important in the preschool years when children are acquiring basic intellectual competencies that can assist in the mastery of academic tasks in school. . . .
> John R. Bergan, Albert J. Neumann II, and Cheryl L. Kapp, "Effects of Parent Training on Parent Instruction and Child Learning of Intellectual Skills," *Journal of School Psychology* 21 (1983), 31–39.

11. A number of studies have indicated that the psychic and economic returns available in an occupation, the psychic and economic costs required to prepare for an occupation, and the labor market conditions affecting the availability of jobs in an occupation are considered by individuals making occupations choices.
> Kenneth G. Wheeler, "Perceptions of Labor Market Variables by College Students in Business, Education, and Psychology," *Journal of Vocational Behavior* 22 (1983), 1–11.

12. These results reveal that the lonelier an adolescent, the more likely he or she was to be anxious, depressed, show an external locus of control, have high levels of public self-consciousness and social anxiety, and exhibit low levels of happiness and life-satisfaction. In addition, loneliness was associated with a reluctance to take social risks.
> "Loneliness at Adolescence," *Journal of Youth and Adolescents* 12 (2) (1983).

13. Factor-analytic studies of instruments used to measure hypochondriasis . . . and clinical descriptions of the syndrome have identified the following set of behaviors that are typically associated with hypochondriasis: excessive concern with health and bodily functioning, exaggerated attention to an unrealistic interpretation of physical signs or sensations, multiple bodily complaints, and belief or fear of having a serious illness. Description of an individual as hypochondriacal usually is based on the presence of one or more of these behaviors.

> Timothy W. Smith, C.R. Snyder, and Suzanne C. Perkins, "The Self-Serving Function of Hypochondriacal Complaints: Physical Symptoms as Self-Handicapping Strategies," *Journal of Personality and Social Psychology* 44 (4) (1983), 787–797.

* 14. Ah, make the most of what we yet may spend, Before we too into the Dust descend; Dust into Dust, and under Dust, to lie, Sans Wine, sans Song, sans Singer and—sans End!

> —*The Rubaiyat of Omar Khayam*, Trans. Edward Fitzgerald.

15. A plausible research orientation of those who wish to study the meaning of social phenomena is to conduct investigations amongst individuals who are directly experiencing such phenomena. Yet a difficulty, or limitation, of this approach is that personal descriptions of experience are obviously confined to the order of things of which individuals are readily aware. Taken-for-granted, or 'latent' meanings cannot be easily expressed because they are locked into patterns of feeling and behavior in an unquestioned, tacit form.

> Stephen Fineman, "Work Meanings, Non-Work, and the Taken-For-Granted," *Journal of Management Studies* 2 (1983).

EXERCISES 2.2B

Now try the process involved in Exercises 2.2A in the reverse direction. For each of these simple declaratives, express the same statement with a long sentence. However, do not add any new information to it. (An asterisk indicates problems that are answered in the back of the book.)

1. Normally, wealth is desirable.
* 2. Exercise can lower your chances of heart disease.
3. Tom can't dance.
4. Whales are big animals.
5. Fred is dead.
* 6. Reagan was president in 1982.
7. Bill Evans was a great pianist.
8. Whistler liked to attack those who criticized him.

    9. Nuclear war is frightening.

*   10. Live jazz is more exciting than recorded jazz.

2.3 The Concept of an Argument

Having defined the concept of a statement, we are now in position to define the concept of an argument. While the term 'argument' usually means a heated or emotional exchange, we have a more technical meaning in mind. An **argument** is a set of statements, some of which (called 'the premises') we (the logical observers) take to be possible evidence for another of the statements (called 'the conclusion'). Arguments are thus structured sets of statements. We (the assessors) view them as representing a process of mental inference in someone's brain—starting with those premises and ending up with the conclusion. But that is only our psychological conjecture. The argument is considered on its own.

    We will follow the practice of putting the premises above a line, with the conclusion below it. This is the common or standard format for arguments, so we will call it 'standard form.'

    An 'argument,' as we define the term, has at least one and possibly many premises. There are one-premise arguments.

    1. People are greedy.

    ∴ Any social system that requires altruism is doomed to failure.

There are two-premise arguments.

    1. Senator Jones is a crook.
    2. We don't want crooks in office.

    ∴ We should vote against Senator Jones.

There are three-, four-, five-, even hundred-premise arguments. Consider:

    1. Unemployment is high.
    2. Inflation still rages.
    3. Our enemies are pushing us around.
    4. Our relations with our allies have deteriorated.
    5. The energy crisis is getting worse.

    ∴ The president is doing a bad job.

By itself, a statement is neither a premise nor a conclusion, any more than a person itself is a sibling or a cousin. Being a premise, like being a cousin, is being related in a distinctive way to something else. The same statement may be a premise in one argument and a conclusion in another. Thus "Nixon was good at foreign policy" is a conclusion in:

1. Nixon opened the door to China.
2. Nixon strengthened our alliances.
3. Nixon ended the Vietnam War.
4. Nixon achieved some measure of peace in the Middle East.

∴ Nixon was good at foreign policy.

Yet the same statement is a premise in:

1. Nixon was good at foreign policy.
2. Nixon was good to little children.
3. Nixon was good at domestic policy.

∴ Nixon was a good president.

Put another way, the main feature of an argument is not so much what makes it up (statements), but how they are structured. Thus the arguments:

1. Billie Mae is a real gambler.

∴ She will take the gamble of dating Fred.

1. Billie Mae will take the gamble of dating Fred.

∴ She is a real gambler.

are quite different. In the first, we view Billie Mae being a gambler to be a known fact, and then infer that she will try dating Fred. In the second, we take it as known that she will date Fred, and infer from that fact that she must be a gambler.

2.4 Identifying Single Arguments

Two themes recur repeatedly in this book. First, as we noted in our discussion of heuristics in Chapter 1, any task becomes easier if it is broken down into parts or subtasks. This is a fact well known to every teacher and computer programmer. Second, some tasks can be done in a mechanical, grind-it-out way, while other tasks cannot. Or, to introduce a couple of technical terms, some tasks are *algorithmic*, while others are *nonalgorithmic*. Nonalgorithmic tasks—like learning how to make a good apple strudel or how to sell used cars—cannot be completely explained by precise rules. You simply have to learn by doing and by watching.

Most logic students are surprised when they discover that many tasks in logic are not algorithmic. This includes the task we now address of identifying arguments contained in passages from books or magazines. So, to handle the job effectively, let us break down the general task of identifying arguments in passages into components:

1. Faced with a passage, we must learn to figure out whether any arguments are in it at all.
2. Suppose we know for a fact that the passage in front of us contains exactly one argument. We must learn how to identify it, that is, to determine what are the premises and what is the conclusion. We must learn to put the argument in standard form.
3. What if the passage contains (as often happens) several arguments interwoven in complicated ways? We must learn to diagram such multiple arguments.

The central subtask is (2). Knowing how to put single arguments in standard form is the key to determining both how many arguments are present (if any) and how they are interrelated. Let us accordingly start with (2). Task (3) will be addressed in Section 2.5, while (1) will be addressed in Section 2.6.

The problem of identifying single arguments—spotting premises and conclusions—is that the order, type, and number of sentences occurring in a passage usually bear little resemblance to the logical structure of the statements in the argument.

Consider first the order of occurrence. The sentence that expresses the conclusion may occur first in the passage:

The Swiss must be loved. After all, they are a lucky people, and people love lucky people.

On the other hand, the conclusion may occur somewhere in the middle, as in:

> The resources of this planet are finite. So there must be a limit to population growth, since an infinite population would require infinite resources.

Here, the conclusion is that there must be a limit to population growth; the two premises flank the conclusion.

Consider next the type of sentences. Remember that in natural languages, sentence form and function do not always coincide. For example, rhetorical questions are not questions; they are statements. This can obscure the logic of the passage:

> Should we be unconcerned about the Russian menace? They threaten to subvert stable governments, they are growing in military strength, and they cause a shortage of vodka!

Here there are no declarative sentences, only an interrogative and an exclamation. But an argument is here:

> 1. Russians cause a shortage of vodka.
> 2. Russians subvert stable governments.
> 3. Russians are growing in military strength.
> ∴ The Russians are a menace we should worry about.

To add to the difficulty, the number of sentences in the passage and the number of statements in the argument are almost never equal. One reason why is that a statement may take two or more sentences for its expression, or may be repeated by different sentences (recall that markedly different words can express the same thought). For example:

> Nope. I can't buy that. Not at all. I'm sorry, I just don't agree that those workers should be fired for striking. They didn't have a "no-strike" clause in their contract, nor is their strike a threat to national security. No threat at all.

In this passage, eliminating the repetition, we have the simple two-premise argument:

> 1. The workers' strike does not threaten national security.
> 2. The workers did not sign a "no-strike" contract.
> ∴ They should not be fired.

A second reason the number of sentences in the passage may not equal the number of statements in the argument is that the passage may contain material that is not actually a part of the argument in the strictest sense, but instead amplifies, clarifies, or explains some of the premises. We are ignoring for now the common situation in which the passage contains a central argument and other material that proves or backs up premises of that central argument. Such a passage really contains multiple arguments, a topic that we will consider in Section 2.5.

As an example of a passage containing explanatory material, consider:

> Bess James was 67 years old when she ran her first race. Now she holds many marathon records. If Bess can run races, Mom, so can you, because you are only 43 years old.

Probably the best way to put this argument in standard form is as follows:

> 1. You are younger than Bess James.
> 2. Bess James can run marathons.
> ∴ You can probably run marathons.

Generally, it is not easy to distinguish which material in the passage is strictly part of the argument and which amplifies or explains.

A third reason (besides repetition of statements and the presence of amplificative material) why the number of sentences in the passage may be greatly different from the number of statements in the argument is that the passage may not include an explicit statement of a premise or the conclusion. An argument (or more exactly, a passage presenting an argument) is an **enthymeme** if part of that argument is unstated but the speaker takes it for granted.

Here is an example of an enthymeme (the scene is that of a man, who has borrowed money from a loan shark, facing two of the loan shark's enforcers):

> Look, pal. We don't like guys who don't pay on time. They wind up looking funny. Get the picture?

The argument stated explicitly in standard form:

> 1. If you don't pay on time, you will be physically injured.
> 2. You don't want to be physically injured.
> ∴ You should pay on time.

Usually, arguments are expressed enthymematically to save time—the arguer does not have to state a large number of things that are obvious to the listeners. This can add to the rhetorical power of the passage.

To sum up, a passage is a sequence of sentences and an argument is a set of statements (one of which is the conclusion, the others being the premises). There is no necessary match-up between type, order, and number of sentences in the passage and the structure of the argument.

How then can we identify premises and conclusions of arguments? It takes practice (which you will get in a moment), but a couple of heuristic guides can be followed.

The first heuristic rule is: English has indicator words in its vocabulary that allow the observer to 'decode' the argument contained in the passage. Some words are **premise-indicators**, which signal that the clause which follows expresses a premise. On the other hand, other words are **conclusion-indicators**, which signal that the clause which follows probably expresses the conclusion. Table 2.1 is a list of the most common indicator words.

This rule is only heuristic, not absolute. Indicator words may not signal an argument. For example, in

> Sue is so angry she could explode.

the word 'so' does not indicate that any kind of inference is involved, but it is used to express the degree of Sue's anger. And in the passage

> Kim Soon paid the bill, then he left.

the word 'then' merely indicates a temporal link between the two events (that is, Kim Soon paying the bill and leaving).

A second heuristic rule is: look to the context of the passage for any indications of relative certitude. By 'context' we mean the passages surrounding the one in question, or—if we are dealing with part of a conversation—the rest of the conversation. As a general rule, what the writer of the passage takes to be more certain is probably what the writer will offer as evidence, because the usual strategy in argument is to argue from what is more certain to what is less certain.

Suppose the following exchange takes place between two fans in a Pittsburgh bar:

Sue: Why do you say the Pirates should try to get Fernando from the Dodgers?
Barbara: Look, winning requires good pitching. We're weak there.

Table 2.1

Premise-Indicators	Conclusion-Indicators
For . . .	Thus . . .
Since . . .	Therefore . . .
Because . . .	So . . .
Due to . . .	Hence . . .
As . . .	It follows that . . .
Inasmuch as . . .	Ergo . . .
Otherwise . . .	Accordingly . . .
As indicated by . . .	We may conclude that . . .
After all . . .	Which entails that . . .
For the reason that . . .	Which means that . . .
Considering that . . .	You see that . . .
In view of the fact that . . .	As a result . . .
As shown by . . .	Which implies that . . .
Follows from . . .	So it is obvious that . . .
Being as . . .	In this way one sees that . . .
Being that . . .	Demonstrates that . . .
In the first place . . .	Bears out the point that . . .
Seeing that . . .	Should lead you to believe that . . .
Firstly/secondly . . .	Allows us to infer that . . .
Assuming that . . .	Which shows that . . .
May be deduced from . . .	This proves that . . .
May be concluded from . . .	This establishes that . . .
May be inferred from . . .	So it is undeniable that . . .
May be derived from . . .	

By the very context, we can assume that Barbara believes:

1. The Pirates do not have good pitchers.
2. Good pitchers are needed for a winning team.
3. It would be nice if the Pirates were a winning team.
4. Fernando is a good pitcher.

And her conclusion (the truth of which Sue had doubted):

∴ The Pirates should try to get Fernando from Los Angeles.

Considerations of context are an important feature in handling inductive arguments. For now, let us adopt the policy of stating premises and conclusions in such a way that nothing is left to context, that things are explicit rather than implicit.

A third heuristic rule is: Some words or punctuation marks, which we shall call **balance-indicators**, signal that the elements on either side—be they premises, conclusions, or arguments—are of the same kind. Some balance-indicators are given as follows:

However	And
Nevertheless	Moreover
What's more	In addition
; (semicolon)	

For example, the statement:

> DePaul, ranked number 2, won; however, UCLA, ranked number 1, lost. I guess DePaul will move into first place in the rankings.

clearly expresses the argument:

1. DePaul was ranked number 2.
2. DePaul won.
3. UCLA was ranked number 1.
4. UCLA lost.

∴ DePaul will be ranked number 1 in the rankings.

EXERCISES 2.4A

The following passages contain exactly one argument. Put each argument in standard form. (An asterisk indicates problems that are answered in the back of the book.)

1. I think that tax-indexing is the best cure for inflation. It would give Congress less of our money to waste, and it would create confidence in the money market.

* 2. The Padres won't win this year. After all, they have weak pitching. Also, their best hitter has been injured.

3. The unemployment rate for blacks is almost double what it is for whites. So any unemployment program should be aimed at blacks. After all, government programs should be aimed at the heart of the problem they are designed to address.

4. Unemployment is down. The stock market is up. I believe this shows that the recession is over. Why, even the budget deficit is shrinking.

5. Fernando is injured. Steve is injured. Ron is injured. And those are the best players on the team. And you say the team will win?

* 6. Why are you surprised that many jazz musicians have heart disease? They are constantly performing, they don't get much sleep, they drink and take drugs, and they eat high-cholesterol food in greasy dives.

7. Why not simply cut off all military aid to Central America? Didn't we learn in Vietnam that military aid leads to direct military involvement? And don't we want to avoid direct military involvement?

8. No, I don't think it would be a good idea for you to go to that rock concert! Don't the kids use drugs? And they drink like fish!

9. If you don't shut up I'm going to smack your face! You don't want your face slapped, do you? Then shut up!

* 10. Treason is worse than murder, for the murderer only kills one person, while the traitor, if successful, kills a whole nation.

11. Whoever preaches violence should be arrested. And whoever advocates universal suffrage is, in fact, preaching violence. So Ted should be in jail, because he advocates universal suffrage.

12. I was so happy to hear Professor Larch tell us that the bacteria she has developed to eat the oil in oil-spills are harmless! I'm glad that I don't have to worry about them attacking the oil in fish or on human skin, or about them attacking oil-drilling platforms, or surviving on the microscopic layer of oil on the ocean's surface! I'm so relieved!

13. No, I don't believe it. I can't believe it! I can't believe it! Are you really suggesting that we give Jason a job? Are you kidding? Don't you realize that he is a hopeless incompetent?

* 14. Left on her own, that kid would burn the house down, so we'd better take her with us, since I sure don't want to lose my house. Yes, we had better take the little brat with us.

15. The life expectancy of women is about five years greater than that for men. Since women live longer than men, any life-insurance system that pegs premium rates to life spans will discriminate against women.

16. Laura has known Steve for a long time. Gosh, I think they were going together even in high school. And, you know, that wasn't exactly yesterday. Anyway, it is surprising that they've broken up, since relationships that have lasted for years tend to continue.

17. The patient led a sedentary life, was grossly over weight, and smoked. It's even reported that he smoked four packs a day, and was known to have loved pizza. These are risk factors in heart disease. So his heart attack comes as no surprise, in the face of his attempt to run 20 miles. What a foolish thing to do.

* 18. Don't get excited. Don't get so excited. If it was Mark who was in that car crash, the police would surely have notified us by now. Just relax, okay?

19. If America continues to neglect our educational system, it will cease to be a great power. If that happens, our enemies will destroy us. Do you want that to happen? Then draw your own conclusion.

20. If you come any closer, I'll shoot.

EXERCISES 2.4B

The following passages contain exactly one argument. Put each argument in standard form. (An asterisk indicates problems that are answered in the back of the book.)

1. Science is based on experiment, on a willingness to challenge old dogma, on an openness to see the universe as it really is. Accordingly, science sometimes requires courage—at the very least the courage to question the conventional wisdom.

—Carl Sagan, *Broca's Brain*.

* 2. As positive proof that the earth is flat, Voliva cites statements from engineers that no allowance was made for curvature of the earth in building canals and railroads. In addition, the issue of the periodical includes a picture taken at Lake Winnebago in Oshkosh, Wisconsin. The opposite shore, 12 miles away, is clearly visible, "proving beyond any doubt that the surface of the lake is a plane, or a horizontal line."

—Daisie and Michael Radner, *Science and Unreason*.

3. Although the studies differ in their approach, their results are consistent: Computer-assisted instruction works. First, it seems that CAI gives students the opportunity to practice what they've learned, and to get better at it. . . . Second, in some circumstances CAI may teach students better study habits.

—Trudy Bell, "My Computer, My Teacher," in *Personal Computing* (June, 1983).

4. When an individual is sleeping and resting comfortably, the intervertebral discs reabsorb fluid and become plumper—to use a common expression—and the spine thus increases in length.

A point, then, to remember is—if you are going to take a physical examination and you want to be at your tallest—appear for the examination in the morning.

—Charles Linart and August Blake, *How to Increase Your Height*.

5. If you can get together a college degree—two years are OK, four are preferred—do so, even if you are a dozen years behind. You will be paid back

lavishly in higher earning potential, increased mobility, and more interesting jobs. It is a kind of passport.

—Tom Jackson, *Guerrilla Tactics in the Job Market.*

*   6. Kelli Stewart, Instructor of Dance for the Department of Physical Education, has requested that the division consider the inclusion of courses in Dance as partial fulfillment of the area of concentration requirement for the Associate of Arts in Humanities and Creative and Performing Arts. The argument for this inclusion is that Dance is a performing art and, at many universities, is housed in the fine or performing arts area.

7. The crucial aspects of any possible future world which might develop from the present situation are now seen in terms of only two variable quantities: the growth of world productivity as a whole, and the distribution of the fruits of that productivity between the rich and poor countries. Each variable may go in either of two directions compared with the present situation: more or less growth, more or less equality between nations. So we are left with only four kinds of possible future world to consider:

   a. High growth, high equality (a big cake fairly shared).
   b. High growth, low equality (a big cake hogged by the rich).
   c. Low growth, high equality (a small cake fairly shared).
   d. Low growth, low equality (a small cake hogged by the rich).

—John Gribbin, *Future Worlds.*

8. Where before winning meant total victory without regard to one's adversary, we know today that it's no longer appropriate for one person to walk away with all the rewards and love and self-esteem. . . .

The most vivid proof of this I have, perhaps, is the kind of people who are enlisting in my seminars on negotiation. Increasingly, they are people who have found that the win-it-all manner in which they had always functioned isn't working for them any longer.

—Tessa Albert Warschaw, *Winning by Negotiation.*

9. It came to me today, walking in the rain to get Helen a glass of orange juice, that the world exists only in my consciousness (whether as a reality or as an illusion the evening papers do not say, but my guess is reality). The only possible way the world could be destroyed, it came to me, was through the destruction of my consciousness. This proves the superiority of the individual to any and all forms of collectivism. I could enlarge on that, only I have what the French call "rheumatism of the brain"—that is, the common cold.

—James Thurber, *Selected Letters,* edited by Helen Thurber and Edward Weeks.

* 10. Inequality may even grow at first as poverty declines. To lift the incomes of the poor, it will be necessary to increase the rates of investment, which in turn will tend to enlarge the wealth, if not the consumption, of the rich. The poor, as they move into the work force and acquire promotions, will raise their incomes by a greater percentage than the rich; but the upper classes will gain by greater absolute amounts, and the gap between the rich and the poor may grow.

—George Gilder, *Wealth and Poverty.*

11. Enemy-making of men would ultimately subvert the whole dream of a women's culture based on mutuality and altruism. The very process of projecting the negative part of their own psychic potential onto males, and failing to own these themselves, would tend to make such women's groups fanatical caricatures of that which they hate. The dehumanization of the other ultimately dehumanizes oneself. One duplicates evil-making in the very effort to escape from it once and for all, by projecting it on the "alien" group.

—Rosemary Radford Reuther, *Sexism and God-Talk.*

12. Many therapists would admit the possibility that some people have a subconscious cause for their agoraphobia and need psychoanalysis. I do not close my mind to this possibility, but it is interesting that over thirty years of practice and after curing many hundreds of agoraphobic people, it was not necessary for me to use psychoanalysis on any of my patients.

—Claire Weekes, *Agoraphobia.*

13. Real winners—people who are self-actualizing and who respond authentically to life—do not need losers. Women winning does not mean that men must lose, any more than men winning should mean that women must lose. If women have more options, so too will men. More choices for women means choices for men. For example, it has long been acceptable for a married woman to quit her job and stay home, explaining, "I'm tired of working. I'm going to relax and do something different for a while." Yet how many married men—no matter how tired or pressured—would be greeted with the same degree of acceptance if they announced, "I'm tired of working. I'm going to stay home with the family for a while"?

—Dorothy Jongeward and Dru Scott, *Women as Winners.*

* 14. Without perceiving some sacredness in human identity, individuals are out of touch with the depth they might feel in themselves and respond to in others. Given such a sense, however, certain intrusions are felt as violations—a few even as desecrations. It is in order to guard against such encroachments that we recoil from those who would tap our telephones, read our letters, bug our rooms: no matter how little we have to hide, no matter how benevolent their intentions, we take such intrusions to be demeaning.

—Sissela Bok, *Secrets: On the Ethics of Concealment and Revelation.*

15. On the whole, then, the task of determining who is and who is not religious in the USSR is far from complete. Soviet sociologists have not yet developed completely satisfactory criteria to answer this question, nor have they achieved general agreement, even among themselves, concerning what criteria should be used. The introduction of a considerable variety of classification schemes confuses the picture further. A great deal of data on this subject has been collected; but much work remains to be done before Soviet sociology will be able to differentiate clearly between the religious and the nonreligious citizen.

—William Fletcher, *Soviet Believers.*

16. Fifty years ago there were 435 congressmen and 96 senators. Today there are 435 congressmen and 100 senators. The number of legislators is

fixed; the legislature's workload has grown exponentially. So the legislature's work is increasingly done elsewhere, no matter what you read in your high-school civics text.

—George F. Will, "More Government, Less Control," *Newsweek* (July 4, 1983).

17. Since biological determinism possesses such evident utility for groups in power, one might be excused for suspecting that it also arises in a political context, despite the denials quoted above. After all, if the status quo is an extension of nature, then any major change, if possible at all, must inflict an enormous cost—psychological for individuals, or economic for society—in forcing people into unnatural arrangements. In this epochal book, *An American Dilemma* (1944), Swedish sociologist Gunner Myrdal discussed the thrust of biological and medical arguments about human nature: "They have been associated in America, as in the rest of the world, with conservative and even reactionary ideologies. Under their long hegemony, there has been a tendency to assume biological causation without question, and to accept social explanations only under the duress of a siege of irresistible evidence. In political questions, this tendency favored a do-nothing policy." Or, as Condorcet said more succinctly a long time ago: They "make nature herself an accomplice in the crime of political inequality."

—Stephen Jay Gould, *The Mismeasure of Man*.

* 18. The constitution states that the Supreme Soviet of the USSR is to exercise supreme power in the country: it issues laws and forms and dissolves the government—that is, the Council of Ministers, which is responsible to it. But in real life the Supreme Soviet of the USSR merely rubber-stamps the decision of the top Communist party organs. The purely decorative Supreme Soviet is necessary to give to the Soviet state a semblance of democracy.

Everything attests to the decorative, propagandistic functions of the Supreme Soviet: it is in session for a mere six days a year, and in the course of its forty-three years of existence there has never once been so much as a single vote against a motion submitted by the government, or even an abstention, either in the Supreme Soviet of the USSR or in the supreme soviets of any of the fifteen Union republics that make up the federation that is the USSR. The fictitious nature of the power of the Supreme Soviets is further borne out by the fact that never in any of these Soviet "parliaments" has any question been raised relating to lack of confidence in the government; no member of the government has been questioned on any subject.

—Konstantin Simis, *USSR: The Corrupt Society*.

19. As an academic and professional discipline, economics lives with a fundamental internal contradiction: what is taught in conventional micro-economics is incompatible with what is taught in macro-economics. In the former, every market is a price-auction market that clears based on competitive bidding within a framework of supply and demand. Accordingly, any market is always in equilibrium, having no unsatisfied bidders, and every individual is a maximizer in his decisions to consume and produce. Macro-economics, on the other hand, is basically the study of markets that do not

clear and are not in equilibrium. Such contradictions, of course, are not peculiar to economics. Physics uses both particle theory and wave theory to describe electromagnetic phenomena. But the contradictions in economics are perhaps more severe than in other disciplines.

—Lester Thurow, *Dangerous Currents*.

20. Meanwhile the search was in full swing. Seven people were busily examining chests of drawers, carefully tapping the sides of my antique writing desk in an attempt to discover hiding places (some eluded them, but, alas, only the empty ones), thumbing through hundreds of books on our shelves, and rummaging in our laundry. One of the agents picked up from my desk the mosaic frame containing a picture of my wife, looked at it front and back, and replaced it. But the thin, sharp-nosed, respectable girl who had been appointed official witness in the search—actually just another KGB agent like the rest—scolded him, whereupon he began carefully taking the frame to pieces, trying to find a secret place.

Yet despite all the fuss they were making, my wife and I had a nagging feeling that the whole thing was a sham, that they were only going through the motions of a search. We soon understood why: we realized that they had already been to our apartment over the weekend while we were out of town at our rented dacha. They picked our lock, entered the empty apartment, and carried out their real search, finding everything that might be of interest to them.

Before long we had evidence that this surmise was correct. Part of the manuscript they had taken from my writing desk now lay in front of Investigator Borovik. Turning to one of his people, he ordered, "Bring the rest to me." Without hesitation the man walked straight to my wife's room (where no one in the search party had been yet that day) and fetched the last seventy pages of the manuscript, which had been there since Friday.

—Konstantin Simis, *USSR: The Corrupt Society*.

2.5 Interwoven Arguments

In this section,[2] we will refine our abilities to represent, in the fairest and most accurate way, the logical structure of a given passage. We must first distinguish between single support and linked support among premises. **Single support** is the support of a conclusion by a premise taken singly, as in the following arguments.

Fred is rich and famous, therefore he is rich.

All human beings are mortal, so if Kim is a human being, then Kim is mortal.

[2] The diagram technique employed in this section was first explored in Monroe C. Beardsley, *Practical Logic* (Englewood Cliffs, N.J.: Prentice-Hall, 1950).

The square root of 2 is an irrational number, thus there is at least one irrational number.

In contrast, **linked support** is the support of a conclusion by two or more premises that work together. Consider this example:

All frogs are green. Kermit is a frog. Therefore, Kermit is green.

Knowing only that all frogs are green does not automatically justify the inference that Kermit is green; for all the arguments say, he might be an anteater. Again, knowing only that Kermit is a frog does not justify the inference that he is green; for all the argument shows, frogs might all be pink. Only taken together do those premises support that conclusion.

Two points must be understood about linked support. First, any number of premises can link together to support a conclusion, as the following examples should make clear.

All people are mortal. All mortal beings are lonely. Therefore all people are lonely.

All people are mortal. All mortal beings are lonely. All lonely beings are frightened. Therefore, all people are frightened.

All people are mortal. All mortal beings are lonely. All lonely beings are frightened. All frightened beings are anxious. Therefore all people are anxious.

Clearly, we can generalize the previous argument to any number of premises.

All people are P_1. All P_1 beings are P_2. . . . All P_n beings are P. Therefore all people are P.

The second point to note is that many arguments in which a premise appears to singly support the conclusion are, in fact, enthymemes, that is, have premises omitted. For example:

Zaphod has two heads. Therefore Zaphod needs two hats.

seems to be a case of one premise singly supporting a conclusion. But, in fact, that argument omits the premise "any being with two heads needs two hats," which is not altogether obvious (can't one big hat fit over two small heads?). Telling single from enthymematically linked support is not always easy, even after considerable thought. For instance, it is hotly debated whether Descartes' famous aphorism "I think, therefore I am" is an argument with one premise, or an enthymeme.

We can represent single support of a conclusion by the diagram:

and linked support of a conclusion by the diagram:

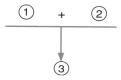

We will call any group of mutually linked premises (including any one single-support premise) a **premise cluster**. All of the examples so far given in this section have been simple arguments, in which one premise cluster supports one conclusion. But we often encounter complex arguments, in which one or more premise clusters support one or more conclusions. Many patterns of complex argumentation are possible; however, four patterns are especially common and deserve our special attention. Those four patterns are listed in Figure 2.1.

Figure 2.1

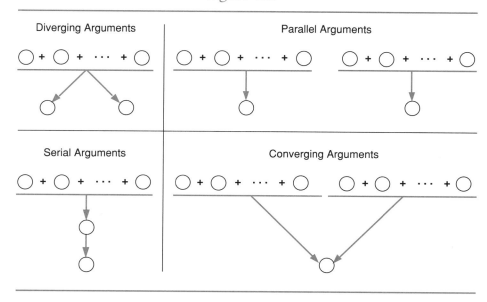

We shall consider each of the four patterns given in Figure 2.1 in turn. Consider first *diverging arguments*, which begin at the same premise cluster, but draw out different and quite often complementary conclusions. For example:

> The American dollar is very strong in comparison with the European currencies. This means that European goods here will be real bargains, but it also means that our exports abroad will be at a competitive disadvantage.

We can see that the previous argument fits the pattern listed earlier by numbering the statements in it and then diagramming the argument.

1 = The American dollar is strong in comparison with the European currencies.
2 = European goods here will be a bargain.
3 = American goods abroad will be at a competitive disadvantage.

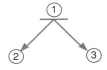

Parallel arguments move alongside one another to establish contrasting conclusions. A case in point is the following:

> While I'm doing poorly in physics, I'm really doing well in chemistry. My physics grades are F, D, and D⁻, whereas my chem grades are A, A⁺, and B⁺.

Here again, since it is easier to put numbers in circles than to put in whole sentences, we number the statements first.

1 = I'm doing poorly in physics.
2 = I'm doing well in chemistry.
3 = My physics grades were F, D, and D⁻.
4 = My chem grades were A, A⁺, and B⁺.

We then construct the diagram:

2.5 / Interwoven Arguments

The third common pattern of multiple or complex argumentation is serial argumentation. In *serial argumentation,* the conclusion of one argument becomes the premise cluster for another. Typically, serial arguments involve the successive unfolding of implications from some initial facts. For example:

> Jazz is improvisational music. This tells us that jazz is better appreciated live than on record. But that, in turn, means that a true jazz lover should go out to jazz nightclubs.

Again, we label the different statements and construct the appropriate diagram.

    1 = Jazz is improvisational music.
    2 = Jazz is better appreciated live than on record.
    3 = Jazz lovers should go to jazz nightclubs.

Converging arguments involve the elaboration of two different lines of attack on the same ultimate conclusion. An example will make this pattern clear.

> Evidence for the existence of the Great Pumpkin comes from two quite different directions. First, there is the undeniable unanimity of testimony of small children. When testimony is unanimous, it is very convincing. Second, fossils reveal pumpkin seeds the size of tennis shoes. The only way to account for such mammoth seeds is to postulate the existence of the Great Pumpkin.

We number the different statements and diagram them as follows:

    1 = The Great Pumpkin exists.
    2 = Small children are unanimous in their testimony about the Great Pumpkin.
    3 = Unanimous testimony is especially convincing.

4 = Fossils reveal giant pumpkin seeds.
5 = Postulating the existence of the Great Pumpkin is the only way to explain those seeds.

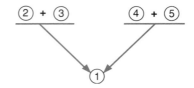

These four basic patterns (diverging, parallel, serial, and converging) can be combined in several ways. The following are a few of the possibilities.

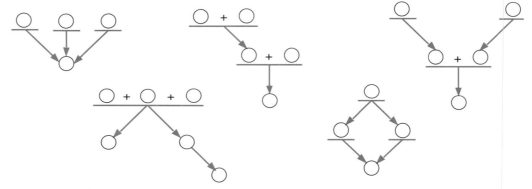

As you have seen before, any complex problem is easier to solve if it is broken down into simple parts. The following sequence of steps will help you diagram passages more easily.

Step 1: Read through the passage, and number every different statement stated in the simplest words. Eliminate redundancies, as well as irrelevant or extraneous material. Fill in any unstated premises.

Step 2: Go back to the passage and focus your attention on the indicator words. Attempt a first pass at diagramming the passage.

Step 3: Return to the original passage, and double check to verify that you have diagrammed the structure correctly.

Remember to fill in all premises that were unstated.

Let us work through several passages from a book by Melvin Konner.[3]

---

[3] Melvin Konner, *The Tangled Wing: Biological Constraints on the Human Spirit* (New York: Rinehart and Winston, 1982), p. 34.

> ... I would be amazed if the creature in question [a direct primate ancestor of the human race of five million years ago] did not have, as part of its reproductive life, competition among males for access to females, occasionally erupting into violence, infants kept in contact with their mothers and nursed freely; and males that, after maturity, rarely attempted to mate with their own mothers. If it violated any of these generalizations, it would be the most freakish higher primate that ever lived in the Old World; so unusual, in fact, that I would not be inclined to believe that it existed.

Step 1: Number the essential statements.

1 = Our primate ancestors probably had competition among males for access to females.

2 = Our primate ancestors probably had infants in close contact with their mothers.

3 = Our primate ancestors probably had males that did not try to mate with their mothers.

4 = All Old World higher primates have competition among males for access to females.

5 = All Old World higher primate infants are kept in close contact with their mothers.

6 = All Old World higher primate males do not attempt to mate with their mothers.

Step 2: Construct a provisional diagram. We seem to have three parallel arguments.

Step 3: Recheck. Have we left out an unmentioned premise? It seems so: we need to add to our original list a statement to the effect that what is true of all known Old World higher primates would probably be true of any primate ancestor of ours.

7 = What is generally true of present Old World primates is likely to be true of any primate ancestor of ours.

We can then refine our earlier diagram as follows:

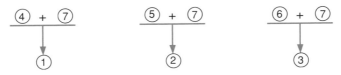

Note that the same premise links with each of the others.

EXERCISES 2.5

Diagram the following multiple arguments. I constructed those passages for you so as to have a minimum of extraneous material, but some passages contain enthymemes. (An asterisk indicates problems that are answered in the back of the book.)

1. Surely everyone loves money. So I feel confident that Ed will respond to the money we have offered him, and thus take the job.

* 2. People hate and fear snitches. So it stands to reason that people hate Richard and fear him.

3. Franklin is a dog. Most dogs bark, so Franklin probably barks, so he will probably disturb the neighbors.

4. Andrea likes cheese. Thus Andrea likes something. Thus somebody likes something.

5. Since everybody loves everybody, Sonja loves everybody. Since Sonja loves everybody, she loves Juan.

* 6. Rhonda just bought a Pomeranian dog, and Pomeranians bark like crazy. So Rhonda's dog will bark a lot, and since old man Crandall hates noise, he's going to hate Rhonda's dog.

7. Politicians are crooks, and Walter is a politician, so he's probably a crook. But, politicians are also sly, so he's sly. Face it—he's a sly crook.

8. Billie is a loner, and loners are distrusted. So we can expect that Billie will be distrusted. But people who are distrusted are disliked, so we can expect further that Billie will be disliked.

9. Unlike the American economy, the Japanese economy is diversified. So while the American economy can be expected to stagnate, the Japanese economy will probably move ahead.

* 10. People are basically greedy. So Communism is bound to fail, since Communism demands that people be unselfish. And so capitalism will win in the end.

11. Either Smythe, Tammy, or the butler killed Lord Louse. Since Smythe was in Blythe, we can rule him out. And Tammy was in Miami, so she couldn't have been the killer. The butler hated Lord Louse, and since the

other suspects (Smythe and Tammy) have been ruled out, the butler must have done it.

12. Food shortages in the Third World are increasing in frequency, and the high oil prices have taken their toll. The Third World is, therefore, in desperate economic straits; and so the Western nations must greatly increase their aid. The communist nations are themselves nearly bankrupt, and hence can't afford to give away any more aid.

13. Since no third-party candidate has won the presidency in the last 50 years, it's obvious that either the Republican or the Democrat will win. But if either the Republican or Democrat win, then no third-party candidate will win in the immediate future, because when any party is in office, it tends to become entrenched.

* 14. The use of computers in education is to be welcomed. Computers are endlessly patient, they force the student to interact rather than just passively watch, and they remind students of video games. Since they are to be welcomed, we should be putting more computers into the schools. Also, we need to start training more teachers in computer science.

15. Since people fear crime, the number of police officers should be increased. Since any increase in government workers requires an increase in taxes, we should increase taxes. After all, police officers are government workers. Since we should increase taxes, we should make the tax system more fair. Let's face it, if the tax system is unfair, people will be unwilling to pay the increase.

16. Given the fact that many people will be crowding the beach this summer, we should open a hot-dog stand there. Also, it really gets hot out there on the sand, and people get thirsty in the heat, so we should sell cold drinks. We should open a hot-dog stand that sells cold drinks.

17. Since people are basically dishonest, and since they are basically greedy, you can expect any legal system to be open to abuse. But given that, no society will ever be free from crime. Also, no society will be free from lawyers. Thus (since crimes mean criminals) every society will have lawyers and criminals.

* 18. Computers really do have the power of creative thought. We can conclude two things. First, computers will eventually be able to replace the highly paid workers, such as doctors and lawyers. Moreover, we should no longer treat computers as mere inanimate objects.

2.6 Telling Arguments from Other Things

We are now able to identify and diagram arguments, both single and multiple. We next want to address the fact that ordinary language contains things that resemble arguments, but that are not. These include: positive assertions, repeated assertions, conditionals, announcements, rephrasals, examples, descriptions, and summaries.

Let us begin with positive assertion. **Positive assertion** is the emphatic assertion of a claim without any warrant. For example:

> Everybody knows about the tremendous problems unemployment causes. You know about it, I know about it. Isn't it about time we acted? Folks, my unemployment bill will do the job. Yes, it will.

Here, the person is not giving any reason to adopt the bill, but merely asserts dogmatically (positively) that it will.

Related to this is repeated assertion, also called the "Big Lie." **Repeated assertion** is the technique (perfected by Hitler's propaganda minister Goebbels) of cleverly repeating one's claim until the listeners take it as proven fact. Hitler and his gang repeated the Big Lie that Jews were responsible for Germany's problems, and by repeating the canard so often and in so many different forms, many people came to accept it as fact. Advertisers use a similar tactic with "spot ads," constantly repeating a slogan, such as "Drink Ratz beer—it's the best!" until it is drummed into people's heads.

A third thing often confounded with an argument is a type of statement called a conditional. A **conditional** is a compound statement in which the truth of one component is asserted as contingent upon the truth of another. For example, the statement "If President Jones gets the support of blue-collar voters, she will win" is a single statement. It is not an argument, because the statement "President Jones gets the support of blue-collar voters" is not claimed, nor is the statement "she will win." All that is claimed is that if such-and-such happens, then something else will happen; that is, one event will happen on condition that another happens first. No premise, no conclusion, no argument. On the other hand, the sentence "Because President Jones is getting the support of blue-collar voters, we can be certain that she will win" does express an argument. Two claims are made—that President Jones has the support of blue-collar voters, and that she will indeed be elected. The argument here is probably an enthymeme.

1. President Jones has won the support of blue-collar voters.
2. She was already close to having the votes needed to win.

∴ She will win.

We will discuss conditionals in Chapter 9.

Announcements are often confounded with arguments. An **announcement** is a statement about where a person or group stands on some issue.

> We, the members of the South State Republican Women's Club, have come out against the new tax law. Ours is a club of 55 members, and our decision was unanimous.

The writer is only announcing her club's opinion, not justifying it by laying out evidence for it. Another thing often confounded with argument is rephrasal. To **rephrase** a point is to restate it in other words. We rephrase to make something clearer. The expressions 'that is' and 'i.e.' usually signal to the listener that we are rephrasing a point, as in the following examples.

> Intraspecific aggression is rather rare; that is, animals rarely attack others of their own kind.
>
> But don't people really seek domination, i.e., don't they really want someone to tell them what to do.

Similar to rephrasal is exemplification. To **exemplify** a claim is to give an example or illustration of it to make it clearer. General statements especially call out for examples. But to cite an example of some general claim is not to prove that claim.

> Kids do the darndest things. For example, my little Flukey just set fire to her brother's hair.

This text usually follows a general claim or definition with one or more examples.

Do not confound description with argument. To **describe** is to simply give a set of propositions that characterize a situation. For example, the following passage describes Fred's financial situation:

> Fred is the owner of a large house in Los Angeles and a small vacation home in Hawaii. He has a large savings account and a modest portfolio of stocks. He also has a retirement income from the company for which he used to work.

Compare this with the following argument intended to prove Fred is not poor.

> Fred can hardly be considered poor. He owns houses, has a savings account, owns stocks, and has a steady income. Poor people don't own much or have good incomes, now, do they?

Finally, summaries should not be confounded with arguments. A **summary** is a set of statements that highlight or repeat points made earlier in a passage. For example:

> We have seen the following: first, that Fred is rich; second, that he is dishonest; and finally, that he is probably a criminal.

Summaries are not arguments, since they don't set forth some of the statements as evidence for another of the statements.

One last complication is that even where an argument is present, it can be unclear whether the author is giving that argument, reporting that argument and agreeing with it, or reporting that argument and disagreeing with it. Compare the three examples that follow:

1. Pigs can't dance. Their legs are too short, and they aren't very coordinated.
2. I doubt that pigs can dance. The well-known pig expert Professor Porcine has pointed out in this regard that pigs are rather uncoordinated and their legs are altogether too short to dance properly.
3. Some have argued that pigs cannot dance. For instance, a certain Professor Porcine has claimed that pigs are uncoordinated and have legs too short for dancing.

In (1), the writer simply gives the argument. In (2), the writer reports and evidently agrees with an argument that Professor Porcine gives. In (3), the writer reports Porcine's argument, but uses phrases ("some . . . have argued," "a certain professor . . . has claimed . . .") that suggest the writer does not agree with the argument. To better determine whether the writer accepts the argument reported, we would look at the context of the passage.

EXERCISES 2.6A

Some of the following passages contain single arguments, while the rest contain no arguments. For each passage, state whether it contains a single argument, and, if it does, put it in standard form. If it does not, tell what kind of technique is being used. (An asterisk indicates problems that are answered in the back of the book.)

1. I've heard this story a thousand times before. I don't care where you heard it, or who told it to you. The whole story is a lie. Do you hear? A lie!

* 2. Dear Editor:

Regarding Lou Salomé's review of Wagner's latest play *The Big Man,* I can only express my utter contempt. Such sycophantic twaddle. Is she kidding, or what?

3. Why can't he get a date? Because no woman is totally devoid of taste!

4. If he had brains, he'd be dangerous. But because we know he isn't dangerous, that tells us he has no brains.

5. Get away. I don't want you around. You hurt me, going out with that other guy. So go date him. I don't care. Get lost. I don't want to be around people who hurt me.

* 6. Health insurance is a must. But that means we'll have to put that in our contract, because the company won't volunteer anything. Not a damn thing. It never does.

7. I think he got cancer because he worked with brake linings, which contain asbestos, and studies have shown that asbestos causes cancer.

8. My friends, you heard my client say on three occasions that she didn't murder her husband. I know she didn't do it. You should know that. She simply couldn't have done such a terrible thing.

9. I know you think otherwise, but Juanita couldn't have been the one who stole the carpet, since Juanita hates Persian carpets.

EXERCISES 2.6B

Some of the following passages contain single arguments, while the rest contain no arguments. For each passage, state whether it contains an argument, and if it does, put that argument in standard form. If it doesn't, classify the passage. (An asterisk indicates problems that are answered in the back of the book.)

1. Since the late 1960's America's economy has been slowly unraveling. The economic decline has been marked by growing unemployment, mounting business failures, and falling productivity. Since about the same time America's politics have been in chronic disarray. The political decline has been marked by the triumph of narrow interest groups, the demise of broad-based political parties, a succession of one-term presidents, and a series of tax revolts.

These phenomena are related. Economics and politics are threads in the same social fabric. The way people work together to produce goods and services is intimately tied to the way they set and pursue public goals. This link is perhaps stronger today than at any time in America's past because we are moving into an era in which economic progress depends to an unprecedented degree upon collaboration in our workplaces and consensus in our politics.

—Robert Reich, *The New American Frontier.*

* 2. American values have been transformed during the 1960's and 1970's, leaving us with a legacy of noble concerns. The society we have created is indeed more compassionate for those who are old, in trouble, or are yet unborn. We have become more cautious, perhaps more responsible. But these good deeds and intentions are not without a price. We shall pay for them through slowth (slow growth). That price can be expressed not only monetarily but also in our feelings of well-being, emotions that we still measure in very concrete and individual ways. For us as individuals, slower growing material standards arising from these values will mean disappointment, a loss of confidence, and an eclipse of hope. The way rising living standards have become intertwined with our identities makes this an unavoidable consequence of any protracted economic slowdown.

—Martin Kupferman, *Slowth: The Looming Struggle of Living with Less.*

3. Japanese still tend to think in terms of personal relationships and subjective circumstances in their business dealings. Thus an agreement between a Japanese and a foreign businessman should be reduced to its basic elements, and each point thoroughly discussed, to make sure each side understands and actually does agree to what the other side is saying.

—Boyne De Mente, *The Japanese Way of Doing Business.*

4. Although the United States remains the largest single market for telecommunications equipment, with approximately $15 billion in sales in 1980, the non-U.S. market is now almost twice as large. Moreover, the non-U.S. market is growing more rapidly. Unless AT&T becomes a greater force in world exports, its global market share will steadily decline.

—Hunter Lewis, *The Real World War.*

5. Most of the behavioral "traits" that sociobiologists try to explain may never have been subject to direct natural selection at all—and may therefore exhibit a flexibility that features crucial to survival can never display.

—Stephen Jay Gould, *The Mismeasure of Man.*

* 6. Before starting this diet (or any diet) get your doctor's approval. And after you've begun, use your good sense in choosing the foods you eat. Protect your heart by watching your intake of saturated fats (dieting or not dieting, you should do that); protect your liver by including many protein-rich foods (eggs, cheese, fish and meat); protect your overall health by including foods rich in vitamins and minerals, by supplementing your diet with a multivitamin every day. Finally, bear in mind that if you remain on a low-carbohydrate diet more than two weeks, you should be sure to include limited but daily quantities of milk and citrus fruit.

—*The New Carbohydrate Gram Counter* (Dell Original).

7. On the whole, concentration is a natural state which can easily be reproduced by simple methods. It is only supposed to be exceptional because people do not try and, in this, as in so many other things, starve within an inch of plenty. Those who do try have never been disappointed in the process but have sometimes experienced disappointment in themselves.

—Ernest Dimnet, *The Art of Thinking.*

8. I never cease to be amazed at the public unawareness of elementary legal procedures and techniques, and of the widespread suspicion and hesitation about going to a lawyer. Because of this, I felt a real need existed for this book.

—Edward Siegal, *How to Avoid Lawyers.*

9. There are so many books on writing as an art or as a craft that one naturally asks some questions about any new book. How is this book different from others? Is it for me? Will it help me to write more effectively and with more joy?

How to Write and Publish differs from most books on writing because it does not deal exclusively with any one type of writing; rather it gives a writer an opportunity to take a guided tour of the wide writing field. Second, from the first chapter it encourages the writer to think, feel and believe for himself, and promotes writing as a natural form of communication growing out of what he has already discovered about himself.

In my own writing and in years of teaching classes, conducting workshops and lecturing to writers' groups I have found that some advice is particularly helpful to both beginners and those who have already written for publication. This is the advice I have included in this book.

—Helen Hinkley Jones, *How to Write and Publish.*

* 10. This book is an attempt to redress the balance, to begin to integrate the rational and intuitive approaches to knowing, and to consider the essential complementarity of these two modes of consciousness as they are manifest in science in general, in psychology in particular, and within each person psychologically and physiologically. A growing body of evidence demonstrates that each person has two major modes of consciousness available, one linear and rational, one arational and intuitive.

—Robert Ornstein, *The Psychology of Consciousness.*

11. The purpose of Freud's lifelong struggle was to help us understand ourselves, so that we would no longer be propelled, by forces unknown to us, to live lives of discontent, or perhaps outright misery, and to make others miserable, very much to our own detriment. In examining the content of the unconscious, Freud called into question some deeply cherished beliefs, such as the unlimited perfectibility of man and his inherent goodness; he made us aware of our ambivalences and of our ingrained narcissism, with its origins in infantile self-centeredness, and he showed us its destructive nature. In his life and work, Freud truly heeded the admonition inscribed on the temple of Apollo at Delphi—"Know thyself"—and he wanted to help us do the same.

—Bruno Bettelheim, *Freud and Man's Soul.*

12. I was on my own in Cambridge, but I was not left alone for long. One evening in November, when I was hunched over my desk and the mice were nibbling at the crumbs I had set out for them, our doorbell rang. Mrs. Roff allowed as how there were two young gentlemen come to see me. Two students in black gowns stood in my doorway. One had a birdlike head and manner; his name was James Klugman. The other had black, curly hair, high cheek bones, and dark, deep-set eyes. His entire body was taut; his whole being seemed to be concentrated upon his immediate purpose. His name was John Cornford.

—Michael Straight, *After Long Silence*.

13. It should be apparent by now that this book will deal with certain physical disorders that are capable of mimicking emotional disturbances and mental disease. These sicknesses can start with tension, with vitamin deficiency or an imbalanced diet, with a disturbance in body chemistry, with allergy, even with a viral infection. All the symptoms overlap, even though the causes are so different, for the brain and the nervous system have limited ways in which to express reaction to insult.

—Carlton Fredericks and Herman Goodman, *Low Blood Sugar and You*.

* 14. Stick to your subject. Beware of introducing irrelevant matters. They distract readers. In writing for any public whatever, use no words, phrases, or thoughts which turn the reader's attention from your main point, even for an instant. Every diversion, however slight, tends to make him lose interest in what you have to say.

—Walter Pitkin, *The Art of Useful Writing*.

15. But enough of that. In my effort to tell you about the facts of educational life, I seem to be getting awfully moralistic. Let's pass on to something else—to the underlying question of why you should improve your reading.

There is a simple answer to that question: You should improve your reading so that you can easily read *Silas Marner* by George Eliot. No, I am not joking. I am just drawing a logical inference from well-known, established facts. These are the facts:

1. Educational experts agree that the highest purpose of reading is the study of literature.
2. The most important works of literature are required reading in our schools.
3. The book most widely required in our schools is *Silas Marner*.

So, in essence, you learn to read so that you can read *Silas Marner*.

—Rudolf Flesch, *How to Make Sense*.

16. More than twenty-five million people, ranging from employers to relatives and from friends to co-workers, are directly or indirectly involved in the problem [of alcoholism]. It affects three percent of our workers, and that costs industry at least three billion dollars yearly. This means that anything even remotely helpful to alcoholics is worthy of investigation.

—Carlton Fredericks and Herman Goodman, *Low Blood Sugar and You*.

17. The Bible is the most-read book that has ever existed, and there are uncounted millions of people in the world who, even today, take it for granted that it is the inspired word of God; that it is literally true at every point; that there are no mistakes or contradictions except where these can be traced to errors in copying or in translation.

There are undoubtedly many who do not realize that the Authorized Version (the "King James Bible"), the one with which English-speaking Protestants are most familiar, is, in fact, a translation, and who therefore believe that every one of its words is inspired and infallible.

Against these strong, unwavering, and undeviating beliefs, the slowly developing views of scientists have always had to fight.

Biological evolution, for instance, is considered a fact of nature by almost all biologists. There may be and, indeed, are many arguments over the details of the mechanics of evolution, but none over the fact—just as we may not completely understand the workings of an automobile engine and yet be certain that a car in good working order will move if we turn the key and step on the gas.

There are millions of people, however, who are strongly and emotionally opposed to the notion of biological evolution, even though they know little or nothing about the evidence and rationale behind it. It is enough for them that the Bible states thus-and-so. The argument ends there.

Well, then, what does the Bible say, and what does science say? Where, if anywhere, do they agree? Where do they disagree? This is what this book is about.

—Isaac Asimov, *In the Beginning*.

* 18. If we read on in *Mein Kampf* we find that Hitler gives us a description of child's life in a lower-class family. He says: "Among the five children there is a boy, let us say, of three . . . When the parents fight almost daily, their brutality leaves nothing to the imagination; then the results of such visual education must slowly but inevitably become apparent to the little one. Those who are not familiar with such conditions can hardly imagine the results, especially when the mutual differences express themselves in the form of brutal attacks on the part of the father toward the mother or to assaults due to drunkenness. The poor little boy, at the age of six, senses things which would make even a grown-up person shudder. The other things the little fellow hears at home do not tend to further his respect for his surroundings."

In view of the fact that we know that there were five children in the Hitler home and that his father liked to spend his spare time in the village tavern where he sometimes drank so heavily that he had to be brought home by his wife or children, we begin to suspect that in this passage Hitler is, in all probability, describing conditions in his own home as a child.

—Walter C. Langer, *The Mind of Adolf Hitler*.

19. Twenty years after arriving at Harvard as a graduate student, I returned as Professor of Psychology. I had spent five of those years as a postdoctoral fellow at Harvard and the rest on the faculties of the University

of Minnesota and Indiana University. In Minnesota I had married Yvonne (Eve) Blue, and in 1938 our daughter Julie was born. Later that year *The Behavior of Organisms* was published. During the war I taught pigeons to guide putative missiles, and in the Guggenhim year that followed I worked on my book on verbal behavior, built the "baby-tender" for our second daughter, Deborah, and wrote *Walden Two*. An invitation to give the William James Lectures brought me back to Harvard, and while there I was asked to join the department. *Walden Two* was published a few months before I returned to Cambridge.

—B.F. Skinner, *A Matter of Consequences*.

20. Individual Japanese industries are efficient, but not to the degree the world has been led to believe. Much of their advantage is based on unsavory practices. Japan has "borrowed" or copied foreign technology, or acquired it through joint-venture agreements which it has later disavowed. When this has failed they have resorted to bribery, industrial espionage and outright theft. Its industries often act in concert, as did the prewar Japanese cartels, the zaibatsu, targeting their competitors in other nations and dumping their products at a temporary loss in order to win larger and larger shares of the world's markets and eventually achieve monopoly positions. The Japanese educate their scientists and engineers in American and European universities; they then return home to use their new skills in a trade war against those who educated them. Japan, it is now becoming clear, is winning the trade war because it refuses to play by the rules.

—Boyne DeMente, *The Japanese Way of Doing Business*.

2.7 The Uses of Argument

In ordinary life, arguments are used to do various things. We would do well to devote attention to how people use arguments. For what purposes are arguments given?

Dozens of different uses occur for arguments, but let us focus on five: (1) to establish belief, (2) to explain, (3) to justify action, (4) to cause action, and (5) to amuse. First, the central use of arguments is to establish a *belief*, which means to prove that some statement is true. We might, for example, put forward the argument:

1. Very few third-party candidates have even come close to winning the presidency.

∴ The third-party candidate will not win in this election.

with the intention of establishing the conclusion as true. Normally, a person will construct an argument to convince an audience of a point, by selecting premises agreeable to that audience.

A second use of argument is to explain. *Explaining* involves using an argument, but not to establish a conclusion. Instead, the conclusion is as accepted by the speaker and audience as are the premises. The audience wants to see the connection between what they already believe and the thing to be explained. For example, someone might explain why a college team lost the game by saying, "They lost because their best quarterback was not in the game." The argument is:

1. The college team played without its best quarterback.
2. (Enthymeme) A team that plays without its top quarterback is very likely to lose.

∴ The college team lost.

Here, both the premises and the conclusion are already believed—the person succeeds in explaining by showing the connection between accepted beliefs. To distinguish explaining from establishing is a task for which no precise rules exist. An heuristic rule is this: "*A* because *B*" probably expresses an establishing argument if the context of its occurrence indicates that *A* needs proof and *B* does not. On the other hand, if *A* is presented as at least as clear as *B*, then it probably expresses an explaining argument. We can only judge, and we need to judge according to context.

People may give multiple arguments, where some are used to explain, others to establish. Scientific research involves this kind of complicated arguing. Someone might, for example, marshal evidence to establish that some phenomenon occurs, and then explain it in terms of currently accepted scientific theories. Indeed, in a scientific revolution, a scientist may have to establish a major new scientific theory first, and then use it to explain a puzzling phenomenon.

A third use of argument is to *justify behavior*, that is, to give reasons for some action one has taken. Suppose, for instance, that Nicole has chosen to major in computer science, and someone asks her why, that is, asks her to justify her choice. She gives this argument:

1. Computer scientists earn very good wages and will continue to be in great demand for the indefinite future.
2. My goal is high pay and good job security.

∴ Majoring in computer science is my best choice.

That justifies her action, in the sense that it shows how her action is appropriate to her goal. Her goals may be wrong or immoral—but that is a different matter.

The argument should not be put in this form:

1. Computer scientists earn very good wages and will continue to be in great demand for the indefinite future.
2. My goal is high pay and good job security.

∴ I chose computer science.

This would not be the best representation for two reasons. First, there is a difference between reasons and causes. Specifically, Nicole's reasons may not be the cause of her action—perhaps she impulsively chose computer science and only later thought up good reasons for it. Second, even if she had recognized that given her goals, computer science was the best choice for her major, she still might have chosen otherwise.[4] For these reasons, we will get in the habit of representing action-justifying arguments as being of this form:

1. The person involved has goals 1, 2, 3 . . . .
2. Action A would best fulfill those goals.

∴ The person ought to do A.

The fourth use of arguments is to *persuade*, that is, to cause people to act in desired ways. Someone may well knowingly use an illogical argument to persuade someone else to do or believe something. This situation should be distinguished from the case where someone sincerely tries to prove a point but unknowingly uses an illogical argument. Logically good arguments can be used in morally bad ways, and logically bad arguments can be used in morally good ways.

A fifth noteworthy use of arguments is to *amuse*. We can offer an argument as a joke or as a sarcastic device. For example:

Why are you so mad that somebody stole your Porsche? Surely your mother taught you to share your toys with others.

Using arguments as jokes is derivative from their use to establish belief. Part of getting the joke is seeing how the argument is illogical, that is, how it would not be a genuine proof of the conclusion. This point was well put by the famous nineteenth-century logician Richard Whately.[5]

---

[4] This last claim has been disputed: some have argued that if a person truly believes that a course of action is in that person's best interest, that person will do it. This whole issue falls under the heading of "the problem of free will," and is beyond the scope of this book.

[5] Richard Whately, *Elements of Logic* (London: Mawman, 1826), pp. 202–203.

[I]t may not be improper to mention the just and ingenious remark, that Jests are Fallacies; i.e., Fallacies so palpable as to not be likely to deceive any one, but yet bearing just the resemblance of argument which is calculated to amuse by the contrast; in the same manner that a parody does, by the contrast of its levity with the serious production which it imitates. There is indeed something laughable even in fallacies which are intended for serious conviction, when they are thoroughly exposed. There are several different kinds of joke and raillery, which will be found to correspond with the different kinds of Fallacy: the pun (to take the simplest and most obvious case) is evidently, in most instances, a mock argument founded on a palpable equivocation of the middle Term: and the rest in like manner will be found to correspond to the respective Fallacies, and to be imitations of serious argument.

It is probable indeed that all jests, sports or games, properly so called, will be founded, on examination, to be imitative of serious transactions; as of war, or commerce. But to enter fully into this subject would be unsuitable to the present occasion.

EXERCISES 2.7

The following passages contain an argument. Put each in standard form, and indicate for what purpose it is used. (An asterisk indicates problems that are answered in the back of the book.)

1. After years of research and thousands of documents declassified with the aid of the Freedom of Information Act, Eisenhower's shabby place in history was explained. He worked largely undercover. America's most popular hero was America's most covert President.

—Blanche Wiesen Cook, *The Declassified Eisenhower.*

* 2. My father told me Buffalo Bill could tell the damnedest stories you ever heard, entertaining his troops of performers for hours with Old West blood and guts make-believe. He was admired by all, including the hundreds of Indians he took along on tour. Indians love a man who can tell good stories and every tribe has its favorite yarn spinners.

—Iron Eyes Cody, *Iron Eyes: My Life as a Hollywood Indian.*

3. Years later in the dark depression years when he was trying to earn a buck on the road as a shoe salesman, [my father] checked into a small-town hotel. "Fine," said the clerk, reversing the register and reading his name. "You'll like it here, Mr. Reagan. We don't permit a Jew in the place."

My father picked up his suitcase again. "I'm a Catholic," he said furiously, "and if it's come to the point where you won't take Jews, you won't take me either." Since it was the only hotel in town, he spent the night in his car in the snow. He contracted near-pneumonia and a short time later had the first heart attack of the several that led to his death.

—Ronald Reagan, *Where's the Rest of Me?*

4. The two skinny, illiterate kids behind the palm trees had obviously been given the job of following us everywhere. We'd be walking down a street and suddenly Henry would pivot around and snap their pictures, while they would break out into peals of laughter. They hid behind bushes from us, waved through the leaves, and flirted with me.

We would walk in one side of the cathedral with appropriate dignity, then dash out the side door and lose them.

It was all endlessly diverting, but it was a deadly, defeating diversion. I understood their tactics all too well: no one from the guerrillas would dare to contact us so long as the skinny ones were about. That was their final card in this strange game.

—Georgia Anne Geyer, *Buying the Night Flight.*

2.8 Arguments vs. Dialogues

You may have found yourself puzzled by the way we have defined the term 'argument.' It sounds as if arguments are what a person gives, rather than what two or more people have or do. But does it make sense to say "My father and I argued about the draft yesterday"? It certainly does. The term 'argument' is ambiguous (that is, has more than one meaning). Our definition in Section 2.1 captures (or reflects) only one of those meanings. For the other meaning, the notion of an argument as an interactive or interpersonal sort of thing, let us choose the word 'dialogue.' By a **dialogue** we mean an array of statements and questions, structured in rounds. Normally, people mean by 'logic' the logic of arguments. But it is also correct to talk about the logic of dialogues. To get a clearer idea of what dialogues look like, consider the following:

Son: I just don't think the draft is moral or legal, and I doubt that I will register for it.

Father: Why do you say it's illegal?

Son: Because the Constitution forbids involuntary servitude, and slavery is what the draft is all about.

Father: But every court that has considered the matter has ruled that the draft is not involuntary servitude.

Son: (silence)

2.8 / Arguments vs. Dialogues

Father: And I don't think the draft is immoral at all.
Son: Why?
Father: Because you owe your country service. After all, it has protected and educated you.
Son: But did I ask it to?

This dialogue can be structured as a matrix (an array of rows and columns), each entry being a question or a statement:

	Son	Father
First Exchange (Round)	The draft is immoral. The draft is illegal. I won't register.	Why is the draft illegal?
Second Round	The draft is slavery, so it is unconstitutional.	No court has ruled it unconstitutional, so it is constitutional.
Third Round	Pass [No reply].	The draft is not immoral.
Fourth Round	Why?	Our country has educated and protected you, so you owe it service.
Fifth Round	I did not ask it to.	Pass.

Understanding the nature of dialogues will be important when we discuss fallacies in Chapters 4 through 7. To better understand dialogues—which are arrays of questions and statements—we must first discuss questions. Let us be as careful in defining our terms here as we were with statements. A **question** is the kind of thing that calls for a statement as an answer. Now let us distinguish sentence from question. Different interrogative sentences can be used to put the same question. We speak of 'putting a question' in the same way we speak of 'making a statement.' We usually use a declarative sentence to make a statement, and an interrogative sentence—a sentence ending with a question mark—to put a question. Thus, the sentences:

Does Heather love John?
Is Heather in love with John?
Is Juan loved by Heather?

all put the same question.

Moreover, interrogative sentences from different languages can put the same question. For example:

Where is my money?
¿Donde es mi dinero?

put the same question.

Finally, we do not have to use an interrogative to put a question. For example, the declarative:

I guess a car like that probably costs a lot.

is often what we would say to ask the owner tactfully what it costs. In fact, we do not need to use any sentence at all to ask a question—a raised eyebrow can do the job.

The two basic aspects to any question are what it presupposes and what counts as a direct answer to it.

A **direct answer** to a question is a statement that completely answers the question but gives no more information than is needed to completely answer the question.[6] Table 2.2 gives some examples of questions and their direct answers.

A **presupposition** of a question is any statement that has to be true if that question is to have any true answer. Presuppositions are important, since, if the presupposition of a question is false, trying to answer the question may waste time.

Consider the classic question, "Have you stopped beating your wife?" Two direct answers are possible:

(Yes) I have stopped beating my wife.
(No) I have not stopped beating my wife.

Neither of these possible answers can be true unless I have a wife and have beaten her in the past. We call questions that have false presuppositions 'loaded.'

---

[6] The terminology and account given here follows that in Nuel D. Belnap and Thomas B. Steel, Jr., *The Logic of Questions and Answers* (New Haven, CT: Yale University Press, 1976).

2.8 / Arguments vs. Dialogues

Table 2.2

Question	Some Direct Answers
Who is the president of the United States today?	Clinton is president today. Perot is president today. Mickey Mouse is president today.
Do unicorns exist?	(Yes) Unicorns exist. (No) Unicorns don't exist.
When did Lorraine say she would get here?	She said she would arrive at 1:00 p.m. She said she would arrive at 1:15 p.m.
What is an example of a sexy car?	A Corvette is an example of a sexy car. An Oldsmobile is an example of a sexy car.

Table 2.3 offers some examples of questions and their presuppositions.

In recognition of the possibility that one of the presuppositions of a question may be false (which means that the question has no true direct answer), we need to introduce the notion of a 'corrective answer.' A **corrective answer** to a question is an answer informing the questioner of a false presupposition. For example, a corrective answer to the question "Have you stopped beating your wife?" would be "I have never married." Another corrective answer would be "I have never beaten my wife at all."

Table 2.3

Question	Some Presuppositions
Who is the president of the United States today?	The United States exists and has a president today.
Do unicorns exist?	No presuppositions!
When did Lorraine say she would get here?	Lorraine exists; Lorraine said she would get here.
What is an example of a sexy car?	There are sexy cars.

EXERCISES 2.8

For each of the following: (a) put the question in the simplest interrogative you can; (b) state an obvious presupposition if any; (c) give an example of a direct answer (make it complete); and (d) give an example of a corrective answer. (An asterisk indicates problems that are answered in the back of the book.)

Example: "How about those Dodgers? Aren't they fantastic?"
   a. Aren't the Dodgers a good team?
   b. Presupposes the Dodgers are a team.
   c. (Yes) The Dodgers are a good team.
   d. The Dodgers were all killed this morning in a plane crash.

1. When, oh when, are you going to learn to dance?
* 2. What movie shall we see?
3. You remember Suzy—she walked with a limp because of the alligator incident. Where is she living now?
4. Have you stopped stealing cars?
5. Where is the best spot for us to open our lemonade stand?
* 6. Why are Americans so unhappy, Mr. Sociologist? Just answer that.
7. The killer must have hidden the gun somewhere in this house—but where, Watson, where?
8. Was it through stupidity or dishonesty that this administration has destroyed the independence of the Supreme Court?
9. What I am curious about is this: do any sea-dwelling mammals like cheeseburgers?
* 10. Dear Gas Company: Can I have instruction on how to light the pilot light in my furnace?
11. Mr. Jason, you have stalled long enough. When can we expect to see this bill paid?
12. Daddy, does Santa Claus really exist?
13. Do Martians exist?
* 14. When did Mom say the train would arrive?
15. Can you keep quiet?

2.9 Identifying Dialogues

We now have a better understanding of questions, on top of our earlier grasp of statements. Basically, the trick to identifying multiple arguments is learning to see past the rhetorical surface of sentences to grasp the underlying evidential web of statements. Similarly, the trick to identifying dialogues is to discern in a given passage the underlying web of questions, answers, and arguments.

To do this we need a technique, a mathematical tool. We will employ grids, with one column for each participant in the dialogue, and one row for each round in the dialogue. Let us try an example:

A: When will Mr. Godot arrive?
B: At midnight.
A: Then he'll be alone, since none of his friends are awake at that hour, and he never travels with his enemies.
B: Are you eager to see him?
A: Yes.
B: Then you are hoping midnight comes quickly.

Step 1: Set up the grid.

	A	B
Round 1		
Round 2		
. . .	. . .	. . .

Step 2: Work through the dialogue, putting each statement and question (reworded as simply as possible) in its appropriate place. When any statement is inferred from prior statements, put the whole argument in standard form, with arrows drawn from the earlier statements. In our example (on page 62):

EXERCISES 2.9

Diagram the following dialogues. (An asterisk indicates problems that are answered in the back of the book.)

1. *Mike:* When will Mom get home?
   Bill: About five o'clock.

	A	B
Round 1	When will Godot arrive?	Godot will arrive at midnight.
Round 2	1. None of Godot's friends are awake at midnight. 2. Godot will only arrive at midnight. 3. Godot never travels with enemies. ∴ Godot will be alone at midnight.	Are you eager to see him?
Round 3	Yes.	1. You are eager to see him. 2. He arrives at midnight. ∴ You are eager to see midnight.

    Mike: Then we had better go now.
    Bill: Yeah, I see what you mean. The game we're going to will take 3 hours, it's now one o'clock, and it's a 20 minute drive to the park. And Mom would kill us if we weren't here when she got home!
    Mike: Exactly. And I don't want to die.
    Bill: Neither do I.

* 2. *Vida:* I'm going to have to work harder if I want to get into medical school.
    Gary: You didn't do well this semester?
    Vida: No, my grades were lousy.
    Gary: I hate to remind you, but without good grades a person doesn't stand a chance of getting in.
    Vida: That's my whole point.

  3. *Stewart:* I'm concerned about our nation's wilderness areas, Mo.
    Mo: So am I. Did you know that every day that passes another squirrel dies?
    Stewart: I know, I know. And every squirrel's death diminishes us all.
    Mo: Do we really want to be diminished?

* 4. *Fan A:* I don't think the Yankees are going to do it this year.
    Fan B: Ah, you doubting-Thomas types bug me. We got hitters, we got pitchers, we got the best catcher! What more do we need to get the pennant?
    Fan A: Something you apparently haven't even thought of.

Fan B: What?
Fan A: The Force.
Fan B: What? What are you talking about?
Fan A: The Universal Force. You know, *Star Wars*. You know.
Fan B: (edging away nervously) Pal, you're weird.

5. *Ginger:* This is a great day for the beach!
   Ducky: Yeah—let's get there quick, 'cause it's going to be packed.
   Doreen: I don't think so. It's a workday.
   Ducky: Yeah, but it's summer.
   Ginger: And it's going to be really smoggy and hot!

(In this conversation, as in most conversation, many premises go unstated.)

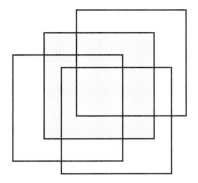

3

The Nature of Logical Rules

3.1 Conversational Orientation

In Chapter 2, we defined the key notion of an argument, and we saw that arguments can be used for different purposes: to establish, to explain, to justify, to persuade, and to amuse. Indeed, an argument may be used to accomplish several of these purposes at the same time.

More generally, conversations (which are composed of arguments, questions, jokes, statements, and so on) serve many purposes and have many orientations. The four most common conversational orientations are information, action, emotion, and aesthetics.

Conversations are **information-oriented** when they are directed primarily at exchanging or acquiring information. Some obvious examples are giving the details of a crime to a police officer, discussing the latest sales report with a coworker, delivering a lecture on an academic subject, and discussing the poor performance of your car with a mechanic. Some less-than-obvious examples of information-oriented dialogues are: a scientist carrying on a research program directed at solving a problem, your inner dialogue directed at resolving a problem about which you are undecided, and a judge deciding a case.

Conversations are **action-oriented** when they are directed at persuading people to act. The people being moved to action need not be participants in the conversation. Here are a few examples of action-oriented conversations: a political candidate working a crowd, a salesperson trying to make a sale, and an advertiser on TV pitching a product.

Other conversations are **emotion-oriented**, that is, aimed at expressing and inducing emotions. Examples of emotion-oriented conversation are expressing condolences and reciting love poetry.

Finally, some conversations are **aesthetics-oriented**, that is, intended to entertain, amuse, or perform artistically. Comedy routines, poetry recitals, and plays fall into this category.

Sometimes what we say or write is focused in one or two of these areas, excluding the others. Thus, a medical report, however informative, is rarely pretty prose. Love poetry is often uninformative (beyond telling who loves whom). A political harangue is primarily intended to get the audience to act in some way. A crossword puzzle does not move us to tears.

Quite often, however, what we say includes all four dimensions. For example, a good speech surely aims to inform, but it also aims to amuse, to inspire, to stir—and to get people to act. To get people to fight for their country, we must do more than dryly state the facts. To convince yourself, reread Lincoln's Gettysburg Address, and restate it in purely factual terms.

A focus throughout this book will be on information-oriented conversations. We will see that such conversations impose rules upon the participants.

EXERCISES 3.1

In each of the following passages, the discourse is focused in one of the conversational orientations discussed in this section: information, action, emotion, or aesthetics. Identify the primary orientation or purpose of each. (An asterisk indicates problems that are answered in the back of the book.)

   1. My argument is that some thinkers long dead have defined the tasks of today's government, and their definition is inadequate and, in the long run, dangerous.

—George Will, *Statecraft and Soulcraft: What Government Does.*

*   2. Graduation is not automatic on the completion of requirements. Students who intend to graduate must take the initiative. When they believe that they are eligible, they should file an application with the Evaluations Office, Administration Building. The Class Schedule each semester specifies the exact date. An application fee of $19.00, which is non-refundable, is required. After the degree is granted no changes can be made in the undergraduate record.

—San Diego State University General Catalog.

3. One guy sees a drunk looking for something on the ground. The guy asks the drunk, "What are you looking for?" "My keys," the drunk replies. So the guy gets down on his knees, too, and starts looking. Pretty soon he asks the drunk, "Where exactly did you lose them?" The drunk replies, "In my house." "Then why are you looking here?" the guy asks, getting mad. The drunk answers, "You stupid jerk—there's more light here!"

4. The Lord is my shepherd; I shall not want. He maketh me to lie down in green pastures; he leadeth me beside the still waters.

He restoreth my soul; he leadeth me in the paths of righteousness for his name's sake.

Yea, though I walk through the valley of the shadow of death, I will fear no evil: for thou art with me; thy rod and thy staff they comfort me.

Thou preparest a table before me in the presence of mine enemies; thou anointest my head with oil; my cup runneth over.

Surely goodness and mercy shall follow me all the days of my life; and I will dwell in the house of the Lord for ever.

—Psalm 23.

5. Pretzel (pret's l) n. A glazed biscuit, salted on the outside, usually baked in the form of a loose knot or stick.

—The American Heritage Dictionary.

* 6. The Little Girl and the Wolf

One afternoon a big wolf waited in a dark forest for a little girl to come along carrying a basket of food to her grandmother. Finally a little girl did come along and she was carrying a basket of food. "Are you carrying that basket to your grandmother?" asked the wolf. The little girl said yes, she was. So the wolf asked her where her grandmother lived and the little girl told him and he disappeared into the woods.

When the little girl opened the door of her grandmother's house she saw that there was somebody in bed with a nightcap and nightgown on. She had approached no nearer than twenty-five feet from the bed when she saw that it was not her grandmother but the wolf, for even in a nightcap a wolf does not look any more like your grandmother than the Metro-Goldwyn lion looks like Calvin Coolidge. So the little girl took an automatic out of her basket and shot the wolf dead.

Moral: It is not so easy to fool little girls nowadays as it used to be.

—James Thurber, The Thurber Carnival.

7. One backlash effect on Affirmative Action is more rights for males. For instance, in December 1973 a three-judge Federal court in Newark, New Jersey, ruled that a widower is entitled to collect widow's Social Security benefits. So if in doubt, keep asking questions and raising objections

for you may, like the immigrants of old, have rights you are not aware are yours.

—Robin Higham, *The Compleat Academic.*

8. Now let us consider what happened to different kinds of bivalves during the Paleozoic-Mesozoic transition. For our purposes we may consider a bivalve as an animal with three main components: (1) a pair of siphons, one of which takes up food-laden water, while the other serves for discharge; (2) a pair of gills, provided with cilia, which serve as both filter and pump; (3) palps, to either side of the mouth, which help sort and transfer food, and which can also take it up in some forms. Other important organs include the well-known bivalved shell, closed by a pair of strong adductor muscles, and a foot, used in burrowing, crawling, and attachment.

—Michael Ghiselin, *The Economy of Nature and the Evolution of Sex.*

9. In the high-resolution graphics mode, an image is generated on the screen as an arrangement of dots. The term **high resolution** stems from the fact that greater visual fidelity can be obtained through dots than with small rectangles.

In the high-resolution graphics mode, the screen is regarded as a matrix 280 dots wide and either 160 or 192 dots high. The rows are numbered 0 through 159 or 0 through 191, and columns are numbered 0 through 270. High-resolution graphics consists of lighting up the appropriate dots to create the desired visual image. As with low-resolution graphics, screen positions are numbered from left to right and from top to bottom.

—Harry Katzan, *Microcomputer Graphics.*

* 10. Helen, thy beauty is to me,
like those Nicean barks of yore;
That, gently, o'er a perfumed sea,
the weary, way-worn wanderer bore,
to his own native shore.

—Edgar Allen Poe, *To Helen.*

11. The recommendations in this chapter are designed to strengthen every link in the chain of educational authority: parent/student; parent/school board; school board/superintendent; superintendent/principal; principal/teacher; teacher/student. Wherever you, the reader, find yourself in this chain, I urge you to do your part to help America's students receive the education they need, and deserve.

—Paul Copperman, *The Literacy Hoax.*

12. As a black hole cannot be seen directly, its detection depends upon spotting secondary effects. Having the mass of a star, it will produce intense gravitational disturbance among its neighbors. This is especially true of a black hole in a so-called binary system. A great many stars in our galaxy are not wandering about individually, but are grouped in twos or even threes. Two stars can remain close together, without falling into one another, by orbiting around their common centre of gravity. Binary star systems are

known with orbital periods varying from hours to years, depending on the separation between the components.

—Paul Davies, *The Edge of Infinity*.

13. To extend this analogy, the perspective from which we wrote the last chapter is in many ways similar to that of the mapmaker. The facts and figures we included prove that our perspective has some validity. But they provide a less than adequate picture of the modern world simply because their statistical accuracy does nothing to flesh out the description. If you are not accustomed to seeing the world as an information environment, we may have to work harder to change your perspective.

For that is our aim. We want to take your individual perspective on the modern world, and lay over it, like an extra filter, the concept of the information environment. The information environment does not skulk in a corner of the modern world. It is enmeshed with almost every aspect of it, and understanding its importance should change your entire concept of the world.

—Gordon Pask, *MicroMan: Computers and the Evolution of Consciousness*.

* 14. It is a common experience to fear that the admirations of youth will wear thin, and precisely because *1984* had so enormous an impact when it originally came out more than thirty years ago, I hesitated for a long time before returning to it. I can still remember the turbulent feelings—the bottomless dismay, the sense of being undone—with which many people first read Orwell's book. My fear now was that it would seem a passing sensation of its moment or even, as some leftist critics have charged, a mere reflex of the cold war. But these fears were groundless. Having reread *1984*, I am convinced, more than ever, that it is a classic of our age.

—Irving Howe, *1984 Revisited: Totalitarianism in Our Century*.

15. The fundamental physics governing the structure of the sun is a balance of forces. Any two particles are attracted to each other by gravity, so the mutual gravitational attraction of all the material making up the sun pulls inward, trying to make the sun collapse on itself. But gravity is not the only force at work in the sun. Any gas at a temperature above absolute zero exerts pressure. Therefore, the hot gases that make up the sun exert pressure, which tends to make the sun expand. These two forces, the gravity pulling in and the pressure pushing out, balance, and the sun remains stable. It is, in fact, a self-regulating system.

—Robert Rood and James Trefil, *Are We Alone? The Possibility of Extraterrestrial Civilizations*.

3.2 Two Types of Evidential Relations

In Chapter 2 we defined the term 'argument' much more narrowly than often is done. We said that an **argument** consists of a set of statements, of which one is the conclusion, and the rest are the premises.

We learned to identify arguments, both singly and in webs. But the main task of logic is to evaluate arguments: to figure out which are good (and why) and which are bad (and why). How can this evaluation be done? What exactly are we looking for?

Consider this argument:

1. All frogs are streetfighters.
2. All streetfighters are devoted to nonviolence.

∴ All frogs are devoted to nonviolence.

Is this argument 'good'?

In one sense, it is not. From a factual point of view, the previous argument is ridiculous. Frogs can hardly be streetfighters, carrying knives and such. Similarly, a streetfighter can hardly be considered a person devoted to nonviolence. But in another sense, the argument is good: the premises would establish the conclusion if they were, in fact, true.

As logicians, we will be interested in the structure rather than the factual content of arguments. Given an argument, we look to see not if the premises are true, but whether they would, if true, support the conclusion. We want to know what sort of *evidential relation* (if any) holds between the premises and the conclusion of the given argument. **Logic** is the study, not of the factual content of arguments, but of the evidential relations between premises and conclusions of arguments.

A question arises here. How many types of evidential relations exist? One? Two? Several? Infinitely many? The consensus among logicians is that just two types of evidential relations exist. Consider these two arguments:

1. All frogs are green.
2. There is a frog in my briefcase.

∴ The frog in my briefcase is green.

1. Almost all frogs are green.
2. There is a frog in my briefcase.

∴ The frog in my briefcase is green.

Are both these arguments 'good'? The question again is not whether the premises are true, but rather whether those premises, if true, would support their respective conclusions.

We may have doubts about the second argument. Clearly, the first argument is logically acceptable; but the second? Consider it further.

3.2 / Two Types of Evidential Relations

Granted we would not stake our life on the truth of the conclusion, even if we were convinced of the truth of the premises. Still, if we were betting, we would bet that the conclusion is true. We would probably win the bet.

This distinction—between arguments whose conclusions follow absolutely from their premises and those whose conclusions follow only probabilistically from their premises—is what we support by the following definitions.

An argument is **deductively valid** if it is impossible for the premises to be true and the conclusion false.

An argument is **inductively strong** if it is not impossible, but is unlikely, that the conclusion would be false given that the premises are true.

Since logic is the study of the evidential relations between premises and conclusions of arguments, and since there are exactly two such relations (validity and strength), then we have two branches of logic: deductive logic and inductive logic. **Deductive logic** is the study (including analysis and assessment) of arguments from the point of view of validity. The deductive logician is concerned to figure out what validity is and how to detect it. **Inductive logic** is the study (again including analysis and assessment) of arguments from the point of view of inductive strength. The inductive logician is concerned with the notion of probability, what it is and how to measure it.

Let us consider several useful points regarding deductive validity and inductive strength. Truth and validity are different notions, although they are related. It is a mistake to say (as people often do), "so-and-so's argument is true" or "so-and-so's statement is valid."

To illustrate the point, consider these arguments:

1. 1. All cats are mammals.
    2. All mammals are animals.
    ∴ All cats are animals.

2. 1. All cats are good poker players.
    2. All good poker players are gamblers.
    ∴ All cats are gamblers.

3. 1. All cats are fish.
    2. All fish are huge.
    ∴ All cats are huge.

4. 1. All cats are fish.
    2. All fish are mammals.
    ∴ All cats are mammals.

Each of these four arguments is valid. Of each, it is correct to say that if the premises were true, then the conclusion also would have to be true. Indeed, they all share the same structure:

1. All A are B.
2. All B are C.

∴ All A are C.

But the truth values of the statements in each vary. In the first argument, both the premises are true, and so is the conclusion. In the second argument, the first premise is false but the second is true, while the conclusion is clearly false. In the third argument, both the premises and the conclusion are false. In the fourth argument, while the premises are false, the conclusion is surely true.

Thus our first point is that an argument can be valid without consisting of true statements. In that important sense, the notions of 'truth' (which applies to statements, not arguments) and of 'validity' (which applies to arguments, not statements) are different. But they are related: by the definition of validity, every combination of truth values is possible in a valid argument, except the premises being all true and the conclusion false.

Put another way, the fact that an argument is valid does not guarantee that the conclusion is true. Validity only guarantees that we will not be taken from true premises to false conclusions. Thus we prefer our arguments to be more than just valid: we want them to be sound. An argument is **sound** if it is valid and, in fact, has true premises.

Notice how easy it was to describe those valid arguments by using the symbols A, B, and C to characterize their structure. This is the crucial insight that underlies symbolic logic. **Symbolic logic** is the study of the evidential relations between premises and conclusions of arguments, by means of symbolic techniques. We will discuss symbolic logic in Chapter 9.

The second point to make is that no correlation exists between the nature of the evidential relation and the degree of generality of the statements in any given argument. At one time, logic books spoke of inductive and deductive reasoning. 'Inductive reasoning' was defined as something that allegedly goes on in the empirical sciences, and 'deductive reasoning' as something that supposedly goes on in mathematical work. It was believed that inductive reasoning was reasoning from particular statements to general statements. Thus the paradigm of inductive reasoning was generalization, as in this example:

3.2 / Two Types of Evidential Relations

1. Frog no. 1 is green.
2. Frog no. 2 is green.
   •
   •
   •
n. Frog no. n is green.

∴ All frogs are green.

On the other hand, it was said that deductive reasoning was reasoning from general rules to particular statements. The following example is typical.

1. All frogs are green.
2. Fred is a frog.

∴ Fred is green.

In the previous example, the first premise, at least, is 'general.' Unfortunately, this neat account is wrong. For one thing, it confounds the questions about **logic** (the normative study of evidential relations between statements) and questions about **scientific method** (the study about how scientists should do their work). But even from the point of view of logic, the traditional account is mistaken. It is mistaken because no connection exists for the nature of the evidential relation between statements and their generality or particularity, as the examples in Table 3.1 show.

Logic studies evidential relations, and these are of two types, having nothing to do with the degree of generality of the statements involved.

One last point. Inductive strength is a matter of degree. The premises can make a conclusion quite probable, somewhat probable, or only slightly probable. Validity is an all-or-nothing relation: an argument either is or is not valid. But the premises of an argument may offer inductive evidence ranging from near zero to near certain, as summarized in Figure 3.1.

Figure 3.1 The Scale of Evidential Strength

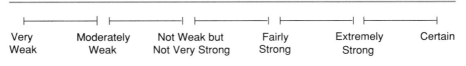

Table 3.1

Valid Arguments	Strong Arguments
Deductively valid, particular to particular:	Inductively strong, particular to particular:
1. Fred is rich and ugly.	1. The man stood over the corpse with bloody knife in hand.
∴ Fred is rich.	∴ The man may have killed the victim.
Deductively valid, particular to general:	Inductively strong, particular to general:
1. Kermit is the only talking frog in the universe, and he is talented.	1. Frog no. 1 is green. . . . n. Frog no. n is green.
∴ All talking frogs are talented.	∴ All frogs are green.
Deductively valid, general to particular:	Inductively strong, general to particular:
1. All frogs are green.	1. Almost all frogs are green.
∴ If Fred is a frog, then he is green.	∴ If Fred is a frog, then he is green.
Deductively valid, general to general:	Inductively strong, general to general:
1. All rats are rodents. 2. All rodents are animals.	1. Mammals and marsupials occupy similar ecological niches.
∴ All rats are animals.	∴ There are species of each that are morphologically similar.

EXERCISES 3.2

Each of the following arguments is already in standard form, and none are enthymemes. Relying on logical intuition, determine of each whether it is valid, strong, or neither valid nor strong. (An asterisk indicates problems that are answered in the back of the book.)

1. 1. All mice love cheese.
    ∴ All mice love crackers.

*2. 1. All mice love cheese.
    2. Anything that loves cheese loves wine.
    ∴ All mice love wine.

3. 1. Almost all mice love cheese.
    2. Anything that loves cheese loves crackers.
    ∴ Almost all mice love crackers.

4. 1. If Nicole were a money-grubber, she wouldn't be dating Gary.
    2. Nicole is dating Gary.
    ∴ Nicole is not a money-grubber.

5. 1. Kelli and Nicole are sisters.
    ∴ Kelli and Nicole are nice to one another.

*6. 1. If Juan were lonely, he'd call Nicole.
    2. He called Nicole.
    ∴ Juan is lonely.

7. 1. Either Kelli or Nick stole the book.
    2. Kelli didn't steal it.
    ∴ Nick stole it.

8. 1. We pulled 50 slips of paper from that bucketful of slips of paper, and 95% of them were green.
    ∴ The next slip of paper will be green.

9. 1. All of the slips of paper drawn were green.
    ∴ The first was green.

*10. 1. I have rolled these dice 12 times, and number 7 has come up each time.
    2. The dice are not loaded.
    ∴ The next roll will probably not come up as number 7.

3.3 Is There a Third Evidential Relation?

In Section 3.2, we defined the two evidential relations of deductive validity and inductive strength, and claimed that these were the only evidential relations. However, how do we know that no other types of evidential relations exist? In this section, we address only the question of a third evidential relation (besides validity and strength).

We might ask, then, whether a third relation needs to be distinguished from validity and strength. Two candidates with imposing names have been mentioned: deontic sufficiency and retroductive strength. We will examine each in turn.

The best way to introduce the notion of *deontic sufficiency* is to imagine the following situation.

> Kim is visiting her brother, who lives in the forests of Colorado, where large bears can be found. She decides to take a hike, and her brother tells her that a hiker's path is the most scenic walk. He also informs her that the path continues for about a mile but then has a branch. One branch leads back to town, and the other leads to a dead end. He does not remember which branch is which, however.
>
> Kim decides to try the path. As she walks along, she hears something following her. Turning around, she sees a large bear, which runs toward her. She turns and runs, very quickly reaching the fork in the path her brother had told her about. She does not have much time to decide which path to take. A glance at the paths reveals that the grass on the path to the left is more trampled. She decides to take it.

The advocate of the third evidential relation called 'deontically sufficient' would say that the argument behind Kim's choice was this:

1. Kim has to choose one of those two branch paths.
2. The grass is more trampled on the left path.
3. Grass being trampled may indicate that it is the path to town.

∴ She should take the left path.

We might accordingly define an argument as being **deontically sufficient** if, and only if, it is reasonable for a person to adopt the conclusion, given that those premises are true. The information in the premises of a 'deontically sufficient' argument would be enough to enable us to choose the course of action indicated in the conclusion, even though not logically sufficient to warrant asserting the conclusion as certain or even probable.

However, we might ask whether the advocate of this "new" type of evidential relation is not confounding the use of an argument with the strength of evidence in it. We have an argument here (expressed enthymematically):

1. The grass on the left path is more trampled.
2. Grass is usually trampled down by people's feet.
3. People tend to walk to towns rather than dead ends.

∴ The left path leads to town.

Now that argument has some degree of inductive strength, which presumably can be measured. Recall that inductive strength ranges from comparatively weak to comparatively strong. The choice of an alternative is based on the best evidence, even if the best evidence is weak. But the use of the argument does not alter the nature of the evidential relationship between the premises and the conclusion.

Consider next the notion of *retroductive strength*. If deductive validity involves certainty, and inductive strength involves probability, should we not distinguish a third evidential relation that involves mere plausibility?

Another example will make clear the motivation for such a proposal.

1. This mold culture plate exhibits an unusual structure.
2. This structure could be explained if we postulate the existence of a bacteria-killing substance given off by the mold.

∴ The mold does give off such a bacteria-killing substance.

The arguer does not put the conclusion forward as certain or even probable, but instead as 'plausible,' that is, worthy of test or of further investigation.

Thus (it might be argued), we should define an argument as being **retroductively strong** if and only if the conclusion is plausible, given that the premises are true.

Again, however, we should distinguish the use of an argument—here, to get people to investigate a hypothesis, as opposed to accepting it unreservedly—from the nature of the evidential relation involved.

Even if the strength of evidence in retroductive arguments is generally weak, it would not follow that the nature of the evidential relation in a retroductive argument is of a special sort. Why not just say the obvious: retroductive arguments are inductive arguments, but comparatively weak ones? Naturally, it takes more evidence to prove a hypothesis than to justify taking it seriously (considering it worthy of testing), but the nature of the evidential relation is the same in proposing as in testing.

Consider this example: suppose Fred's statement that he just saw Bo at home suggests to Ivan that Bo indeed is at home. Ivan may verify his hypothesis (in the ordinary sense of 'verify') by asking several other individuals who have just returned from Bo's house. If they testify that Bo is at home, Ivan's evidence for his hypothesis consists not only of their testimony, but of Fred's as well. The initial evidence (testimony) continues to be evidence for the claim, and forms part of the decisive proof (the totality of all the witnesses' testimony).

To sum up this section, no reason appears to deny what most logicians take for granted, namely, that, at most, two different types of evidential relations occur between premises and conclusions of arguments. We will adopt this view.

3.4 Modalities

Logic texts often define a 'deductive argument' as one that purports to be deductively valid, and an 'inductive argument' as one that purports to be inductively strong. But these are not useful definitions. Arguments do not "intend" or "claim to be" anything (only people can make claims). Since people may not even know the difference between inductive strength and deductive validity, they may not make any claim about the strength of evidence in their arguments.[1]

But you can see why some texts cling to those definitions. How else are we going to take account of the undeniable fact that people consciously intend to state valid arguments in situations that call for valid arguments (such as mathematical proofs)?

We will take the view that while arguments do not include claims about the evidential relations holding between the premises and conclusions, people often (though not always) offer arguments together with modalities. A **modality** (for a statement) is a claim about the degree of confidence the speaker has in that statement. Regarding an argument, the modality attached to the conclusion usually indicates the arguer's belief about the strength of evidence the premises furnish for the conclusion. Modalities, in effect, are statements about arguments, in that they describe the strength of argument.

How can you identify the modality given (supposing one is given at all) with an argument? Again, our language contains indicator words

---

[1] Bryan Skyrms makes this point in *Choice and Chance,* 2nd ed. (Belmont, CA: Dickenson, 1979), p. 12.

3.4 / Modalities

that signal how the argument-giver views the strength of evidence in his argument. Table 3.2 lists the most common modality indicators.

As an example, the passage:

> I think it is very likely that Kelly is ill. After all, she is cross, won't eat, and keeps crying.

contains the argument:

1. Kelly keeps crying.
2. Kelly is cross.
3. Kelly won't eat.

∴ Kelly is ill.

and the modality:

Table 3.2 *Modality Indicator Words*

Words		Meaning
Be sure, surely Be obvious, obviously Be evident, evidently Be certain, certainly Be clear, clearly Has to be, has to follow Must, automatically follows	→	Indicate that the speaker has no doubts about the claim
Think Suppose Believe Probably, likely Presumably Supposedly, chances are Should	→	Indicate confidence yet show some doubt
Guess, conjecture Seem Can, could May, maybe, might Possibly Perhaps Conceivably Plausible that	→	Indicate more doubt

This argument is inductively strong.

Beware of taking a heuristic rule (about some class of indicator words) as an absolute hard-and-fast rule. Modal words can be used for several purposes besides expressing modalities. For example:

You really must try this cake!

is simply a strong encouragement to try the cake, and

God necessarily exists.

attributes a special property ('necessary existence') to God.

EXERCISES 3.4

Each of the following passages contains an argument, and most of the passages also contain modalities. Put each argument in standard form, and if the argument is accompanied by a modality, state it explicitly. (An asterisk indicates problems that are answered in the back of the book.)

1. The health hazards arising from such air pollution have been less clearly established. . . . Nevertheless, the last two or three years have also brought powerful new evidence of the risks to health occasioned by unwanted exposure to cigarette smoke, or 'passive smoking' as it is popularly termed. Acute exposures to smoke-filled rooms have reduced the effort tolerance of patients with angina pectoris; chronic exposure to smoke-polluted air has apparently caused respiratory disease in young children and lung cancer in adults; while placental transfer of tobacco products has led smoking mothers to bear premature, under-weight infants with continuing abnormalities of growth and development.

—Roy Shephard, *The Risks of Passive Smoking.*

* 2. The kind of structural psychological shifts that we think necessary cannot be sufficiently implemented in one generation. Women and men currently carry with them deep feelings of misogyny and unconscious sexism. Even with changes in child-rearing arrangements, these influences will have their impact on at least the first generation raised by two parents.

—Luise Eichenbaum and Susie Orbach, *Understanding Women.*

3. Feminists in particular may rebel at the thought of looking to the science of biology for information that bears on the human condition. They

may be put off by the fact that among our nearest relations, the other primates, the balance of power favors males in most species.

—Sarah Blaffer Hrdy, *The Woman that Never Evolved*.

4. Perhaps the only sensible course is to treat computers as we treat our fellow human beings, and work on the assumption that they mean us no harm until the opposite is proven.

There are grounds for optimism. Our experience so far tells us that a reasonably well designed computer system makes a complex process which has to be kept within strict limits safer than it would be if it were controlled by a human alone. Of course computers can break down and programs can exhibit unforeseen bugs. But human beings are even less trustworthy; their attention wanders, they like to think all is well when it manifestly is not, and they react slowly and irrationally in moments of crisis.

—Gordon Pask, *Micro Man: Computers and the Evolution of Consciousness*.

5. Estimates put the number of supernova events per galaxy at around three per century, though the last to be recorded in our own galaxy was as long ago as 1604. It is not known how many such events produce black holes rather than neutron stars, but it would be surprising if it were not a fair proportion. It therefore seems reasonable that, over the ten billion years that our galaxy has existed, millions of black holes have formed from the death throes of old stars.

—Paul Davies, *The Edge of Infinity: Where the Universe Came from and How It Will End*.

* 6. Since an open universe is less dense than a closed one, more of the deuterium emerges from the fireball unscathed. If the universe is closed, then we should expect to observe only a fraction of the 20 to 30 parts per million that we do see. Unless there is some other way to explain how this deuterium managed to survive, we must conclude that the universe is open.

—Richard Morris, *The Fate of the Universe*.

7. The infinite, open model is predicted to show smaller change in the galaxy recessional velocities, for in this case the deceleration of the Universe is less.

What do the data indicate? Is there any evidence for a faster recessional velocity among the more distant galaxies? In a nutshell, the most distant galaxies do seem to show a substantially greater recessional velocity than those nearby. The accuracy of these observations is rather poor, however. The most distant galaxies are obviously faint, and observations of them are notoriously hard to make. Nonetheless, this second cosmological test suggests that the Universe is closed and finite.

—Eric Chaisson, *Cosmic Dawn: The Origins of Matter and Life*.

8. If we read on in *Mein Kampf*, we find that Hitler gives us a description of a child's life in a lower-class family. He says:

> Among the five children there is a boy, let us say, of three. . . .
> When the parents fight almost daily, their brutality leaves nothing
> to the imagination; then the results of such visual education must
> slowly but inevitably become apparent to the little one. Those who

are not familiar with such conditions can hardly imagine the results, especially when the mutual differences express themselves in the form of brutal attacks on the part of the father toward the mother or to assaults due to drunkenness. The poor little boy, at the age of six, senses things which would make even a grownup person shudder. . . . The other things the little fellow hears at home do not tend to further his respect for his surroundings.

In view of the fact that we now know that there were five children in the Hitler home and that his father liked to spend his spare time in the village tavern where he sometimes drank so heavily that he had to be brought home by his wife or children, we begin to suspect that in this passage Hitler is, in all probability, describing conditions in his own home as a child.

—Walter C. Langer, *The Mind of Adolph Hitler.*

3.5 Rules of Dialogue

Discourse, in the broadest sense of the term, is any oral or written expression between people. In this broad sense, arguments are but one type of discourse, as against questions, expressions of sympathy, promises, and a thousand other types. We are primarily interested in arguments as opposed to other forms of discourse.

But any argument takes place within a broader interaction between people. We are about to discuss rules of logic, and the proper way to begin is with the rules that more broadly govern discourse between people.

Recall from Section 3.1 that conversations (discourse between people) can be oriented toward or can be concerned with any number of purposes, including amusement, garnering information, moving people to action, expressing emotion, and any combination of those purposes. Let us focus on just one sort of conversation, namely, conversation that is oriented solely at acquiring and exchanging information. We shall use the term **dialogue** to mean an exclusively information-oriented conversation. What then are the logical rules governing dialogues?

The rules governing information-oriented conversation are easy to discover. They are common sense. One philosopher[2] has given the following examples of rules of dialogue:

1. Make your contribution to the dialogue as informative as required.
2. Do not make it more informative than is required.

[2] H.P. Grice, "Logic and Conversation," in *The Logic of Grammar,* edited by Donald Davidson (Encino, CA: Dickenson Publishing Co., 1975).

3. Do not say what you believe false.
4. Do not say that for which you lack adequate evidence.
5. Be relevant.
6. Avoid obscurity in what you say.
7. Avoid ambiguity.
8. Be brief.
9. Be orderly.

We could add many other such rules, such as:

10. When you use a modality that leads people to expect a deductively valid argument, do not substitute one that is merely inductively strong.
11. Do not raise a question that has already been answered adequately.
12. Do not ask a question that presupposes that which you know is false.

Rules of dialogue fall into two groups, 'global' and 'local.' *Local rules* are rules that govern the immediate response to someone's comment, or to use terminology introduced in Chapter 2, that govern the acceptability of responses in a given round. The *global rules* (to put it roughly) govern how the rounds fit together. Rules 10–12 are examples of global rules of dialogue.

In practice, the local rules of dialogue are of most interest. The most important of these rules are those governing the acceptability of arguments. Formulating these rules is the main business of the deductive and inductive logician. Since formulating logical rules will be crucial in what follows, we need to discuss how logical rules are discovered, and what form they take.

3.6 Refutation by Counterexample and Logical Analogy

Let us take up the question of how rules of logic can be discovered or formulated. Suppose someone says:

1. Women who choose elementary school teaching as a career love children.
2. But women who love children want to have children of their own.

∴ All women want to be mothers.

This argument can be refuted by pointing to a counterexample. A **counterexample** is a realistic (although possibly imaginary) case for which the premises are true, but for which the conclusion is false. In the previous example, we might point to a woman who is a research scientist and quite happy to forgo motherhood. Since the first premise only speaks of women teachers, and the second only of those who love children, they may be true but yet don't apply to our case. But the conclusion does, since it speaks of all women, and the conclusion is not true of our case. So our case is indeed a counterexample.

A more flexible method for refuting arguments is refutation by logical analogy. **Refutation by logical analogy** is the technique of giving an argument of the same logical form as the one you are trying to refute and that clearly has true premises and a false conclusion. For example, suppose you favor rent control, and your opponent argues:

1. You favor rent control.
2. Communists favor rent control.

∴ You are a Communist.

You can overturn the argument by giving this logical analogy:

1. You oppose rent control.
2. Fascists oppose rent control.

∴ You are a Fascist.

The key to refutation by logical analogy is finding an argument of 'the same form.' But two problems here call for a more scientific approach. First, what if you cannot devise an analogy or counterexample? These techniques are limited by our power of imagination—something that may be in short supply when dealing with a complex argument. Moreover, the use of logical analogy depends upon an opponent conceding that your refutation has the same logical form. But what exactly is logical form?

Considerations such as these lead to the view that logic should assess arguments by focusing upon form rather than content, and this means getting clear on the concept of form, what it is and how to represent it. In addition, the logician seeks to assess arguments by using some system of rules, some *organon* (a Greek word that means "tool"), which eliminates the need for less dependable techniques such as 'imagination.' The extent to which we can replace adventitious imagination by automatic rules is a great question for logic, and it is not an easy one to answer. But we should be clear on why logicians seek rules: rules make assessments more achievable and more consistent.

EXERCISES 3.6

Refute the following arguments first by counterexample, then by logical analogy. (An asterisk indicates problems that are answered in the back of the book.)

   1. Dogs are friendly. Ronald is friendly. So Ronald is a dog.

* 2. If Joni loved Roberto, she would act friendly toward him. But she does not love Roberto. So she will not act friendly to him.

   3. Kim, Kelly, or both went to the game. Kim went. So Kelly must not have gone.

   4. If Andrea were rich, she would own a car. So if she owns a car, she must be rich.

   5. If Andrea were rich, she would live in Hawaii. If Andrea were rich, she would drive a Porsche. So either Andrea will move to Hawaii or else buy a Porsche.

* 6. Kim is going to study either medicine or law. So she is going to study law.

Refute the following arguments by logical analogy. Why can't they be refuted by counterexample?

   7. All mammals are animals. All cats are animals. So all cats are mammals.

   8. Some dogs are friendly. Some dogs are smart. So some dogs are smart and friendly.

   9. All dogs are territorial. Some dogs are vicious. So all vicious things are territorial.

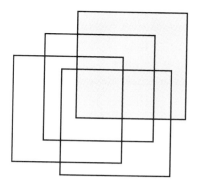

4

Fallacies of No Evidence

4.1 The Concept of a Fallacy

In this chapter, we begin our job of learning how to assess arguments and dialogues. We do this in an indirect fashion, by learning the most common ways people reason badly. Sometimes people do so deliberately, sometimes unconsciously; but irrespective of their motives, we should be able to detect their errors in reasoning.

We are going to examine fallacies. By the term 'fallacy,' people often mean a mistaken belief—as when someone says that it is a 'fallacy' to think that women do not make good police officers. But what we will mean by **fallacy** is a type of incorrect reasoning. The fallacies we shall examine have an added feature: they are often plausible. Some fallacies are so obvious that nobody could ever be fooled by them, and therefore they are not worth spending time on.

To judge an argument requires the same kind of fine discrimination as is required to judge a movie or a person. To judge arguments, movies, food, or people is an exercise in practical reasoning. The most striking fact about practical reasoning is that no automatic and precise rules govern it. We cannot state in precise terms what makes a person good (how much generosity is good? How much thriftiness? Friendliness?). Similarly, telling fallacy from good argument requires fine discrimination.

Good practical reasoning thus requires considerable practice. We judge on the basis of past cases, precedent being as important as rule.

For this reason, each fallacy we discuss is illustrated by several examples. The examples you meet in Chapters 4–6 are genuine: they were taken from newspapers and magazines. The only changes we have made were altering the names (to avoid giving unnecessary offense), eliminating profanity, and filling in the context surrounding the controversies.

How many fallacies are there? How many standard mistakes do people commonly make in reasoning? Hundreds, and some books cover dozens of them. But we will be much less ambitious, focusing upon only the most common fallacies.

In the list of fallacies we cover, each label will be defined and illustrated in such a way as to make it distinct from the others. But, as with the terms 'blue' and 'gray,' cases will occur in which we could apply correctly more than one label. The point is not to get overly concerned with labels, but to use the labels to help detect flaws in reasoning.

To categorize fallacies, perhaps the easiest division is into three broad groups: fallacies of no evidence, fallacies of little evidence, and fallacies of language. *Fallacies of no evidence* are fallacies in which no evidence at all is presented. We cover such fallacies in this chapter. *Fallacies of little evidence* are fallacies in which inherently flawed evidence is given. We discuss these in Chapter 5. Finally, *fallacies of language* are fallacies involving some abuse of language. These we discuss in Chapter 6. We consider an extended example in Chapter 7.

The fallacies of no evidence we discuss in this chapter are (1) pooh-poohing; (2) shifting the burden of proof (including appeal to ignorance); (3) attacking the person; (4) appeal to fear; (5) appeal to pity; (6) appeal to the crowd; (7) ignoring the issue; (8) loaded question; (9) begging the question. These fallacies are grouped together and defined in Sections 4.2 through 4.5.

4.2 Fallacies of Complete Evasion

The first two fallacies on our list describe ways in which people can completely evade or run away from their obligation to prove their point.

The fallacy of **pooh-poohing** is the fallacy of dismissing, simply brushing aside, someone else's point of view. To pooh-pooh a point is to put it down without logically addressing it.

A person can pooh-pooh a point in two ways. One way is to dismiss the point outright, as in these dialogues.

Reporter: Mr. President, would this law that you propose discriminate against the poor?
President: So what? The poor have a raw deal anyway.

Student: Dean, we are hereby presenting you with our list of grievances.
Dean: Being told what you don't like about the university is about as interesting to me as being told you don't like strawberries!

Opponent of ERA: The Equal Rights Amendment would require the elimination of separate bathrooms for men and women.
Proponent: That's the stupidest thing I have ever heard! I won't even bother to answer that nonsense.

A more subtle way to pooh-pooh a point is to agree with it in general, but then to disagree with the specific case at hand. For example:

Senator: Our automobile industries need protection from the Japanese.
Reporter: But doesn't that conflict with the whole idea of free trade?
Senator: Now look, young lady, nobody is more in favor of free trade than I. But I repeat, our companies need protection!

Politicians often pooh-pooh embarrassing objections to their proposals in this way.

The second fallacy on our list is also a way of running away from our obligation to prove our assertions. The fallacy of **shifting the burden of proof** means trying to make the other person prove what you should prove. For example:

A: Midwesterners are really uncreative.
B: Why do you say such a thing?
A: Well, can you name one creative midwesterner?
B: Uh, no . . .
A: Well, then, . . .

One variety of shifting the burden of proof deserves special mention. **Appeal to ignorance**, called in Latin **argumentum ad ig-**

norantiam (it is prudent to learn the Latin labels for the fallacies because those labels are still widely used and you will not be mystified when you encounter them), is the fallacy of arguing that something must be true because nobody can prove it false (or alternatively, that something must be false because nobody can prove it true). Such arguments involve the illogical notion that we can view the lack of evidence about a proposition as some kind of evidence for it or against it. But lack of evidence is lack of evidence, and it supports no conclusion. An example of an appeal to ignorance:

> Styrofoam cups must be safe; after all, no studies have implicated them in cancer.

This argument is fallacious because it is possible that no studies have been done on those cups, or that what studies have been done have not focused on cancer (as opposed to other diseases).

Here are a few examples of arguing from ignorance that Richard Robinson[1] gives.

1. You believe in immortality?—I have not sufficient data not to believe in it.
2. To say that punishment does not always cause psychic damage is to evade the issue, for we do not know what reaction the punishment will cause in later years.
3. Although this hypothesis leads to somewhat improbable conclusion, there is no reason for rejecting the possibility that it comes more or less near to the reality which is so hard to reconstruct.

In (1), the arguer concludes that he ought to believe in immortality because he does not have the data to refute it. In (2), the arguer concludes that punishment is always harmful, because we have no evidence that it does not cause later harm. In (3), the arguer concludes that the hypothesis is plausible, because we have no reason to reject its possibility.

Let us see why appeal to ignorance is fallacious. When we are ignorant of the truth of proposition p, we should not conclude that not-p is true. Instead, we ought to ask whether the balance of evidence favors p or not-p.

Two seeming 'exceptions' regarding appeals to ignorance are often mentioned. First, if the FBI investigates a person and finds no evi-

---

[1]Richard Robinson, "Arguing from Ignorance," *The Philosophical Quarterly* 21(83) (April 1971), 97–108.

dence that the suspect is a Communist, is it not right for them to conclude the person is not a Communist? Second, in a court of law, if no evidence is presented that proves the defendant guilty, is not the jury obligated to return a verdict of not guilty?

Neither of those cases is an exception; that is, in neither case is it correct to say that the argument form:

1. There is no evidence regarding statement S.

∴ S is true (or false).

is held to be logically acceptable. In the FBI case, the FBI is not arguing:

1. There is no evidence that Jason is a Communist.

∴ Jason is not a Communist.

Instead, it is arguing:

1. If Jason were a Communist, he would probably belong to organization X, Y, or Z.
2. Jason does not belong to organization X, Y, or Z.

∴ Jason is probably not a Communist.

Again, in a court of law, from the lack of evidence proving Jason is guilty, the jury does not conclude he is innocent (that he positively did not commit the crime), but only that he is not guilty in the narrowly legal sense that his conviction is not justified by the evidence presented.

You may have noticed that it was possible to state the general form of the fallacy of appeal to ignorance as an argument pattern:

1. There is no evidence regarding statement S.

∴ S is false.

But we gave no such pattern with pooh-poohing or shifting the burden of proof. Why?

Recall from Chapter 2 the distinction between *arguments* and *dialogues*. Some of the fallacies we discuss are best viewed as types of incorrect (neither valid nor strong) arguments. For those, we will put forward a general description in standard form, using a variable S instead of a particular statement. But other fallacies are better viewed as illogical moves in a dialogue.

In the case of pooh-poohing, for instance, we expect in any rational exchange that the other person will answer questions honestly put. That is an unwritten rule of rational discussion. In the case of shifting the burden of proof, the unwritten rule that gets broken is the rule that the arguer should answer a request for evidence for the claim either by giving evidence or else by ceasing to make the claim. It is illogical to deflect a request for proof by making a counterrequest for proof.

4.3 Emotional Appeals

The next group of fallacies are emotional appeals of various sorts, including appeals to hatred, fear, and pity.

The first of these fallacies is **attacking the person (argumentum ad hominem)**, which is the fallacy of criticizing a person who puts forward a proposal or claim rather than giving evidence to logically refute the person's point of view. Arguing against the person is illogical even if the attack is factually correct. It is illogical because even bad people can be correct in what they say, and to figure out whether a statement is correct we have to look at it, not the person who originated it.

We can distinguish several varieties of personal attacks. One form is the *abusive form*, in which the person's character is attacked. Dismissing a person's claim on the basis of being a 'fascist,' 'pinko,' 'nut,' 'creep,' 'thief,' or any other (alleged) defect in character is to commit the abusive form of this fallacy. Some examples:

> Dear Editor:
> Regarding Fred Boar's claim (see his letter to this paper May 13) that the 55 mph speed limit doesn't save lives, I have this to say: Boar, you are the stupidest jerk I have ever run across. I would expect more smarts from a clump of fungus!

Logically speaking, we may decide whether the 55 mph limit saves lives by looking at the statistics concerning accident rates (among other things). Fred Boar's character is irrelevant to the issue.

> Ramsey Clark's only saving grace as an American citizen is that he can be used as the ultimate bad example of same. After a four-day whirlwind visit to Nicaragua earlier this

month, America's modern-day version of the pusillanimous Neville Chamberlain advises us that the Sandinistas are not communists nor have they been supplying arms to leftists guerrillas in El Salvador.

If you believe that spiel give me a $10,000 deposit and we'll open escrow on the Brooklyn Bridge.

If Clark's contention about Central America is so obviously wrong, then it should be easy to give us some evidence that proves it. The evidence should be about Central America. The reason why personal attacks often work is the natural psychological tendency to transfer hatred of a person (or group) to hatred of the position that person favors (or they favor).

To see why personal attacks are illogical, try putting an argument against the person in standard form:

1. The people who accept statement S have qualities X, Y, and Z (say, 'stupidity,' 'greed,' and so on).
2. X, Y, and Z are bad traits.

∴ S is wrong.

As it stands, this clearly is logically incorrect. The best we can do is treat it as an enthymeme and add the missing premise:

1. The people who accept S have qualities X, Y, and Z.
2. X, Y, and Z are bad qualities.
3. All or most statements believed by people who have X, Y, and Z are false.

∴ S is false.

But premise 3 is clearly false.

The second form of personal attack is the *circumstantial variety*. Here, we do not so much attack the other person's character, as accuse the person of being biased. Again, it does not matter whether the accusation is correct, because even biased people can be right. Some examples of this fallacy:

Having Judge Callister, a member of the Mormon Church, rule on the fate of the most significant piece of women's rights legislation since the 19th Amendment 60 years ago is akin to having an executive of the National Rifle Association decide on the constitutionality of gun control legislation.

If the writer wants to prove the judge's decision is incorrect, she should talk about that decision rather than the judge's religion.

Consider this reply by someone who was accused of severe mismanagement:

> Leonard J. Hansen, *Senior World's* founder, publisher and editor, says that while he has recently experienced "severe cash problems," a reorganization has cut overhead and put the newspaper on the road to good health. Hansen dismisses the allegations as being from "a couple of disgruntled former employees who are going around trying to assassinate me."

Has Hansen proven those charges false by accusing those who made them of being prejudiced against him?

The third form of this fallacy is **tu quoque** ("you also"), where a person's point of view is dismissed because of his (alleged) hypocrisy. But even hypocrites can be right. It is thus illogical to dismiss your father's warning about the addictive use of drugs merely because he drinks addictively. Even if he is a hypocrite, his warnings may be right.

Consider this example of tu quoque, which took place during a debate between former California Governor Jerry Brown and former San Diego Mayor Pete Wilson. Brown asked Wilson to explain a $70,000 loan the mayor had gotten that allowed him to invest in a tax shelter and escape federal income taxes in 1980. Wilson responded that it was:

> . . . the ultimate in brass even for Jerry Brown to come up with a comment on taxes. . . . Over the last three years, sir, I have paid more taxes than you. So if I have not paid a fair share of taxes, neither have you, brother.

A fourth form of personal attack is *poisoning the well*. This is the fallacy of attacking the other person before he has a chance to speak—discrediting the speaker in advance. The attack here involves planting doubt in the listeners' minds, which prevents the speaker from getting a fair hearing. As an example, imagine this comment in a political debate.

Politician: My opponent is going to speak to you in a moment about what he thinks is our need for more defense spending. As you listen to his tired old rhetoric, keep in mind that he worked for many years as a defense industry consultant, and indeed has been a big player in the military-industrial complex that President Eisenhower warned us about so many years ago.

The politician is trying to discredit his opponent in advance by insinuating that his opponent is biased and perhaps even part of some sinister conspiracy.

We need not use words to commit the fallacy under discussion. *Caricature* involves using pictures or cartoons to attack your opponent. This technique is commonly employed by some photojournalists and political cartoonists. (See Figure 4.1A, B.)

One final variety of personal attack deserves mention. Often, an idea (theory, practice, or proposal) will be attacked on the basis of its origins (its 'genesis'), but the people who originated it are not specifically named. We call this a **genetic fallacy**. For example, such a fallacy would be committed by someone who argued against the idea of a 4-day workweek by saying it was a "Communist" idea.

Do not confound the genetic fallacy with guilt by association. *Guilt by association* is the discrediting of a person by pointing to the group to which that person belongs, or the person's friends or associates. The following is a case of guilt by association.

Figure 4.1A

Reprinted by permission: Tribune Media Services.

Figure 4.1B

Reprinted by permission: Tribune Media Services.

Senator Jason has advocated this civil-rights bill. But how dubious the bill is will become clear to you when I point out Jason's associates: Bill Marko, well-known communist organizer; Ted Wylong, a member of the Wallaby Communist Cell; and Sharon Blank, a left-wing activist.

In the genetic fallacy, an idea is attacked on the basis of the group that originated it; in guilt by association, an idea is criticized on the basis of the person who advocates it, but that person's character is attacked on the basis of the group to which that person belongs.

Personal attacks are sometimes thought to be acceptable in a court of law, as when an attorney attacks the credibility of a witness. This, however, mistakes the case. There is a difference between the activity of testifying and these activities of proposing, theorizing, arguing, suggesting, and speculating. To **testify** is to ask people to accept a claim on our own say so. In such cases, it is quite logical for others to examine who we are and what our character is like. But in situations where we are not testifying, other logical standards apply—and argumentum ad hominem attacks should not be indulged in.

A last note about arguing against the person: it is equally illogical to argue for the person. An argument like:

> Freda is so kind. She is against President Goodman, so I guess we should be, too.

is as fallacious as any argument against the person. Freda's kindness, smartness, integrity, or other good qualities tell us nothing about her stand on President Goodman. Again, we are not talking here about cases of testimony: if Freda is a political scientist, her testimony regarding, say, Goodman Supreme Court appointments may be worth credence. We discuss expert testimony in Chapter 5.

The next fallacy is **appeal to fear**, called in Latin **argumentum ad baculum**, which means an argument directed to the 'rod.' 'Rod' here is a stick for beating someone, as in "spare the rod and spoil the child." A person can commit this fallacy in two ways. We can directly threaten to use force, as in these examples:

> "Politicians who do not deal with the Equal Rights Amendment in the Virginia legislature are playing a dangerous game if they plan to stay there." Those were the words of Barbara Lomax, one of the Virginia State Coordinators for LERN (Labor for Equal Rights Now), speaking at a massive pro-ERA demonstration in Richmond on January 22, as ERA lobbyists shifted their tactics from cool persuasion to outright threats.

We must be careful here. Typically, appealing to fear is done to get somebody to do something, not to believe something. As such, it is odd to talk about an argument here: vote for ERA or else we will vote you out of office! But the point is that, even if we give in to a threat, we should still realize that no evidence has been given to justify the action. In a better world, the force of evidence might be the only force that makes people act.

> (From a punk-rocker.)
> Your rag says we're all rich kids spoiled with all kinds of money from our parents. *Esquire* says we're children from broken homes, left on our own for years by alcoholic and drug-addicted parents. The hell with you. We live where we live and we do what we want. And if any of you tourists got anything to say,

come down to the beach and we will make you wish you never came to California.

Sometimes the threat is not physical.

> Associated Students President Harriet DeMarco walked out of yesterday's council meeting and threatened to resign after two business council representatives accused her of withholding important information from A.S. Council.

A more subtle method of appealing to fear is to use 'scare tactics,' wildly implausible claims about what will happen if such-and-such is or is not done. "If this proposition passes, the schools will be closed down within two weeks!" "If this man gets elected, there won't be a free America after the election!" Examples:

> LOS ANGELES—President Reagan arrived in California last night for a weeklong working vacation after a cross-country trip in which he warned of a "larger and increased possibility of war," if Congress fails to approve the military spending he wants. President Reagan is resorting to scare tactics.

> WASHINGTON [AP]—The Soviets could knock out the entire U.S. military command system with two or three well aimed nuclear blasts disrupting communications, a senior defense offical said yesterday. At the same time, this official said, the U.S. satellite-based system for warning the United States against nuclear attack is vulnerable to sabotage because a relatively small number of "guys with wirecutters" could disable cables leading from satellite ground stations to command posts. The official gave these graphic statements about the vulnerability of the U.S. command and control system to a group of reporters in a session apparently aimed to generate support for the Reagan administration's costly plan to correct such weaknesses.

In the previous examples, the arguments are not offered by the writers, but are reported as being given by someone else. Here again, scare tactics are used to garner support for a defense program.

A third type of emotional appeal is to pity. Often, a speaker will try to persuade her listeners to do what the speaker wants by appealing to their sense of pity, by 'pulling at their heartstrings.' The speaker

4.3 / *Emotional Appeals*

persuades them by making them feel sorry for some person or situation. This is called **appeal to pity (argumentum ad misericordiam)**. We may represent an appeal to pity as having the form:

1. Persons A, B, and C believe statement S.
2. A, B, and C . . . deserve pity because of their circumstances.

∴ So S is true.

This form is clearly fallacious.

Often, appeals to pity make reference to small children or animals. Figure 4.2 illustrates this practice.

Figure 4.3 gives a parody of appeal to fear (courtesy of *The National Lampoon*). Recall the point we made in Chapter 2, that an argument may be used as a joke.

The point here is not that a logical person is a person without sympathy for his fellow beings. Instead, the point is that it is not a sufficient reason to donate to a particular charity that we want to help small children and cute animals; we should demand genuine evidence that the money we donate will be put to good use.

Moreover, as with appeal to fear, we have to be careful to see what is being 'argued.' Consider, for example, appeals to pity by defense lawyers. For example, here is the summation to the jury that Samuel Leibowitz offered in defense of a man accused of killing his own son.[2]

> In his summation, Leibowitz made a frankly emotional appeal, and when, overcome by his own emotion, he stopped, "the sobbing of the women spectators was the only sound in the courtroom," said the New York Post. "You talk about tragedy. You talk about purgatory. If that child had only been born blind or crippled or deaf and dumb—or even an invalid confined to a life in a wheel chair—the blow would not have been so terrible. But this child was just a lump of flesh.
>
> "This mother would have taken her eyes out," Leibowitz declared. "She would have cut her arms off. This man would have cut his heart out—if that child could have taken just one step, could have said one word. This man's mind was worn down bit by bit. It was like a drop of water," he said. "Drop by drop, wearing down the stone. Hour after hour, week after week, and year after year, every moment—awake or asleep. Human flesh could not stand it! His mind could not stand it!" Before reviewing the actual killing, when Greenfield chloroformed his son, Leibowitz said, "Suppose you were walking along the street

---

[2]Quentin Reynolds, *Courtroom: The Story of Samuel S. Leibowitz* (New York: Farrar, Straus and Company, 1950), pp. 176–177.

Figure 4.2

Reprinted with permission of the Associated Humane Societies (Forked River, NJ; 609/693-1900).

and saw a dog lying helpless in the gutter, his body torn by pain, after being run over by the wheel of an automobile—just a poor, yellow mongrel dog, lying there in agony? You would say, 'I wish some policeman would come along and put that poor thing out of its misery.' So, Greenfield saw his boy in agony that day, as he had seen him so many times before. The boy couldn't tell him where he was being hurt. Then something dragged him to the closet, where he had kept the chloroform hidden for two months. He took it out. He put it on a handkerchief and placed it on his son's face. And life went out of that lump of flesh. No more torture at the hands of doctors. No more suffering for this poor woman." Here he pointed to Mrs. Greenfield. "If what he did

4.3 / Emotional Appeals

Figure 4.3

Reprinted with permission of *National Lampoon* (Los Angeles, CA).

was the moral thing to do, you can't find him guilty. How much more suffering does Louis Greenfield deserve?"

Indeed, the jury found the defendant Greenfield innocent. Taken as arguments that the defendant did not commit the crime, such appeals are logically irrelevant. But taken as attempts to remind the judge or jury that they ought to be merciful, that mercy is a virtue, such appeals need not be fallacious. Once again, determining fallacy from legitimate reasoning requires judgment.

The next fallacy to be discussed is **appeal to the crowd (argumentum ad populum)**. This involves appealing to your audience's feelings of group loyalty. Such appeals come in two varieties, the 'bandwagon argument' and 'mob appeal.'

In a *bandwagon argument*, the arguer asserts that because most people believe some proposition P, P must be true. In standard form, we can represent it as:

1. Most people believe P.

∴ P must be true.

which is clearly illogical, or we can represent it as an enthymeme:

1. Most people believe P.
2. Whatever most people believe is true.

∴ P is true.

But the problem then becomes premise 2, since the majority of people often believe false things.

Illogical as it is, the bandwagon argument is common. For example, one car company argues in its commercials that its cars must be the best because they sell more than any other model.

The bandwagon argument frequently takes the form of saying that you should believe or do something because 'the winners,' 'the leaders,' or 'the smart people' do so. We call this 'snob appeal.' In Figure 4.4 (from a British magazine), a camcorder is advertised under a picture of Patrick Macnee, former TV star who embodied classiness.

As we shall see in Chapter 5, to cite the testimony of experts to back your claim can be logically acceptable. But simply saying, "The people in the know buy this product" does not constitute a reasonable citation of expert testimony.

A more subtle (even sneaky) method of ad populum persuasion is to appeal to feelings of patriotism, ethnic or racial pride, religious

Figure 4.4

All good tailors recommend Hitachi Hi 8 camcorder because it is the only one which doesn't deform your pockets.

Lightness is good. But more is necessary to make an exceptionnal camcorder. Stereo sound, digital zoom x 20, three possibilities of fading, red colour filter, sepia or B/W, the WMH 37 shoots in Hi 8 **HITACHI** perfectly. Hand-sewn in fact.

VMH 37 CAMCORDER

Reprinted with permission of Hitachi Corporation (Tarrytown, NY).

clannishness, or hometown sentiment. We call this *mob appeal*, and it is a favorite of demagogues and advertisers. Chevrolet advertises its cars by *appeal to patriotism*: "What does America love? Baseball, hot dogs, apple pie and Chevrolet." (As if only a traitor would buy a Porsche!) Dodges are very often pictured (in ads) in front of "hometown America" scenes, such as family picnics and homecoming celebrations. Brands of televisions are advertised with the TV screens showing pictures of the Lincoln Memorial, the Statue of Liberty, and other national symbols. Such an appeal to patriotic sentiment is all the more effective because it operates on a subliminal level.

One other form of appealing to the crowd is worth mentioning, which, for lack of a better name, we shall call *appeal to sex*. In this fallacy, allusions to sex replace rational evidence. This is not unheard of in advertising.

For example, in Figure 4.5, a bikini-clad woman and words such as "sensual" and "uninhibited" entice travelers to visit a Carribean resort. In Figure 4.6, attractive models in a European setting advertise a liqueur. A young woman is used to advertise car engines in Figure 4.7. Finally, in Figure 4.8, an ad for men's footwear depicts an aura of elegance and excitement.

In summary, arguing against the person, appeal to fear, appeal to pity, and appeal to the crowd all involve substituting emotional manipulation for rational evidence. The first involves appealing to our hatred or distrust for specific groups. Appeal to pity usually involves appealing to our sympathy for specific groups (underdogs, little children, animals), of which we are not (and do not desire to be) members. Appeal to the crowd involves appealing to our desire to belong to a group (the rest of the people, 'the winners,' or 'the beautiful people'), or else our feeling of loyalty to a group to which we do belong (nation, family, race, religion, hometown, and so on).

EXERCISES 4.3

Each of the following passages contains one or more of the fallacies discussed. State what is wrong with the passage in specific terms (do not just repeat the argument, but criticize it), then pick a label that best identifies it. The best heuristic rule to follow is to compare the problem you are working on to the examples given in the book, and the closest match will determine which label to choose. Do not just look at the definitions of the labels—look at the cases to which they were applied. (An asterisk indicates problems that are answered in the back of the book.) (Note: Exercises 4.3 continue on page 109.)

Figure 4.5

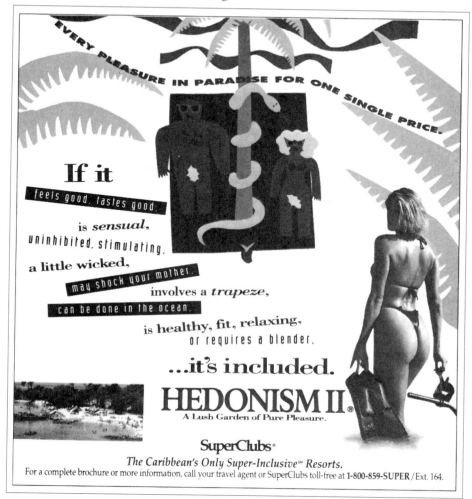

Figure 4.6

Figure 4.7

Figure 4.8

Reprinted with permission of Johnson & Murphy (Nashville, TN).

4.3 / *Emotional Appeals*

(Exercises 4.3 resume from page 104.)

Example: The American Bar Association thinks it would be dandy if Uncle Sam would foot the bills for interested citizens who want to be heard in regulatory proceedings, but lack the wherewithal for that representation. Part of the reimbursable expenses, of course, would be legal fees—a little point that makes the ABA proposal sound considerably less noble and altruistic. In fact, it makes it look like a full-employment scheme for lawyers.

Response: The author doesn't address the merits of the ABA proposal, but instead accuses the ABA of being biased. Attacking the person.

1. McClellan's editorial, "Defects or Risks?" (February) was the most outrageous, calloused and ill-considered example of journalistic tripe ever published on an aviation subject. It should have been preceded by the warning: "Reading and believing what follows will surely be hazardous to your health." The only conclusion readers could draw from the editorial is that, in McClellan's opinion, manufacturers should never be held accountable for injuries caused by their defective products. Apparently, according to his Simple Simon reasoning there are no "defects"—only "risks" that users assume as general aviation participants. That caveat emptor philosophy has been substantially gone from general commerce for decades.

* 2. Because of your recent diatribe against the San Diego medical community and the president of the San Diego County Medical Society, I direct you to cancel my subscription to *San Diego Magazine* effective immediately. I demand a complete refund of my subscription fee paid. Your attack was unfair and one-sided. At the very least, you should provide the San Diego County Medical Society, and/or its president, "equal time" and space in your next issue for a rebuttal, to present its side of the controversy.

If the intended article is published, and if an apology is made for the harangue in this month's issue, I will be happy to continue as a subscriber of your magazine, and I will urge my colleagues, friends, and relatives to do the same. If you proceed to publish and make no apology, my actions will be to the contrary.

3. Bubba Smith's allegation that the 1969 Super Bowl was fixed was called "ludicrous" yesterday by National Football League Commissioner Pete Rozelle.

"I don't feel it merits even a comment. We shouldn't have to justify those things," Rozelle said.

4. Do you miss your mother? You don't ever have to miss her again. You don't ever have to miss any important telephone calls again.

Superfone can automatically answer your phone and take messages when you're not home.

5. We call him Trixie. This puppy was painted with shoe polish and thrown out from a third floor window. He's alive, paralyzed in the back. He wants to live as you can see by the hope in his eyes. We will see that he does, regardless of the cost. We will help Trixie as we do thousands of unwanted, mistreated, and abandoned animals. Will you help us? We know who threw

him out of the window but the witnesses are afraid to testify. Your gift is tax deductible.

* 6. Lack of vitamins can cause you a slow, horribly painful death! Don't take chances. Take a vitamin pill.

7. Dodge—America's driving machine!

8. To Lester Jones, on your article about Jim Morrison: I don't rightly care for you or your damn opinions. You put the man down through the whole article and at the end tried to say you like him. Can't you make up your mind?

At least Morrison had his own style about everything he did. Maybe you're just jealous 'cause you couldn't make it past being a plain old hippie. It takes a lot of guts to be different and not a plastic person who does only what other people want him to do. If anyone's a bozo, you are. I read your opinion of Morrison so you can read my opinion of you!

9. Mayor Pete Wilson of San Diego thinks California ought to have a law requiring "the gubernatorial candidates of the two major political parties to square off in debates."

"Were such a requirement to exist," Wilson said, "I'm confident that TV stations throughout the state would be inclined to broadcast them in prime time."

We suppose Wilson's suggestion ought to be taken seriously, even if it does involve him in a conflict of interest. He is a prospective Republican candidate for governor next year, and he also has a reputation as a rough, tough debater.

* 10. A Million English Women Can't Be Wrong. For over 40 years, the natural ingredient of YEAST has been ending the skin problem blues of English lasses. Blackheads, grease-filled pores, and other acne-causing impurities are drawn out by the natural action of deep cleansing agents fortified with yeast.

11. Phillip Smith's defense of Rose Bird (Letters, Sept. 17) is what one might expect from a resident of Beverly Hills. It simply reeks of that parlor liberal elitism that emanates from our local Mt. Olympus and whistles down the nearby canyons like so much hot air.

12. Mayor Elisa Turkle was outraged that critics have criticized her park program. She said in reply to them, "I challenge you to provide documentation that shows some other program is better than mine!"

13. You can fill the emptiness in Rena's eyes and tummy. Little Rena lives in a hut made of palm branches and mud. She and her sisters sometimes go for days with little more than bread and water. You can help Rena, or a child like her. Just $21 a month will help provide food, clothes, medical attention, school, and maybe a toy. There are so many whose lives and little tummies are empty. Please hurry with your help.

* 14. Do you want people to be offended by your presence? Do you want children to run away when you walk by? No? Then you need Tuffstuff Deodorant Spray.

15. The beautiful people have discovered Magic Cream. The legendary lifestyle you live year round starts with Magic Cream. The ultimate in a radiant complexion can be yours. Enjoy the look of confidence and contentment that comes from looking and feeling your best.

4.4 Ignoring the Issue

The next fallacy on our list may be the most pervasive of all. Often, faced with an issue they cannot logically address, people will ignore the issue at hand and instead talk about something else. We call this **ignoring the issue (ignoratio elenchi**, which means ignorance of the question at hand). This fallacy is also called *irrelevant conclusion,* in that the evidence given supports only a conclusion irrelevant to the discussion at hand. As you might imagine, politicians often commit this fallacy: rather than admit they do not know the answer to a question, they will talk on and on about matters they can address. It does not matter if the irrelevant evidence is good enough to establish the irrelevant conclusion—the point is that the question at hand has been ignored. The only way to represent such a fallacy as an argument form would be as:

1. P is true.

∴ Q is true.

which is not very helpful. We can better view this fallacy as breaking a rule implicit in rational dialogue: always answer the question at hand, instead of changing it.

Varieties of this fallacy have been considered fallacies in their own right. They include: glittering generalities, diversion, red herring, strawman, slippery slope, and apples and oranges. We will discuss, in turn, each variety of ignoring the issue. We begin with *glittering generalities*. It is expected that people try to propose solutions to the problems they face. But reasons need to support these proposals. When a person supports a proposal by speaking in generalities (such as how terrible the problem is) rather than specifics (such as why this particular proposal will solve the problem and solve it in the best way), the person ignores the issue. Politicians commit this fallacy with depressing regularity. Ask a senator to justify his bill on unemployment, and he will likely give you only glittering generalities about how terrible it is to be unemployed, how it hurts the family, how it saps a person's self-esteem, and so on. All true, but all irrelevant to the real issue: why vote for this bill? What makes the generalities "glittering" is their obvious truth and compassionate nature.

Another form of ignoring the issue is *diversion*, which is to change the subject by joking. Two recent presidents stand out for their exceptional ability to evade embarrassing or tough issues by joking: John F. Kennedy and Ronald Reagan. Wit is an admirable quality, but not if it is used to evade the responsibility to justify one's beliefs. An example of diversion:

> The story is told about Wendell Phillips, the abolitionist, who one day found himself on the same train with a group of Southern Clergymen on their way to a conference. When the Southerners learned of Phillips' presence, they decided to have some fun at his expense. One of them approached and said, "Are you Wendell Phillips?"
> "Yes, sir," came the reply.
> "Are you the great abolitionist?"
> "I am not great, but I am an abolitionist."
> "Are you not the one who makes speeches in Boston and New York against slavery?"
> "Yes, I am."
> "Why don't you go to Kentucky and make speeches there?" Phillips looked at his questioner for a moment and then said, "Are you a clergyman?"
> "Yes, I am," replied the other.
> "Are you trying to save souls from Hell?"
> "Yes."
> "Well—why don't you go there?"

The example contains argumentum ad hominem attacks by both Phillips and the clergyman, but diversion is present as well: Phillips jokes his way out of the issue (namely, why he does not give antislavery speeches in the South).

We saw in Chapter 2 that an argument can be given as a joke, but that is not what happens in diversion. In diversion, a joke is used to evade the responsibility for rational argument, to avoid the burden of proof. Indeed, we might have called ignoring the issue 'avoiding the burden of proof,' in contrast to our earlier fallacy of shifting the burden of proof.

The third way to ignore the issue is to *raise a red herring issue*. That is, faced with a difficult issue upon which a person is not prepared to give logical evidence, the person might muddy the waters by raising controversial issues superficially like the one at hand but essentially different. As an example, while a feminist was speaking on a talk show in favor of the ERA, the question was put to her, whether the ERA would not require the drafting of women into combat in times of war. The feminist responded by asserting that the draft was immoral, that in a world run by women politicians there would be no war, that wars are due to capitalism, and that women already serve in the armed forces and deserve equal pay. All interesting issues—a pity they were utterly irrelevant to the issue at hand!

A fourth variety of ignoring the issue is *strawman*, which is the distortion of another person's position. (The name arises from the metaphor of setting up a straw man, a dummy, and vainly trying to prove your prowess by knocking it down.) We can distort our opponent's position by oversimplifying it (leaving out important qualifications and details) or by extending it to situations to which it was never meant to apply.

Examples of strawman:

> Candidate: My opponent wants to increase the number of daycare centers. But do we really want the government to take over child-rearing? I say: let the parents raise the kids!

The candidate has mistated the opponent's position. The opponent only wanted more daycare centers, not for the government to take over child-rearing entirely.

Here is a letter opposing California State Senator Bob Wilson's anger at the use of the taxpayer's money to support unusual forms of art.

> So State Sen. Bob Wilson doesn't like Paul Fericano's poetry. Or he doesn't understand it. And he can't relate to any of a list of California Arts Council special projects. That's unfortunate, but does Wilson really think that lack of appreciation entitles him to decide what a legitimate expression of artistic effort is, and what it is not? I suspect that not even the most jaded art critic would undertake that sort of pretension. One can say, quite honestly, that one likes or does not like this or that. And one can give reasons. But art should not have to shoulder the burden of a public official's approval merely to exist. Perhaps most of the projects on Wilson's list are, in fact, pointless to all but their creators. So what. No one has a God-given right to decide, in advance, what should or shouldn't be created. I suggest that Bob Wilson have an extended conversation with William Wilson, the *L.A. Times'* distinguished art critic. Better yet, he should enroll in an art appreciation class. And he should stop taking himself so seriously—no one else does.

Wilson did not say that those artists should be forbidden to create whatever they wanted; only that the taxpayer's money should not be spent to support it!

Politicians often set up strawmen. During the Carter-Ford presidential campaign of 1976, Carter said that he would not automatically kick Italy out of NATO if the people there voted in a Communist government. During one of their debates, this strawman exchange occurred:

> *Ford:* Mr. Carter has indicated that he would look with sympathy to a Communist government in NATO. I think that would destroy the integrity and the strength of NATO, and I am totally opposed to it.
>
> *Carter:* Now Mr. Ford, unfortunately, just made a statement that isn't true. I have never advocated a communist government for Italy, that would obviously be a ridiculous thing to do for anyone who wanted to be President of this country.

Ford starts out by distorting Carter's position—which was that Carter would not necessarily kick Italy out of NATO if the Communists won there, not that he was sympathetic to that prospect. Carter distorts Ford's claim that Carter is *sympathetic* to the prospect of a communistic government in Italy to the claim that Carter *advocates* such a government.

A fifth way to ignore the issue is *slippery slope*. We shall discuss this fallacy more precisely in Chapter 6, when we talk about vagueness of concepts. For now, let us say that a fallacy of *slippery slope* occurs when the arguer changes the issue by degrees. That is, faced with some issue, the arguer says:

> But if we agree to A, then why not A_1? Or A_2? Or A_3? But A_3 is obviously absurd!

and the conclusion is drawn that the orginal issue, or claim A, is false, too. What allows this fallacy to appear plausible is that the claims A_1, A_2, A_3, . . . , differ by degree only.

Consider this example:

> Dear Editor:
>
> The PTA has asked the television networks if they [the PTA] can have the power to stop programs with a lot of violence in them from being aired. I am outraged by the PTA's request. If the PTA is allowed to determine what shows I can watch, maybe

> they can next determine what I can eat, when I should sleep, and what I read. I suppose they'll be burning books next!

Notice that the writer has shifted away from the real issue ("Should the PTA be allowed to decide if a TV program has too much violence to be aired?") to a much easier one to refute ("Should the PTA be allowed to burn books?")

Another example:

> Some have suggested lowering the voting age to 18. But if we lower it to 18, why not 17? Or 16? Or 15? Or 14? But can you imagine the way fourteen-year olds would vote? My God, some rock star would be elected President!

The real issue ("Should we lower the voting age to 18?") has been ignored. We, the listeners, have been carried down a slippery slope to an entirely different issue.

One caution. Some good arguments superficially resemble slippery slopes. These are arguments from 'precedent': If we pass this law (or allow this exception, or such like), then that will set a legal precedent which will allow bad consequences. Given the fact that laws are indeed applied in a court of law in part on the basis of past decisions, such arguments can be logically quite acceptable.

A sixth form of ignoring the issue is *apples and oranges*. In this form of the fallacy, the speaker lumps issue A in with another issue B, and then proceeds to defend B instead of the real issue A. For example, proponents of welfare programs will often lump together AFDC (Aid to Families with Dependent Children—the program that gives money to women with children whose fathers refuse to give support) with programs to support the handicapped, and defend both programs by focusing their remarks on the second. Another example:

> Still, criticism persists that much of what the National Science Foundation does fails to meet any reasonable definition of spending priorities. Its defenders say scientific advancement and improved technology depend upon the foundation's continued growth. Last March, when Rep. John Ashbrook (R–Ohio) offered an amendment to cut $14 million from the foundation's biological, behavioral, and social science research, some warned that he could be denying money that might lead to breakthroughs in medical research.
>
> "How many people here would vote for $100,000 to study the growth of viruses in monkey kidney cells?" asked Rep. Tom Harkin (D–Iowa). While that foundation-funded research had no immediate payoff, Harkin said, Dr. Jonas Salk a few years later used the study in his own research and came up with a polio vaccine.

But Harkin and Ashbrook, it seemed, were talking about apples and oranges. Ashbrook was not attacking medical research. Instead, he was criticizing studies that he argued were indefensible and simply wasted tax dollars. Like the $83,839 the foundation gave to the American Bar Association to study the social structure of the legal profession.

Ashbrook attacks social science funding; Harkin responds by exploiting the fact that the same agency, which funds social science, funds medical science as well, in order to defend social science funding by defending the irrelevant issue of medical funding.

4.5 Loaded Question and Begging the Question

We saw in Chapter 2 that questions usually presuppose something or other. To ask, "How is your mother doing?" presupposes that the listener has a mother. To ask, "How many miles per gallon does your Chevy get?" presupposes that the listener owns a Chevy. There is nothing wrong with asking a question that presupposes something; what is bad is to presuppose something false or debatable. When we ask a *loaded question*, one that presupposes something false (like, "Have you stopped beating your wife?") or something that needs to be argued for, we are committing the fallacy of **loaded question**. (Some textbooks call this fallacy 'complex question.')

Examples:

Aren't you glad you use Romeo deodorant?
(Who says I do?)

Why is Congress so insensitive to the needs of the elderly, the handicapped, and the poor?
(Who says Congress is insensitive to the needs of the poor?)

The fallacy of loaded question is used to introduce or establish a claim without ever having to argue for it. As such, it is a way of avoiding the burden of proof. A common use of loaded question is in sales.

I can see you're impressed with this coffeemaker. Will this be cash or charge?

The salesperson is trying to avoid proving to the customer that the customer should buy the coffeemaker by raising the issue of manner of payment.

4.5 / Loaded Question and Begging the Question

The last fallacy we examine in this chapter is the fallacy of **begging the question**, or "arguing in a circle," which in Latin is called **petitio principii**.

A *circular argument* is an argument in which claim C is backed up by premises $P_1, P_2, \ldots, P_n$, but where one of those premises is, in fact, equivalent to the conclusion. Remember that the same statement can be made using many greatly different sentences, and this is what makes circular arguments so difficult to spot in practice. Consider these examples:

> To allow every man an unbounded freedom of speech must always be, on the whole, advantageous to the state; for it is highly conducive to the interests of the community, that each individual should enjoy a liberty perfectly unlimited, of expressing his sentiments.

In this passage, the conclusion argued for (that freedom of speech is worthwhile for the state) and the premise offered for it mean the same thing, but this is not obvious because they are expressed by different words. The example just given was devised by the famous nineteenth-century logician Richard Whately, who pointed out that English is especially suited to constructing circular arguments, since it was formed from two distinct languages. In the previous example, the words in the first part are of Saxon origin, while those in the second part are of Norman origin.[3] Another example:

> A person's strongest desires determine that person's actions. For people do just what they want to do most.

Making circular arguments even harder to detect is the fact that many premises may intervene, causing the listener to lose track of what conclusion is to be proved.

He: God certainly exists.
She: How do you know?
He: The Bible states clearly that He does.
She: Maybe the Bible is wrong.
He: Impossible—it is too consistent.
She: So maybe it is consistently wrong.
He: Impossible—it is written by prophets.

[3]Richard Whately, *Elements of Logic* (London: J. Mawman, 1826), p. 181.

She: What do you mean by "prophet"?

He: A person inspired by God.

The premise "prophets inspired by God wrote the Bible" would not be accepted by someone who did not already believe in God's existence.
Another example:

> Papandreou had a simple proposition. It was that the CIA had bank-rolled the colonels and told them to stage the coup d'état.
>
> "That's a hell of a statement for you to make, Mr. Papandreou, how can you prove it?"
>
> "I have seen the documents."
>
> "What documents?"
>
> "They are secret."
>
> "But if they are secret, how is it that you are quoting from them?"
>
> "They cannot be seen by unfriendly eyes."

But the statement "They cannot be seen by unfriendly eyes" assumes that the documents exist.

Some logic books include the Big Lie (which we discussed in Chapter 2) as a type of begging the question. This points out the difficulty of precisely categorizing fallacies. Does repeating a claim count as an attempt at argument? If it does, it is merely begging the question.

4.6 Summary

In this chapter we have discussed fallacies of no evidence. These include:

Pooh-poohing: Dismissing rather than arguing against your opponent's point.

Shifting the burden of proof: Trying to make the other person prove what you should prove. (Including **argumentum ad ignorantiam**—arguing that something must be true or false because nobody can prove it is not.)

4.6 / Summary

Attacking the person: Attacking the person rather than the point. (*Abusive* = attacking the character; *circumstantial* = accusing the person of bias; *tu quoque* = accusing the person of hypocrisy; *poisoning the well* = discrediting someone in advance; and *genetic fallacy* = attacking an idea, concept, theory, or practice by attacking its origins.)

Appeal to fear: Using threats or scare tactics to get your point accepted.

Appeal to pity: Appealing to pity instead of giving evidence to get your point accepted.

Appeal to the crowd: Either arguing that something must be true because everybody believes it (*bandwagon argument*) or else using an appeal to the emotions of the crowd (patriotism, ethnic pride, or hometown sentiment) (*mob appeal*) instead of evidence to get your point accepted.

Ignoring the issue (ignoratio elenchi): Arguing about something other than the point at hand (*glittering generalities, diversion, red herring, strawman, slippery slope, and apples and oranges*).

Loaded question: Asking a 'loaded question,' that is, presupposing in your question something that needs to be argued for.

Begging the question: Assuming during the course of your argument the very thing you are supposed to prove.

EXERCISES 4.6

Each of the following passages contains at least one of the fallacies discussed in this chapter. State in specific terms what is wrong with the argument, and only then find a label that describes it. Several correct labels may apply; if so, select the best one. (An asterisk indicates problems that are answered in the back of the book.)

1. It is one of the most unique works of original jewelry ever created. It contains at least nine magnets. For their size they are unbelievably powerful. In fact, since it was first introduced in Japan just two short years ago, over three million people have purchased it, worn it, and valued its effects. Currently over 100,000 necklaces per month are being sold, and it would not be stretching a point to say that it is the most popular necklace in all of history. Its appeal is universal. It is worn by men and women, young and old.

As more and more people experience the powers of this mysterious necklace, word has begun to spread around the world. Articles about its vast popular acceptance have appeared in leading American newspapers.

* 2. Wally Schirra, an international publicity spokesman for *Realty World*, was asked what an astronaut was doing in real estate. He replied, "Who's seen more real estate, anyway?"

3. Boston's pitcher Bill Lee was asked why baseball doesn't go to a 3-ball walk to speed up the game.
Lee: "Why don't they go to three balls and two strikes? Or just eliminate the pitchers and put the ball on a batting tee? Soon you could have all the managers sit down in Florida with presto boards and they could conduct the season electronically."

4. Question: "Are there any dangers in experimenting with gene transplants on humans?"
Answer: "Experimentation on human patients is done all the time. There are always experiments on human beings—patients with diseases. How else would we learn how to control diseases? Don't you think we experiment on patients with leukemia or cancer? Most patients, or a great number of patients, with leukemia or cancer here are treated with experimental protocols. It is the very basis of medical research. If you're going to do medical research it has to be on people."

5. How do we know that we have here in the Bible a right criterion of truth? We know because of the Bible's claims for itself. All through the Scripture are found frequent expressions, such as, "Thus says the Lord," "The Lord said," and "God spoke." Such statements occur no less than 1,904 times in the 39 books of the Old Testament.

* 6. Give the gift America has been giving since 1842! Reynaldo's Famous Chocolates.

7. After long speaking out against federal bail-outs of troubled industries, Reagan indicated last week he might support federal aid as a last resort to help rejuvenate the nation's steel industry. And he has indicated he might back price supports for farmers.

But ask Reagan and he denies emphatically that he's changed any of his positions—or even shifted closer toward the middle ground—to win favor with organized labor or other voting blocks.

At a planeside news conference in Birmingham, several days ago, Reagan bristled at the suggestion.

"Look, I've been on the mashed potato circuit so long. I was on radio so many years with those five-day-a-week commentaries. I had a twice-a-week newspaper column that was in more than 100 newspapers. How could I have changed my position? I'm still where I was these last 20 years."

8. Haven't you promised yourself a new car long enough?

9. Last June in St. Louis, well-organized conservatives at the annual meeting of the Southern Baptist Convention elected a stemwinding preacher

named Baily Smith, 41, as president of the nation's biggest Protestant group (13.4 million members). Said Smith: "It's interesting to me at great political battles how you have a Protestant to pray and a Catholic to pray, and then you have a Jew to pray. With all due respect to those dear people, my friend, God Almighty does not hear the prayer of a Jew. For how in the world can God hear the prayer of a man who says that Jesus Christ is not the true Messiah? It is blasphemous."

* 10. Attention investors—Why is Hammer Mortgage Company different?

11. Legislators say that raising the drinking age will save lives. Using this logic, they should not stop at 19 or 21 years. Why not raise it to 30 and save more lives, or outlaw drinking altogether?

12. Question: "What would you do as President to counter Soviet-backed subversion in the Caribbean?"

Answer: "There, I think, is one area where this administration has been woefully lacking. There's no question but that the Caribbean is being made—by way of Cuba, the Soviets' proxy—into a Red lake. It is so vital to us with regard to the sea lanes: We forget that the overwhelming majority of minerals essential to our industry are imported and come in by ship and that the Caribbean intersects a great many of those sea lanes.

There is also the Communist move into Central America. While we look at the far stretches of the world, we're long overdue for the United States to really make an effort to align ourselves with the other countries in the Americas. When I announced my candidacy, I proposed that we develop a North American accord—Canada, the United States and Mexico. It shouldn't be done by Big Brother trying to impose something: let's go to them and ask what their ideas are."

13. Why not be part of the excitement that is public television? Make it official—phone with your pledge.

* 14. How can we afford to sell our world-famous soft contacts at this price? $139. The question is, can you afford not to buy them at this price?

15. Proposition 5 [a proposal to restrict smoking in restaurants] is chipping away at people's rights. The government has its hand in too many things already. First it's smoking, then they'll be telling us how many children we can have, or what kind of car we can drive, or the type of food we should eat.

Today it's smoking, tomorrow it's something else. Vote No on Prop 5.

16. Dear Editor: Roger Smith is right when he wrote to protest letting that kid go—you know, the ten-year-old kid who was caught writing graffiti. Smith is right when he says graffiti invades our privacy, but that is the least of it. A strap-hanger's logic tells him that if a ten-year-old can get away with writing on a subway train, an eleven-year-old can kick out windows, a twelve-year-old can snatch purses, and a thirteen-year-old can commit murder and get away with it.

17. Will you open up your heart to lovable little Jamie? More than 100,000 special children in America have never felt the love, warmth, and

security of a permanent family. Their parents gave them up for adoption. But they haven't been adopted because they are very special children with mental, physical or emotional problems. Here is another touching story of a special child who needs a home and a family full of love.

Send money right away.

* 18. The shame and embarrassment of being a constituent of Rep. William E. Dannemeyer is overwhelming. Every time the man opens his mouth, I cringe with horror. However, to compare Nelson Mandela to Willie Horton is sinking to an all-time low, even for Mr. Dannemeyer. I guess since President Bush got so much mileage out of Mr. Horton during the presidential race, Mr. Dannemeyer, not being a particularly clever man, figured it was time to dredge him up again.

Well, Mr. Dannemeyer, you once again have shown the country what kind of person you are. You appeal to the worst element in mankind—ignorance—and my only wish is to suddenly make you black and live in Soweto. I would love to hear some of your thoughts then.

Until then, you sit on your throne in Orange County, comfortable being the elder statesman for right-wing extremist views. You do not represent Orange County, though; you represent intolerance, hatred and bigotry. Those qualities do not have political boundaries but flourish under the leadership of men like you.

—Daniel Maguire, *A New American Justice.*

19. America's #1 Sinus Remedy!! Maximum strength Sinustuff!

20. (Advertisement) Read today's *Daily Inquirer* and find out which TV shows actually make you smarter.

21. America's true colors come through on GE. The rich green of Miss Liberty. The bright red of a football jersey. The vibrant yellow of a harvest moon. These are America's true colors—colors that come through vivid and life-like on GE TV.

GE color TV: It brings America's true colors into your life. We bring good things to life.

* 22. I object to Ted Smith's record review of George Harrison's *Somewhere in England*. Actually, it wasn't a review at all; it was more like a senseless putdown. His comments on the record were as twisted as his own mind with his stupid insults of a truly great artist and musician. With this guy doing this sort of weirdness, it's not doing the magazine any good. He wasn't the right person for the article and yes, it upsets me and I'm sure plenty of others also, seeing how you put him in there and printed all that trash. You ought to be responsible enough to let someone who knows what he is doing write a new article. This is the real truth.

23. "It would be dumb to think we made the wrong choice by starting the season with Haden at quarterback," Rams coach Malavasi said Tuesday at his weekly breakfast with the media.

"Anybody who thinks that is stupid and doesn't know football. I wouldn't waste time discussing it if someone brought it up."

24. Thanks America. We're celebrating our 100th birthday and you've made us your favorite soap.

25. The attempt by Seymour Minot (Letters, Oct. 17) to tie the demands for corporal punishment to this year's desegregation efforts smacks of the type of intellectual flabbiness and dishonesty that I have grown to expect from psychology professors.

* 26. Question: Mr. President, Former President Ford and others have accused your administration of endangering national security by publicly confirming that we are working on the "stealth" aircraft that could avoid enemy radar. How do you respond to that?

Answer: That's a silly charge. And anyone who knows the facts would realize that it's not accurate.

As a matter of fact, the existence of the "stealth" program was well-known before I became President. Its existence was not even classified when President Ford sat in this office. A contract let for development of the aircraft was an unclassified document.

In 1977, under the orders of Defense Secretary Harold Brown and with my approval, the "stealth" program was classified. Since that time, the size of the program has increased more than a hundred fold. There have been no leaks, so far as I know, about the details, the design, the scope of the program. The existence of the "stealth" program is the only thing that has been revealed—done against our wishes through leaks from individuals opposed to this administration—and is no more than what was known before I became President.

27. Question: Just where would you slash the federal budget if you were sitting in the Oval Office today?

Answer: Everyone seems to think that the only way to cut government is to eliminate programs. Of course, there are unnecessary programs. My experience in California as governor showed me that some programs do not benefit the people. But—more importantly—virtually everything run by government has an overhead higher than that of the private sector. I found that our greatest savings in California—and they were tremendous—resulted from the elimination of waste and fraud and abuse rather than elimination of programs.

I think it's significant that in the federal department that used to be HEW—now the Department of Health and Human Services—they recently eliminated waste, fraud, and abuse. They eliminated it by having a meeting of all the department heads in which Secretary Patricia Harris made a rule that henceforth they would no longer use the words fraud, waste, and abuse. Those words won't appear any more; officals will refer now to mismanagement.

28. A certain doctor advised his patient to get his vitamins with his knife and fork. And I would like to show how dangerously wrong this advice is. If you know something about nutrition, then stand and observe people checking

out of super-markets with their purchases. You will see how wrong it is, for how many people today are educated enough to know what a balanced diet is? How many are strong enough to resist such sugary and refined things as ice cream, cakes, sodas, and other emasculated sweets and starches that lack any semblance of nutritional intelligence?

29. If Dallas elected their Council by district, we would have more voter participation. People will vote if they know that it is only those in their district who are voting for a leader to represent them.

* 30. (This ad is a parody of other ads, so it is a fallacy meant as a joke. What fallacy is Volkswagen making fun of?) Volkswagen does it again. VANAGON holds twice as many baseballs, hot dogs and apple pies as the Chevy station wagon.

31. Question: How do you get interest rates down?
Answer: The problem is monetary policy. The literal problem today in the United States is that there's an enormous run on the dollar, and that run is picking up steam. You can see it in the development of money market funds, dollar accounts, gold accounts, etc. How much does it cost you today to hold a hundred dollar bill in your back pocket for a year? It will cost you 15 to 20%, 15 to 20 dollars. People don't hold money to go bankrupt. People hold money to augment their wealth, not to reduce it. Unless people have the belief that the thing called a dollar is going to be worth about the same in the future as it is now, they're going to drop dollars, and you can see the numbers very clearly in the velocity of money. The velocity of money since 1965 has gone up, what, 70%?

32. (From an anti-evolutionary theory pamphlet): Evolution is the faith seldom doubted as the way to a natural good life that will come to man. It says we are growing up from low animals. It says that we are becoming better man-like animals by stretching our bodies and expanding our minds. It says we are getting higher and higher.

It would be nice to be able to believe some of the things evolution is saying about us. That people are becoming nicer, smarter and better adapted to our environment. But it takes blind faith to believe that humanity is evolving. We have to be blind not to see: That instead of adapting to our environment we are ruining it. Instead of behaving wisely and lovingly toward our fellowman we are behaving beastly and we aren't very nice.

People with their eyes closed might see evolution going up but people with their eyes opened see devolution going down. God made us to be like him Gen. 1:26 (you can't be much better than that). We didn't evolve. We decided we wanted our way, not his, Romans 3:23.

Scientists who said they wanted to give us a better mankind did it by giving us better bombs, biological warfare, and atomic pollution. . . . That's what we call Devil-ution.

33. If absolute accident prevention were the reason for having a speed limit of 55, then I am sure that those imposing the limit would have

dropped the limit to 45. But that would not do a complete job of preventing accidents.

Would 30 mph do the job—or 10 mph? I doubt it, from my own experience. In the past 10 years, each of the two auto accidents I've been involved with concerned one auto stopped and moving cars traveling at less than 10 mph.

* 34. Dear Editor:

As a personal injury defense lawyer, I realized after reading "Taking handguns to court" that this type of litigation would be a marvelous form of job security for me, and in fact would permit me to enlarge our defense firm. After all, why stop at handguns? If someone is beaten to death by an assailant using a baseball bat, then sue the manufacturer, as the bat is obviously a dangerous instrument under those circumstances. If someone is run over by a speeding motorist, then sue the automobile manufacturer because obviously the car too is dangerous. Never mind the good intentions of these manufacturers, or the beneficial uses of the products.

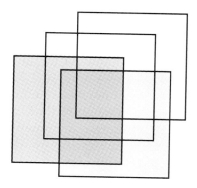

Fallacies of Little Evidence

5.1 Reasoning with Strings Attached

In Chapter 4, we talked about fallacies. We defined **fallacy** as an error in reasoning. We then discussed several complete **nonsequiturs**, which are types of arguments in which absolutely no evidence whatsoever is given for the conclusion. In this chapter, we consider fallacies in which a little rational evidence for the conclusion is given, but that evidence is inherently insufficient.

The best way to view these fallacies is as types of inductive arguments in which logical safeguards or restrictions (which we shall call 'constraints') are violated. Put another way, we shall be looking at kinds of reasoning that have strings attached, and a fallacy occurs when some of those strings are broken.

We can picture the difference between fallacies of no evidence and fallacies of little evidence in terms of the evidence scale given in Chapter 3.

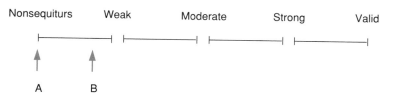

Fallacies of no evidence would be at point A on the scale, for they are total nonsequiturs. Fallacies of little evidence are somewhat farther up the scale, at point B, but are still quite weak.

Remember that, by definition, no inductive argument ever contains absolutely airtight evidence, but such arguments can still offer good evidence by commonsense standards. The inductive argument forms we shall discuss are: testimony as evidence, reasoning by dilemma, reasoning by analogy, generalization, particularization, and inference to cause. In this chapter, our focus is upon how these inductive arguments can go wrong. Later, we will discuss how arguments of these forms can be made strong.

5.2 Testimony as Evidence

Not long ago, I was walking across campus, returning to my office. A young couple came up to me, and the woman asked if the student store was open. I said that it was not (I had just walked by there), and the couple went to their car after thanking me.

Was it illogical for that couple to accept my testimony? Not at all. I was in a position to know whether the bookstore was open and I had no reason to lie. Personal *testimony* can be quite good evidence for a given claim. It can also be bad evidence, and we should get clear on what goes into making testimony logically acceptable.

To begin with, consider personal *credibility*. Is the witness unbiased? Of sound mind? It would have been illogical for that couple to accept my word if I had been drunk or crazy, or were the owner of a competing bookstore (though the couple might not have been able to see that). Testimony stands or falls upon the credibility of the testifier. In saying this, please note that we are not ourselves committing an **argumentum ad hominem** fallacy, because testifying is quite a different activity from arguing (or suggesting, proposing, or explaining). In testifying, people offer themselves as support for their claim ("believe it because I say so"). In such a case, we are entitled to examine their background.

Besides having credibility, the witness must have been in a *privileged* position to observe or know what happened. I was 'privileged' in the sense that I walked by the bookstore, so I was in a position to see whether it was open.

A third requirement for reliability of testimony is that it be *consistent*. If someone tells you that she saw a large brown bear poking around the trash cans on campus, but later you overhear her telling someone else that it was a large timber wolf, you are entitled to be skeptical.

Finally, testimony is more worthwhile if it is *corroborated* by other evidence and testimony by other people. If that couple had doubts about my credibility (say, if they smelled alcohol on my breath), they would have been prudent to ask other people besides me. These four criteria—credibility, privilege, consistency, and corroboration—are matters of degree. But the more credible the witness, the more privileged is that person's position to know, while the more consistent the story, and the better it is corroborated, the more logically worthwhile is that testimony.

These criteria (or 'constraints') apply not only to ordinary eyewitness testimony, but to expert testimony as well. In this age of increasing specialization, we must often appeal to the testimony of experts. We decide whether to stop smoking or stop eating greasy food, on the basis of what our doctor says about these practices. We decide to acquit the defendant as insane on the basis of what the psychiatrists say about the accused. We decide how to vote on nuclear power on the basis of what engineers say about it. Such testimony can be logically worthwhile, but when certain constraints are broken, the testimony should be rejected. We will use the label **bad appeal to authority (argumentum ad verecundiam)** to refer to such a case.

The constraints on the acceptability of expert testimony are obvious, falling under those same broad headings of credibility, privilege, consistency, and corroboration. The authority cited should be clearly identified. Who is this person and with what background? Advertisements for many medicines commit fallacious appeals to authority by using expressions like "experts agree that. . . . ," "doctor-tested," and "studies hail the effectiveness of this product."

A second constraint upon the acceptability of expert testimony is that the 'expert' really must be an expert. Admittedly, spelling out exactly what makes someone an expert in an area is difficult, but just reading a few books on the subject does not suffice. A Ph.D. or other advanced degree from a respected university in the relevant field, publications in respected journals, and membership in relevant professional societies are indications, though not guarantees, of expertise.

The authority cited should be unbiased. For instance, a doctor who testifies that cigarettes are harmless deserves little credence if a tobacco company pays her salary. If the doctor offers objective evidence rather than personal expertise for the claim, that is a different matter.

The proposal or claim under discussion should be in the area of competence of the authority cited. Hard as it is for some people to believe, a philosopher's or a physicist's word on economic matters deserves no more credence than anybody else's, unless that person happens to be an expert in both fields.

The authority cited must base the opinion on evidence, and that evidence (usually research findings) must be open to inspection by other experts. If other experts disagree, they should be mentioned. Lawyers often violate this constraint by "shrink shopping," that is, by going from psychologist to psychologist until they find one who will say that their client is insane, and then using only that testimony in court without mentioning the fact that those other psychologists disagree.

If something that some expert wrote is being used, we should demand that the authority be quoted in full, so that the testimony is not distorted or misrepresented. To argue that morality is relative by 'citing' Einstein as saying "everything is relative" is to commit a fallacy of the silliest sort.

Finally, the authority should be up to date. Aristotle was once a great authority on physics, but using his words in that field now would be rather ridiculous.

Here are a few faulty appeals to authority.

> New university super "crash-loss" diet turns ugly fat into harmless water and it flows right out of your system by the gallons! Works so fast, you shrink your waistline as much as a full size smaller in just 24 hours (a full inch in a single day), and 4 sizes smaller in just 14 days! That's right! Four inches gone in just 2 weeks!

Which university? Who developed the diet?

> SKINNY LEGS—Try this new amazing scientific home method to ADD SHAPELY CURVES at ankles, calves, thighs, knees, hips! Skinny legs rob the rest of your figure of attractiveness! Now, at last, you can try to help yourself improve underdeveloped legs, due to normal causes, and fill out any part of your legs you wish, or your legs all over as many women have by following this new scientific method. This tested and proven scientific course was prepared by a well-known authority on legs with years of experience.

To what well-known authority on legs is the advertisement referring?

By the way, if you take too much of the skinny-leg formula, we have this:

> FAT LEGS—Try this new amazing scientific home method to reduce ankles, calves, thighs, knees, hips, for SLENDERIZED LEGS. Beautifully firm, slenderized legs help the rest of your figure look slimmer, more appealing! Now, at last, you too can try to help yourself to improve heavy legs due to normal causes, and reduce and reshape ANY PART of your legs you wish . . . or your legs all over . . . as many women have by following this new scientific method. This tested and proven scientific course was prepared by a well-known authority on legs with years of experience.

You wonder if it's the same well-known authority on legs!

Actor: My doctor wants me to switch to a decaffeinated coffee, because caffeine makes me nervous, but I only like real coffee.

Robert Young (who played "Dr. Marcus Welby" on TV): Phil, Sanka Brand Decaffeinated coffee is 100% real coffee and tastes it! Try it.

Robert Young is not a doctor, so why accept his prescription?
Richard Whately offers the following observation:[1]

> One of the many contrivances employed for this purpose, is what may be called the "fallacy of references"; which is particularly common in theological works. It is of course a circumstance which adds great weight to any assertion, that it shall seem to be supported by many passages in Scripture: How when a writer can cite few or none of these, that distinctly and decidedly favor his opinion, he may at least find many which may be conceived capable of being so understood, or which, in some way or other, remotely relate to the subject; but if these texts were inserted at length, it would be at once perceived how little they bear on the question; the usual artifice therefore is to give merely *references* to them; trusting that nineteen out of twenty readers will never take the trouble of turning to the passages, but, taking for granted that they afford, each, some degree of confirmation to what is maintained, will be overawed by seeing every assertion supported, as they suppose, by five or six Scripture-texts.

---

[1] Richard Whately, *Elements of Logic* (London: J. Mawman, 1826), p. 185.

The point Whately is making can be generalized. Often, a kind of ad verecundiam fallacy is committed when a point is buttressed by a mass of footnotes, where the authorities cited in the notes do not really agree with the point the author is trying to make. Indeed, scholars who assume that the absence of copious footnotes in an article or book somehow indicates the lack of scholarly merit commit the reverse fallacy.

A similar point can be made with regard to the presentation of numerical data and statistics. Often, the data will be presented by some stern-looking figure in a white lab coat, with elaborate charts and digital displays. All such rigmarole serves to lend an air of authority to the data. Such tactics can also be viewed (as we observed earlier) as a kind of positive ad hominem in that someone is arguing for a point on the basis of the alleged expertise of the person who plumps for it. As we shall see in Section 5.7, such a fallacy borders on special pleading as well.

5.3 Reasoning by Dilemma

Much of the reasoning we do is choosing, that is, reasoning devoted to selecting a course of action from a list of alternatives. When people buy new jeans, they have to select from alternative brands. When they vote in an election, they have to choose from among two or three candidates (unless write-in votes are allowed). When you choose a career, only a limited number of realistic career alternatives are open to you.

Not every choice is difficult. If you are offered the choice of $100 or a punch in the nose, it would be no tough choice. But some choices are difficult. We call these situations *genuine dilemmas*, situations with only a few alternatives, where a choice must be made among those alternatives, yet where each involves unpleasant consequences. As an example, consider the dilemma of a cancer patient who has but two alternatives: take up a painful course of chemotherapy, or else die. The agony is that a choice must be made, and both alternatives are distasteful.

Thus, the form of any reasoning by dilemma is:

1. Either A or B.
2. If A, then C.
3. If B, then D.

∴ Either C or D.

when C and D (the consequences) are both unpleasant, and where A and B (the alternatives) are the only possible courses of action. The argument form is clearly valid. We can use the argument form to select a course of action.

1. Our alternatives are *A* and *B*.
2. *A* has consequence *C*.
3. *B* has consequence *D*.
4. *C* is better than *D*.

∴ *A* is what we should do.

But—and here we shift from thinking about arguments to thinking about dialogues—this pattern of argument is often used as a debating tool. To *place your opponent in a dilemma* is to show that the position puts the opponent in a true dilemma.

Consider this example:

> An avowed atheist is arguing with you about God's existence. He suddenly says, "Look, you say God is all-good, all-knowing, and all-powerful. If He's all good, then He should not want to see suffering. If He's all-powerful, He should be able to stop people's suffering. If He's all-knowing, He should know about the suffering. Yet the suffering of people continues. So God is either not all-good, all-knowing, or all-powerful. That means that God, as you've defined Him, cannot exist."

This is a dilemma so old that it has acquired a name: *The Argument from Evil* (or the problem of evil). If we examine the argument, we see that it involves a dilemma. First, an argument is given that sets up the dilemma.

1. Evil exists.
2. If Evil exists, then God is either not all-good, all-knowing, or all-powerful.

∴ God is either not all-good, or all-knowing, or all-powerful.

Since three alternatives are involved, we call this a *trilemma* (Greek for "three legs"). The dilemma argument then follows:

1. If God is not all-good, God, as you've defined Him, cannot exist.
2. If God is not all-knowing, God, as you've defined Him, cannot exist.
3. If God is not all-powerful, God, as you've defined Him, cannot exist.

∴ God, as you've defined Him, cannot exist.

We use this classic argument not to digress into theology, but to illustrate how one participant in a dialogue can place the other in a dilemma.

How can reasoning from dilemma be fallacious? Here is where the inductive nature of such reasoning enters. Usually, not all of the logically possible alternatives are taken into consideration. This leaves it possible that one overlooked alternative might be the best and even involve no unpleasant consequences. The fallacy of **false dilemma** occurs when the arguer assumes that a genuine dilemma exists when it does not. Typically, a false dilemma is set up when the arguer either deliberately or unintentionally overlooks genuine alternatives. Indeed, this fallacy can shade into *strawman*—an opponent can represent an opinion as being an exclusive "either-or" and rebut it easily, when it really was meant to allow more alternatives.

False dilemmas often arise because people confound **contradictories** and **contraries**. *Contradictory properties* (characteristics) are such that any given thing must have one of them, but cannot have both completely, at the same time. For example, these pairs are contradictories: red/not red; rich/not rich; happy/not happy. *Contrary properties* are properties such that any given thing cannot have both, but may lack both. Examples: red/yellow; rich/poor; happy/sad. To begin an argument from dilemma using contrary properties is to open the way for a false dilemma:

In this country, you're either rich or poor.

ignores the possibility of being middle class.

"Either you're for me or against me!"

overlooks the possibility of someone being for you in most respects, but against you in some other respects.

So, in a dialogue in which you have been placed in a dilemma, the first thing to do is to check to see if it is a false one. This technique is called *going between the horns of the dilemma*, by pointing out overlooked alternatives. As an example, if a student body president asks in a speech before the students, "Should registration fees be increased or should student services be reduced?" we might reply, "Why not have lower fees and increased services through increased efficiency?"

Another way to rebut (reply to) a dilemma is to *grasp one of the horns*, that is, deny that one of the alternatives has unpleasant consequences. We can argue that the consequences are acceptable after all.

For example, suppose you favor government aid to private schools, and your opponent places you in this dilemma:

> Either government aid will help private schools grow or it will not. If it helps them grow, then there will be fewer students in public schools, and so some public schools will be forced to close. If it does not help private schools grow, then the money will be wasted.

The first premise, since it involves contradictories rather than contraries, is unassailable. You might rebut this by arguing that it would be quite all right for some public schools to close, since they might not be particularly good to begin with.

5.4 Reasoning by Analogy

We often *reason by analogy*, as when a person figures that the Toyota Celica she just bought will last over 100,000 miles because she has owned several other Toyota Celicas, and they lasted that long. Argument by analogy has this form, spelling out in detail the analogy between the things compared:

1. $A_1, A_2, \ldots, A_n$ and B share properties $P_1, P_2, P_3 \ldots$
2. The As have property Q.

∴ (Probably) B has Q.

Arguments of this form can be good or bad, depending upon whether certain constraints are met. We will discuss the constraints on reasoning from analogy in detail in Chapter 8. For now, let us observe the most crucial constraint: no major difference should occur between the things compared relevant to the issue at hand.

When someone argues from analogy, yet overlooks a relevant significant difference between the things compared, we say that person has committed the fallacy of **false analogy**. Here are a few examples of the fallacy of false analogy.

> Despite the fact that there is no evidence that the pilot of a small plane at Houston was at fault in any way, the favorite solution, suggested in the media, has been to ban or restrict small plane operations at major airports. This suggestion ignores the fact that airports are public property, paid for by all taxpayers, just like we

pay for highways. (No doubt, there are bus and truck drivers who would like to see private cars banned or restricted on major highways). The solution is for the airlines to build and operate their own private air-transport terminals—just as the railroads once built and operated their own stations.

A key difference between private autos and private planes does, however, exist. Air traffic must be centrally controlled and channeled, whereas motor vehicle traffic need not be.

> Editor: I believe that the fraternity system should be abolished on all colleges in the United States. The reason for this belief is simple. The behavior of many fraternity members is spiritually reminiscent of Nazis of Hitler's Germany, as well as the leadership of the present-day Communist Party in both Russia and China. This may well seem to represent an extreme viewpoint, but I believe it to be a fair and rational conclusion in view of all the facts. For example, you may recall not only the recent destructive behavior of certain pledges in terms of noise and property (palm trees), but also the recent law passed against hazing (an activity that has taken the lives of a number of students). There was also the fraternity member who endangered the lives of several thousand students by flying his plane at nearly ground level a few semesters ago.

But fraternity members have not killed millions of people, nor have they invaded other countries.

> Bumper sticker: "Guns don't cause crime any more than flies cause garbage."

However, flies do not play a role in the creation of the garbage, while guns do play a role in creating crime.

EXERCISES 5.4

Each of the following passages contains one or more of the fallacies discussed in the chapter so far: bad appeal to authority, false dilemma, and false analogy. For each, state what is wrong with it, and only then label the error. (An asterisk indicates problems that are answered in the back of the book.)

1. A doctor in South Dakota had a remedy for rape, which he symbolically attached to a note scrawled on paper from a prescription pad. "All farmers know how to prevent violence in the barnyard," he wrote. "Their method, used for thousands of years all over the world, is easy, cheap and 100% effective. They transform potentially vicious bulls into socially useful oxen. . . . This method would put the fear that is like no other fear into potentially lawless adolescents."

* 2. A famed research team of sensitive psychologists, plus computer technology, create the amazing new "Astral Sounds."

3. "For the first time in 25 years, I passed my driver's test without glasses!" (C. Smith, CT) Exercise your eyes! Without Glasses or Contacts. Doctor-Approved program shows you simple exercises to build up clear eyesight at home. Call for free brochure.

4. Many of my neighbors and I are disgusted with the current practice of towing cars off of private property in Myrtle Beach.

For example, on September Third, at approximately 11:15 a.m., our houseguests parked in front of our garage. Minutes later, the meter maid arrived and ticketed their car. A few minutes later, the tow truck arrived and, against our protestations, towed the car away while we helplessly stood by. The entire episode took less than 15 minutes.

This deplorable act, in a society that espouses justice for all, makes me ask, justice for whom? Perhaps justice is only for those petty tyrants who are, unfortunately, in positions of authority and power, and incidentally, on the public payroll. Our civil servants are hired to serve us, not to harass us. We can do with less Gestapo tactics and fairer treatment for all.

5. Lose Weight and Keep it off with the Physicians' Clinical Diet Plan! It's simple! It's easy! It works! Written by a physician and based on sound medical principles! No pills or protein supplements. Rush $5.95, check or money order, for prompt delivery.

* 6. A married woman in her late twenties says she has been going topless for the last four years when the weather is pleasant, when she is working in her yard, driving her car, or riding a motorcycle with her husband. . . . Sunday, State Highway Patrolman, T.L. Wolfe, stopped her and her husband while she was topless on the back of their motorcycle. . . . Wolfe said he later let her go because there is no law prohibiting her from being topless in public. "I guess it's not legally indecent to do that," he said, "but I still believe it's improper. It could cause accidents." The woman's husband supports her action. "You can't have two sets of moral values, one for men and another for women," he said. In support for her own actions, the woman said, "If a man can go without a shirt, then so can I. There's not much difference between the chest of a man and the chest of a woman. The only difference I can see is there's a little more fat on a woman's, and a little more hair on a man's. . . ."

7. Either you love me or you don't. If you love me, you'll propose to me right now. If you don't love me, you'll want to say goodbye forever.

8. YES—see the amazing proof in your own mirror! With the New TotaLoss Fat Burn-Off Program, just one powerful pill in the morning launches you on the most incredible 24 hour fat burning blitz. Doctors, studies at Universities, and leading magazines hail the awesome effectiveness of the TotaLoss capsule formula.

9. (A classic dilemma.) You should believe in God. If God doesn't exist, then all you've lost is a few Sunday mornings, but if God does exist, then you have gained eternal bliss.

* 10. (Another classic dilemma.) You shouldn't go to college. After all, either you know what will be taught or you don't. If you know what will be taught, then you don't need to go. If you don't know what will be taught, you won't know what to take.

11. You deserve good health—and we want to help you get it by offering you our BEE POLLEN at a reduced rate only with this coupon. Try our BEE POLLEN NOW and save money.

Bee Pollen has been acclaimed by athletes, health experts and scientists all over the world as a fantastic natural food supplement that provides increased energy, extra stamina, and better health. Now you can try it at no risk! Read all the facts below from the Medical Front thoroughly, then order today with this coupon!

—German research indicates that Bee Pollen may be helpful in reducing the bad effects of stress of all kinds, from arguments with your spouse, to the after effects of radiation therapy.
—A Russian scientist reported good results in the treatment of chronic colitis with Bee Pollen. Bee Pollen taken daily has been shown to reduce the number of harmful bacteria in the digestive tract and may work to correct both constipation and diarrhea.
—Bee Pollen is being used effectively in building immunities to 90% of all allergies.
—Swedish physicians regularly give Bee Pollen in the treatment of prostate problems. Hundreds of U.S. doctors are now doing the same.

12. New Hampshire students will be going to the polls this week and voting the same way they cast their ballot for class president, and will be picking the leaders of this country exactly as they choose their student council officers.

But who can fault them?

They don't know any better, because they are not being taught any better. Schools are not inculcating students with the idea that politics is essential in keeping democracy alive.

When teachers choose not to discuss politics in class, they are showing by example that politics is evil and should be avoided. Instead of guiding students through the confusing political maze with a helping hand, schools are leaving young voters in the middle of a dark street, forcing them to cast their first ballot with only the most primitive understanding of the candidates.

13. As the mother of a freshman at Henning High School, my stomach turns every morning when I take him to school and see the demonstration by the "teachers." Whatever happened to the dedicated teacher who was only interested in educating and forming the minds of young people? Does such a creature exist anymore or are they all in the teaching profession for money?

Granted, we all like and need raises from time to time, but can't this be worked out while on the job? The teachers say they have a right to strike. Do police and fire fighters have the right to strike and deny the public their right to protection? Do soldiers have the right to strike? Certainly not!

* 14. Polly Bergen: No Singer machine has ever saved you more and given you so much. This Touch and Sew II machine has a Soft-Touch Fabric Feed for smooth feeding of all fabrics, a Flip and Sew panel for easy sewing of armholes, cuffs and sleeves, and exclusive slant needle that's easy on your fabrics, a two-step built-in buttonholer, and more! Made in the U.S.A.

Small Print: Polly Bergen is a member of the Singer Board of Directors.

15. Stay younger longer!! Kungpao powder helps reverse aging! At last, you can live life to the fullest. Recent scientific research reveals Kungpao powder can:
    - increase sexual potency
    - increase mental ability
    - improve blood circulation
    - help you sleep better

Twenty million people use Kungpao powder daily—20 million people can't be wrong! Send your check today.

16. For a woman, "her world is her husband, her family, her children, and her home ... neither sex should try to do that which belongs to the sphere of the other."

The above quotation sounds like Phyllis Schlafly, who founded the Eagle Forum to lobby against the ERA, but it isn't. Although Schlafly has said the same thing, this quotation is of Adolf Hitler. This is not all these two have in common.

Phyllis and Adolf share the idea that the male-dominant family structure is, in Phyllis's words, "the basic unit of society." Adolf called it the *Keimzelle*, or basic cell of Nazi society. Hitler liked the patriarchal family because it is the perfect model of dictatorship. The Führer was the embodiment of the "Fatherland," and all Germans had to obey his will. And because the male-dominant family crushes any concept of individual rights by submitting the wife and children utterly to the father's will, it was the ideal base for Hitler's government.

5.5 Generalization and Particularization

The next two fallacies on our list, hasty generalization and accident, are converses of each other. Both have to do with moving between general rules (*generalizations*), and particular cases (*particularizations*).

Hasty generalization: We have noted before that not every appeal to authority is fallacious, not every dilemma is false, and not every analogy is bad. Instead, testimony, dilemmas, and analogies (comparisons) must meet constraints—and when they do so, they are logically acceptable.

Similarly, we must acknowledge that people often generalize on the basis of what they see, and they may be quite right to do so. If we see 30 people die after eating pink mushrooms from a patch, we would be illogical if we did not conclude that all those mushrooms are poisonous. But we have to discuss 'constraints'—criteria that any generalization must meet for it to be logically acceptable.

Let us put the form of generalization thus:

1. All observed As have property P.

$\therefore$ All As are P.

The cases we have observed we call 'the sample,' the whole group we call 'the population.' This terminology derives from one especially useful form of generalization: political polling.

We can now state the constraints, the criteria governing the acceptability of any inference from sample to population. First, the sample has to be sufficient—it is a fallacy to generalize on too few cases. Thus, it would be illogical to conclude that all cats are friendly because the two at your friend's house are affectionate.

The second constraint is that the sample (the cases upon which you generalize) must be representative, not unusual or atypical (often termed 'biased'). This is just common sense. It would be illogical to conduct a poll on who will win the next presidential election standing outside Tiffany's Jewelry Store in Beverly Hills or Manhattan. But detecting bias in your sample is not always easy.

In 1936, the (then famous, now defunct) magazine *Literary Digest* conducted a poll, and concluded that Alf Landon would win the presidency over Franklin D. Roosevelt. Roosevelt won with a landslide victory. The sample size was enormous—over two million people—but the magazine editors were unaware that the sample was biased. It was derived principally from telephone directories; thus it completely missed that segment of the population who could not afford telephones. At that time, only fairly well-to-do people could afford phones, and such people tended to vote Republican.

A full treatment of generalization can be given only with the tool of mathematical statistics, but we do not need sophisticated mathematics to see that generalizing on one or two atypical cases is simply fallacious. Consider the following examples:

> ASTOUNDING REPORTS! A 48-year-old man finds relief from constant arthritic pain in only 10 days of treatment. A middle-aged woman can do housework again after three weeks. A grandmother, once totally affected by rheumatoid arthritis, found relief of pain and swelling in one month. "Inflammation vanished in 15 days and pain is greatly reduced," reported an elderly man. Can you imagine yourself or your loved ones enjoying freedom from pain, heat, swelling, and stiffness? It may be yours once again, regardless of your age. PROVE IT TO YOURSELF! It has been proven, without a doubt, that the Blessex formulation is an effective vitamin in relieving most of the symptoms of arthritis.

Or consider:

> Yes, Dear Friend, maybe you'd like to have more money, a better job, a new home, a nicer car . . . or more energy, a healthy body, a serene mind . . . or sincere friends, a great social life, affectionate loved ones, or anything else you want out of life. You can have all of this and more—easily and automatically! You need only whisper your wishes to the MIRACLE AMULET—and they will come true—through the miracle of universal mystic power. Countless others have already achieved incredible results using this sure-fire amulet—and you will too, once you start using it! Here's why. . . .
>
> I first bought my MIRACLE AMULET from a sage when I was overseas—A week later I was back in the U.S., and at that time, I remember being very tired—with a terrible backache to boot. So, just for the heck of it, I did what the sage prescribed, and in 5 minutes, I was bursting with pep, and energy—minus my back pain. A nice coincidence, I thought.
>
> —Two weeks later, I lost my gold watch. I searched high and low for it—everywhere—but no sign of it. Skeptically, I took out my MIRACLE AMULET and mumbled my wish to it. In 10 minutes the phone rang—it was the locker attendant at the health spa I frequent. "Sir," he said, "I found a watch under a towel near your locker. Is it yours, by chance?" —A few days later, I had to get to an important meeting. On the road, my car broke down—I was 10 miles out of town. I pulled out my

MIRACLE AMULET and said to it, "Please get me to my appointment on time." Five minutes later, a new Datsun whizzed up to the curb. "Need a lift, buddy?" a concerned voice asked. Miraculously, I arrived at my meeting five minutes early. Uncanny? Weird? What would you do at this point—believe or not believe?—Well, just to give the amulet a final test, I went to an interview for a job that paid almost twice what I was earning. I tell you, it was mobbed there—almost 200 applicants showed up . . . for just 3 positions. Grinning, I took out my MIRACLE AMULET, made my wish, and just sat back and relaxed. Then, 2 weeks later, I reported back there for my first day's work! There was no doubt in my mind any more—the MIRACLE AMULET worked! So, remember the advice of the sage, I found the greatest artist of the occult and had him copy my amulet exactly. Then I had it produced in quantity, and gave them out to all my friends.

In both advertisements, the paradigm of hasty generalization is committed: only a few positive cases are selected as proof. The reader has no assurance that the company placing the advertisement has randomly selected the cases—indeed, you may well suspect only favorable cases were chosen, thus biasing the sample. The first advertisement cites only three cases, and the second advertisement cites only one case! "It worked for me, so it will work for everybody"—that is hasty generalization with a vengeance.

Accident: This fallacy is the converse of hasty generalization, in that it occurs when a person particularizes, that is, moves from a general rule to a particular case. When we state a general rule, such as "honesty is the best policy" or "people should return what they borrow," we don't state every possible exception or qualification. Indeed, it would be impossible to do so, for as the world changes, new ways arise for rules to fail to apply. When a person either applies a rule to cases it was never meant to apply to, or applies it in a way it was never meant to be applied, we say the person commits a fallacy of accident.

Thus, it would be a fallacy to conclude that you should not report a robbery, say, because you believe that people should not be informers. That is a fallacious argument, because being an informer means intentionally spying on people and reporting what they do. The term 'informer' hardly applies to a person who happens upon a major crime in progress. A similar fallacy would be to decide to allow a friend to drive home after getting drunk at a party, for the reason that you do not like to tell people what to do. That is a laudable sentiment, but hardly applies to someone too drunk to rationally decide what to do and drunk enough to kill others.

The fallacy of accident arises in dialogues in two different ways. First, as in the previous examples, a person illogically can draw a conclusion about an atypical particular case from a loosely stated general rule. More subtle is a second kind of accident, in which a person tries to refute a general rule, that is, to overturn a general rule, by pointing out exceptions, but where those exceptions are atypical. For example:

> *Ms. A:* I doubt that people will readily accept mass transit. After all, private autos give people freedom of movement, something they prize highly.
>
> *Mr. B:* That's absurd. How can you speak about cars giving us "freedom of movement"? Haven't you ever been stuck in a traffic jam?

Here, Mr. B attempts to refute Ms. A's general rule—that cars allow us to move around freely—by citing the atypical situation of being stuck in a traffic jam.

Take care not to confound hasty generalization with the second kind of accident. In hasty generalization, the arguer tries to establish a generalization by looking at atypical cases. In the second sort of fallacy of accident, the arguer attempts to refute someone else's generalization by citing atypical cases, without necessarily going on to make a new (contrary) generalization.

5.6 Inference to Cause

A *causal claim* is any statement to the effect that something *A* caused (created, led to, resulted in, made) something *B*. Such claims are important in daily life—you want to know what caused the dent in your car, what is causing the pains in your chest, what causes hair loss, and so on. Such claims are even more important in science and law.

Establishing any causal claim is no easy task. Basically, it requires coming up with enough information to rule out all other plausible explanations of *B* except that *A* caused *B*. Thus to prove that Fred got his case of food poisoning at Burger Biggie, we need to show that he did not eat at any other restaurants that day, and that those who ate the same food at Burger Biggie also got food poisoning.

Since it takes so much work to prove adequately a causal claim, some people resort to illogical shortcuts. **False cause** is the fallacy of arguing that A caused B merely on the basis of a temporal connection between A and B. We can mention two ways in which the fallacy of false cause is committed.

First, *post hoc ergo propter hoc* ("after this, therefore because of this") is the fallacy of concluding that A caused B merely on the basis of the fact that A happened before B. For example, to assume that Nixon caused the 1972 recession because it occurred after he was elected is to commit this fallacy.

Second, a *correlation fallacy* occurs when we try to infer that A caused B merely on the basis that A and B are correlated (occur together). Other reasons could exist (besides A causing B) why A and B occur together. For example, by sheer accident my increasing loss of hair is correlated with the growth of Mexico City. On the other hand, A and B may occur together because both are caused by some third thing, C. For example, pains in the chest and cold feet may be correlated, not because one causes the other, but because heart disease causes both. Finally, A and B may occur together because they are reciprocal effects, that is, they cause each other. For example, some people conclude, from the indisputable fact that changes in the economic system cause changes in the political and cultural systems, that the economic system is the fundamental determinant of society. But this is fallacious, for changes in the political and cultural systems can cause deep changes in the economic system.

Politicians habitually commit the fallacy of false cause by taking credit automatically for whatever good things happened while they were in office and charging their opponents as responsible for the bad things that happened while those folks were in office. Consider these examples, taken from the Carter-Ford debates of 1976.

> Carter: As a matter of fact, since the late 60's, when Mr. Nixon took office, we've had a reduction in the percentage of taxes paid by corporations from 30% down to about 20%. We've had an increase in taxes paid by individuals. Payroll taxes from 14% up to about 20% and this is what the Republicans have done to us. Which is why tax reform is so important.
>
> Ford: I think the record should show, Mr. Newman, that the Bureau of Census, we checked it just yesterday, indicates that in the four years that Gov. Carter was Governor of the state of Georgia, expenditures by the government went up over 50%. Employees of the government of Georgia during his term of office went up over 25%.

And the figures also show that the bonded indebtedness of the state of Georgia during his governorship went up over 20%.

Carter's claim that the Republicans are responsible for the "bad" changes in the tax system is backed only by a temporal link. He needs to give more evidence—say, by showing that Republicans are the ones who passed the relevant tax laws. (Ford pointed out that the Democrats controlled Congress during that period.) Ford's implicit claim—that Carter was responsible for the increase in the cost and size of Georgia's state government while he was Governor—again rests on too little evidence. Did Carter's party control the Georgia legislature, which controls spending? Does the Governor in Georgia have the power to veto spending programs at will? These questions need to be answered for the temporal linkage to indicate responsibility.

Superstition quite often comes from false cause reasoning. Consider this silly story.

> The Terrifying Dog of Death—Four years ago, Turuzzu (this devil-dog) was sitting outside my aunt's home. Shortly afterward, she collapsed and died from a heart attack. "It is uncanny," Anna Carioto, 17, eldest daughter of bricklayer Pasquale Carioto, said. "On Oct. 26, 1977, my father died in a car crash. That same morning, that dog was sitting outside our house." Antonio Carchivi, chief of the traffic police, said, "Everyone feeds the dog, but nobody wants to see him sitting outside their house. Everybody is afraid, grown men included. Nobody wants to kill the dog, they are too scared. Last year when the dog had eczema, the local veterinarian treated him very carefully with lots of love and attention." The Major, an elementary school teacher, said, "There have been accidents and a murder in which the dog was seen outside the victims' houses beforehand."

Note the correlation fallacy between appearances of the dog and death.

EXERCISES 5.6

In this chapter, we have covered the following fallacies:

Bad appeal to authority: Appealing to authority is fallacious if it fails to meet the criteria given in Section 5.2.

False dilemma: Arguing on the assumption that a genuine dilemma exists, when one does not.

False analogy: Making a comparison in which significant differences between the things compared are overlooked.

Hasty generalization: Generalizing on the basis of too few cases (insufficient sample) or atypical cases (biased sample).

Accident: Applying a general rule to atypical cases.

False cause: Arguing that *A* causes *B* merely on the basis that *A* and *B* are temporally linked. (This includes *post hoc ergo propter hoc* and *correlation* fallacies.)

Each of the following passages contains one or more of these fallacies. For each, state what is wrong with it, and only then find a label for it. (An asterisk indicates problems that are answered in the back of the book.)

1. I can understand why unemployment is a major problem facing us today. And I can understand why it has become a political football. But I can't understand the idiocy of blaming President Clinton for the unemployment. It is the equivalent of blaming the Pope for the increasing crime rate.

* 2. Baldness is caused by tight neckties and collars and by combing the hair improperly, says a barber who has studied the problem throughout his 50-year career.

Fred C. Boor, of St. Louis, Mo., says his theories are proven by the fact that he takes his own advice and still has a full head of hair at the age of 77.

3. More women prefer a male doctor than a female doctor, a surprising poll reveals. Of 150 women questioned in Washington, D.C., New York, Los Angeles, and the Chicago area, 61 said they preferred men physicians, 46 said women doctors, and 43 had no preference.

4. If your skin is bothered by oiliness and acne, then you need BUF kit for Acne. Developed by a leading dermatologist, it has a super cleansing sponge—BUF-PUF—and a soapless cleanser—BUF Acne Cleansing Bar—in a neat little tray for sink or vanity.

5. Malcontent march. On Saturday, we witnessed an unsavory march down the streets of Washington, D.C.—the AFL-CIO malcontents. These unions are determined to make economic slavery a way of life for the non-union workers.

During the 1930's, Rudolf Hess led his Brown Shirts through Berlin in glorious victory marches that resulted in the enthronement of Adolf Hitler. Do our labor leaders have a Hitler they are grooming to attain their goals?

Non-union workers should avoid this bunch like a plague. They do not represent the working people of America, only union members.

* 6. I hope all the pro-choice people read "Board calls child abuse a national emergency" [news, June 28]. The anti-abortion camp has been stating for years that the overall devaluation of human life through abortion on demand is the primary reason for child abuse. But, of course, it seems no one is listening or willing to admit that there is a connection.

Roe v. Wade occurred in January, 1973. The article stated that the number of reported abuses increased by a factor of 40—that's right 40 from 1974 to 1989. Wake up, America! Abortion is having greater effect on our society than just the lives snuffed out before birth—a national tragedy in itself. Let's face it, abortion, in almost all cases, is simply a matter of convenience, an "easy" solution for a problem created by a lack of personal responsibility in the first place.

7. In fact, we have gone to a lot of expense and trouble proving the Bust Expander does what we say it does. During a medical test using six women that were supervised by two prominent doctors, it was discovered that changes caused by the Bust Expander were phenomenal.

8. Most female university students own at least one stuffed animal—and today's college coeds talk over their personal problems with the cuddly creatures.

That's the result of a survey of 245 women students at Florida State University. All but 10 of the women surveyed had a stuffed animal, and 75% of them admitted that they talked to it to relieve tensions, or hugged it for comfort, warmth, and security.

9. Doctors and nutritionists discover certain Vitamins can stop and reverse gray hair! It may be possible for you to restore and retain your natural hair color and prevent further graying without dyes, rinses, or other coloring agents . . . the natural way. Medical doctors and nutritionists claim that it is possible to return hair to its natural color simply by adding special high-potency nutrients to the diet. . . .

Scientists believe it is the inability of the body to process certain vitamins as we grow older that causes hair to turn white.

A medical doctor had an amazing 82% success ratio treating 460 patients with a vitamin supplement returning hair to its natural color, as reported in "Vitamins and Hormones II."

* 10. I just want to come to the defense of the so-called "gas-guzzler," as larger cars are known. Our gas-guzzler gets 17 miles per gallon on long trips and less in the short runs around the immediate area when we have to go to the grocery store. However, we use the car only two or three times a week at the most. Our gasoline bill is only $20–$25 a month. Now, I would like to mention an interesting point which most people have not thought of or observed. The little car that gets better mileage per gallon uses more gas in the long run. This small car is seen with skis on top going to and coming from the snow areas, and in the summer they are seen going to and coming from beach areas. I have asked quite a number of these small-car owners what their monthly gasoline bills are, and they usually tell me it is $90 a month or more.

11. This vitamin helps thicken your hair! Hair thinning? Scientific research shows that Panthenol (Vitamin B-5) may give your hair the fuller look you want. This amazing B-vitamin has the unique ability to penetrate

the hair shaft and moisturize from within—providing a lovely natural sheen and fullness as it helps strengthen each strand. Repairs split ends and brittleness, and makes your hair easier to manage, too. Try Mama's Panthenol "F" Hair Thickener for Women.

12. The research results are overwhelming. New National Smoker Study provides solid evidence that "Enriched Flavor" Dormans offer a satisfying alternative to higher tar cigarettes.

13. Here's my two cents worth: Crossing guards should be used only for the first three grades of school. Who herds the rest of the kids across streets all summer?

Secondly: Free meals should be done away with in schools. Where do the kids eat all summer? Some of these programs are fine, but only when there is an unlimited supply of money.

Thirdly: Busing should not be mandatory but should be for small fry who must walk more than a mile and a half.

Fourthly: No classes need to begin before 9 A.M. for the first three grades. I went through all the public school system this way and I was OK.

* 14. An Expert's approval—Supa Tweez has been clinically tested by a university professor of dermatology and proven to be safe and effective. One of his patients had previously been tweezing hairs from her chin every day for 15 years. After treating herself with Supa Tweez, she has eliminated this time-consuming chore for the rest of her life! Over fifteen thousand instruments in use by doctors.

15. "Fantasy Island's" Hervé Villechaize is telling pals never to trust a raccoon. Hervé offered a tender morsel of chicken to his pet raccoon, and the ungrateful beast not only bit the hand that fed him—he hung on painfully. Hervé suffered a two inch gash on his left index finger. No stitches, but the doc told him it'll be three months before he'll have full use of his hand.

16. Doctor Tested. Here are just a few of the incredible results obtained by actual Slim-Skins users in a special slimming test conducted by a prominent American Physician.

"I lost 4″ from each thigh, 2½″ from my hips, 6½″ from my tummy and over 5″ from my waist in just 3 days with Slim-Skins." —I.Z.

"Actually trimmed my abdomen 8 full inches and my waist nearly 7″ while my hips came down about 3½″ and each thigh 4″ for a total loss of over 26″ all in 3 days with the truly astonishing Slim-Skins." —C.D.

"The final solution in instant figure slimming. I lost 5½″ from my waist, 5″ from my tummy, 2¼″ from my hips, and over 3″ from each thigh—an overall loss of 19½″ in just one 25-minute period with Slim-Skins." —M.M.

These "fantastic" inch losses are now documented fact. The famous "fat-burning" diet claims a 3-inch waist loss in 7 days. In the doctor's Slim-Skins test, the waist losses were over 3-inches in just one day—actually in just 25 minutes.

17. Your boss has a bigger vocabulary than you have. That's one good reason why he's your boss.

* 18. Not long ago a poll of pregnant U.S. high school students revealed that of the 1,000 unmarried girls questioned, 984 had become pregnant with suggestive pop music as a background.

19. Proposition 13 would result in a 30% reduction in the county budget. Could your household budget stand that?

20. Patients with such diseases as heart disease and diabetes can refuse treatment if they wish. A heart patient has the right to refuse digitalis, and a diabetic can refuse insulin. Members of some religious sects refuse blood transfusions. Why, then, shouldn't a psychiatric patient also have the right to refuse medications?

21. I read your editorial about how college students didn't know who Stalin was, what is the Warsaw Pact, etc. Well, when I was in high school, I knew what the Warsaw and NATO pacts were; I knew who Stalin, Churchill, de Gaulle, and Roosevelt were. I knew at least 85 percent of the world's capitals, and if I was given a blank map of the United States or world, I could fill in all the states and most of the world's countries. I enjoyed and was interested in it.

But, unfortunately, I knew very little about math and other subjects. I am 29 years old and have held only menial jobs and am unemployed today. The knowledge I had one can't use. Math, computers, science you can use to become a success. Geography, history, etc., are useless. I found out the hard way.

* 22. Well, I remember when I lost everything we had: my job, our savings, my pension. . . . and we had to borrow money from our relatives to pay the rent and keep food on the table for our little ones. Now I am one of the highest paid executives in my field, my wife has her own business, our kids go to the finest schools, all the bills are paid, and we have money in the bank.

How did this all happen? Just when I was in the depths of despair, my wife read what Jesus said about the mustard seed to me: "If ye have faith as a grain of mustard seed . . . nothing shall be impossible unto you." Matthew 17:20

I thought to myself, "if only I had something to hold on to, that I could see and touch." So, I acquired some mustard seeds from the Holy Land, encased them in a credit card-size gold metallic prayer card, and began carrying it with me as an affirmation of faith. I was amazed at how my life changed. Not only did I begin to prosper, but my mental and physical health improved.

Are you or someone close to you having problems? I know what it's like to be desperate. I know what it's like to have what you need and want. And I know that you can decide which one it's going to be. What does it require? Faith. Faith and commitment enough to carry your mustard seed from the Holy Land and follow a few simple instructions. I guarantee it. If you will send a check or money order for $4.00 plus $0.70 postage and handling, I will send you a mustard seed from the Holy Land encased in the "Seed of Faith" prayer card with instructions for its use. If you return your "Seed of Faith" prayer card for any reason within 30 days after you received it, I will return your check or money order to you. Can your future wait?

23. Researchers have come up with an ensemble for men that will discourage advances from coeds as surely as bug repellent keeps off mosquitoes.

The outfit—an earring, a tank top, a fur coat, bell-bottom blue jeans and tire-tread sandals—topped the list of most repulsive clothing compiled on the basis of interviews with 56 college women.

"Tank tops are worn by men who try to show off muscles they don't have," sneered a shapely senior from Southern Methodist University in Texas. "And tire-tread sandals make a guy look like a holdover from Woodstock."

The women, who came from four Texas colleges, were asked to finish the sentence, "An item I particularly detest is . . ." Items mentioned by more than half the coeds were put on a most-hated list.

24. Talisman Changes Lives for Millions. An extraordinary phenomenon is sweeping the North American continent, it was learned from reliable sources here recently. A well-known but little understood woman who calls herself Madame Labonga is possessed by a power that allows her to dispense most of life's good things to whomever owns her specially designed Talisman.

Money, wealth, happiness, love, and prosperity unfold the moment the famous Labonga's Talisman is worn or carried. What is this Talisman? It is a specially minted coin that mysteriously confers upon the wearer or carrier an almost certain propensity for happiness and success in every venture of life. There is only one stipulation, however. Instructions must be followed meticulously because for some peculiar reason the effectiveness of the Talisman is dependent upon them. A set of these instructions is enclosed with each Talisman.

People who have worn or carried this mystical Talisman have reported an almost immediate reversal of their luck. Most report that sooner or later their lives are radically changed.

Others report a gradual, though definite, change. Madame Labonga is not available for comment though reporters have been dogging her footsteps for months now. Maybe that's the key to her success: Complete secrecy. And inaccessibility.

Now for those of you out there who are straining at the bit to change your luck here's how to do it. Simply send $2.50 for each Aluminum Talisman, $4.50 for each Bronze Talisman, or $9.45 for each Electro-Plated Gold Talisman.

25. Our pets get more attention than we do, says an expert. Pets get more attention and kind words than people in almost half of American homes, according to a prominent family therapist.

Dr. Ann Bolin, a professor of psychiatric nursing at the University of Maryland, surveyed 60 families about their relationship with their pets, and she got some surprising results. "Asked who in the family gets the most touches, words, smiles, and gestures, 44 percent said the pet," Dr. Bolin explained.

* 26. Kids who skip the prom aren't missing very much. I was quietly amused when I read the letter from the teenager who had a date to the prom. She wrote, "everyone had a terrific time." That girl is living in fantasyland. The truth is most people have a lousy time.

I am 48 years old. My prom was in 1951. I can't recall a worse evening in my entire life. Five days before prom time I didn't have a date, so I asked a senior if he would take me. He said OK, but I'd have to "go all the way." (That was how they put it in those days.) I refused and finally got an 11th grade kid to be my escort. (I had to pay his expenses.)

We triple-dated with two couples I didn't know very well. My date couldn't dance and no one else asked me. So I dragged around in agony half the evening and spent the other half in the powder room with six girls who were also having a rotten time.

When our own son asked if he should spend the money to go to his prom, I said it wasn't worth it but he decided to go anyway. Afterward he said, "You were right, Mom. I should have gone bowling."

This is to let the girls who don't get asked know that they aren't missing much.

27. Thirty-six percent of Americans believe they have been reincarnated, an ENQUIRER survey reveals—and some say they've even had flashbacks of their past lives.

In our survey—conducted in New York, Philadelphia, Washington, D.C., Chicago and Los Angeles—we asked a total of 50 men and 50 women this question: "Do you believe you've lived before in a previous life?" The results: 64 percent said "no," while 36 percent said "yes." And as many men as women believe they've been reincarnated.

5.7 The Fallacy of Special Pleading

One other inductive fallacy deserves mention: special pleading. **Special pleading** is biasing the presentation of evidence for a point. Often, special pleading involves not mentioning the negative evidence against a proposition being advocated. For instance, a student might write a history paper entitled, "Kennedy Caused the Vietnam War" by reporting every bit of evidence favorable to this thesis, and by ignoring every bit of negative evidence encountered.

The fallacy of special pleading is often committed when self-interest is strongly involved. An obvious example is a salesperson trying to sell a product to a customer: the salesperson will tend to focus only on the positive features of the product, leaving it to the customer to dig out the negatives by clever questioning or (sadly) bitter experience. Another obvious example is a politician running for office: the politician will tend to describe all the good things that happened while he or she was in office and ignore or downplay the mistakes made during the same period.

But it isn't always obvious how self-interest leads to special pleading. For example, it is not unusual to see someone take a position and specially plead to prove it, because he feels he would "lose face" if he loses the argument. Many a scholar has selectively presented evidence to prove a theory because his or her reputation is identified with that theory.

Besides suppressing evidence, evidence can be biased in the way it is presented or stated. As we shall see in Chapter 6, loaded language can be used to bias the evidence, as can technical terminology and graphs.

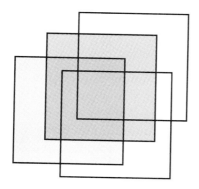

6

Fallacies of Language

6.1 Pitfalls of Language

In this chapter, we discuss fallacies arising from features of language. This is a change from what we have been doing in Chapters 4 and 5, since up to now we have focused primarily upon the ways in which statements can fail to support other statements. But statements are expressed in words; hence the logician must pay attention to the words used in any argument and the manner in which they are used.

We examine fallacies arising out of four features common in both ordinary and technical language: vagueness, ladenness, understatement, and ambiguity. After that, we discuss a tool for combating those fallacies, namely, **definition**. We conclude the chapter by discussing fallacies that arise from technical language (that is, language used to convey numerical and statistical data).

The four features of language under discussion (vagueness, ladenness, understatement, and ambiguity) are not by necessity bad or deceitful. These features must be present in a natural language if it is to facilitate communication and learning. But what is by nature good can be put to use in bad ways, as we shall see in this chapter.

6.2 Vagueness and Slippery Slope

Many concepts have imprecise borderlines. Consider the notion of being rich. Clearly, having an income of $5,000,000 a year qualifies you as rich; it's also clear that having an income of $5,000 a year does not qualify you as rich. But— as the cliché goes—where do you draw the line? Does $50,000 a year make you rich? Most adjectives that deal with matters of degree ('rich,' 'happy,' 'ugly') allow borderline cases. Yet, as the philosopher Ludwig Wittgenstein (1889–1951) pointed out, even common nouns can be open to borderline problems. His example was the word 'game.' Poker, chess, and tennis are clearly games. But how about Russian Roulette? Bullfighting? Dog racing?

A word is **vague** if it lacks a precise meaning. Hence it allows for borderline cases where it is not clear whether the word applied to a given case or not. Many words are vague to some degree. The word 'vague' is itself vague. But this need not be so intimidating. Vagueness allows language to grow in the face of new conditions. The very vagueness of the word 'rich,' for example, enables it to continue to be meaningful in times of inflation (where the real value of currency diminishes at a rapid rate).

However, when a key term in an argument is so vague that participants cannot figure out its application or when vagueness impedes a discussion, the participants have to resort to defining terms more precisely. For instance, a discussion about whether 'the rich' are paying their fair share of taxes will likely be a waste of time, unless the participants get a precise idea of what income level is to be counted as 'rich.'

As we shall see with ambiguity as well, a word can be vague without being used vaguely. The same feature (vagueness, ambiguity) can help or hinder a dialogue, depending upon its use and context.

There is a tie-in between vagueness and slippery slope arguments, which we discussed in Chapter 4. We defined a *slippery slope* argument as the fallacy in which the issue at hand is changed by degrees. What we did not say then (because we had not talked about vagueness) is that slippery slopes often exploit vagueness.

Consider an example:

> *A* argues in favor of allowing political refugees to be granted citizenship without having to wait the normal period of time. *B* replies: "Look, where are you going to draw the line? Any person trying to leave their country can be considered political refugees, since bad economic conditions are the result

of inept government. But we don't have room for all the poor people of this world!"

B has exploited the vagueness of the phrase 'political refugee' to change the issue. The best reply *A* could make is to point out to *B* that the existence of borderline cases does not preclude clear nonborderline cases. Clearly, a person who has opposed a government and is fleeing from that government is a political refugee, while somebody who is looking for a better job is not a political refugee, and whether such economic refugees should be granted asylum is irrelevant to *A*'s point. Borderline cases may exist (such as small shopowners leaving a country just taken over by Marxists), but that does not make *A*'s claim so hopelessly vague that *B* cannot address it rationally.

6.3 Ladenness and Loaded Language

A second feature of language that can be employed to commit fallacies is ladenness. A word or phrase is **laden** if it has theoretical and/or emotional overtones. A word can be theory-laden or emotion-laden, or both. As examples, consider the words 'bourgeois' and 'proletarian.' Both are laden with Marxist economic theory (which posits those economic classes) and with emotion ('proletarian' is laden with positive emotional connotation).

Terms can be theory-laden without being emotion-laden ('electron,' 'ionic bond') or emotion-laden without being theory-laden ('dirty rat,' 'punk').

Ladenness is an extremely pervasive feature of language. Most words are likely to be laden with overtones. That is not per se bad. It is how the ladenness is used that determines whether it is permissible. If a theory-laden term is used in a situation in which the theory at hand is not under question, that is okay. If an emotion-laden term is used in a context (a conversation, say) in which expressing emotion is the central purpose, that is fine. (You should review the discussion of conversational orientations given in Chapter 3.) But if the language is loaded to persuade without proving, that can be bad. Such a situation (in which loaded language is used to slant evidence) is often called **biased description**.

Bertrand Russell's "conjugation of an irregular verb" best conveys the nature of biased description: "I am firm, you are stubborn, he is a pig-headed fool." The same trait (holding fast to one's beliefs) is described favorably or unfavorably by using loaded terms. Biased description is the stock in trade of the "yellow journalist" and the propagandist. For instance, a leftist journalist might describe an incident (a riot by some of the workers at a factory) this way:

> Yesterday, the oppressed proletarians at Smith Clock Works rose up in indignation. The fascistic agents of the capitalists battled the heroic workers for hours.

Yet a rightist could describe the same incident this way:

> Yesterday, union goons and shiftless malcontents tried to destroy private property at the Smith Clock Works. Brave public guardians tried for hours to restore law and order.

All we can safely conclude from reading such slanted accounts is that some violence occurred yesterday at Smith Clock Works. A more neutral description might be:

> A riot occurred yesterday at Smith Clock Works, apparently caused by labor discontent. The police fought the rioters for several hours.

Biased description may be impossible to avoid completely. Ladenness is, like vagueness, a matter of degree. But differences in degree can amount to differences of kind that are important. The following kind of description (given by a writer trying to argue that the National Science Foundation should not have funded a project) is so biased, so loaded, that the argument is worthless.

> Now our government's really gone loco with your tax dollars—blowing $6,710 to study how religion influences life in a small Spanish town!
>
> The nitwit-run National Science Foundation (NSF) has awarded money to a U.S. university student who plans to go to the province of Huelva in Spain for his incredibly useless study.
>
> The student got the grant after convincing NSF's bozos that it's important to examine the influence of religion on the politics and economics of Huelva.
>
> An NSF official insists the project is a "well-thought-out proposal which will help us understand meaning and power in complex societies."

But its meaning for U.S. taxpayers is already clear—they're getting shafted for more bucks, declares Rep. Bill Chappell (D-Fla.). "I doubt whether the NSF should be spending $6,710 on such a personal academic project," he said.

When people use language highly laden with overtones they have no right to employ in the discussion at hand, we say they are committing the fallacy of **loaded language**. Loaded language could be viewed as a type of special pleading. By loading language, people can slant their evidence to make it more favorable to a conclusion. Indeed, if the loading goes far enough, the fallacy of loaded language becomes a case of begging the question (through the use of *question-begging epithets*).

Two points bear repeating. First, vagueness and ladenness are matters of degree, but degree can be very important. Second, something can be vague or laden without being used to commit a fallacy.

EXERCISES 6.3A

(An asterisk indicates problems that are answered in the back of the book.)

1. Think up three "conjugations of irregular verbs" similar to Russell's.

* 2. In the following passage, substitute for the emotion-laden terms words with the same literal meaning that do not have emotional overtones.

> Ah, make the most of
> what we yet may spend,
> Before we too into the
> Dust descend;
> Dust into Dust, and
> under Dust, to lie,
> Sans Wine, sans Song,
> sans Singer and—sans End!
> —*The Rubaiyat of Omar Khayyam*, translated by Edward Fitzgerald.

3. The following words are vague. Explain how.
   a. Intelligent
   b. Wealthy
   c. Happy
   d. Strong
   e. Large

EXERCISES 6.3B

Each of the following passages contains either vague or loaded language. Indicate the words that are excessively vague or loaded. (An asterisk indicates problems that are answered in the back of the book.)

1. Fuzz-brained bureaucrats are actually spending $32,744 of your hard-earned tax money on safari for a study of child-rearing in Africa.

   The ridiculous grant was awarded to a researcher from a New England university by the National Institute of Mental Health. Here's the gobbledygook explanation of the study's aim, as it was listed on the grant request: "The theoretical treatise on the cultural context of child development will serve as a force for the decentralization of current psychological approaches to development as well as an investigation of anthropological and psychological theories."

* 2. Savage cuts—The federal appropriation for public broadcasting has already been drastically cut for the fiscal years 1982 through 1985, and now further savage cuts are proposed for this period.

   This is compounded by the targeting of cuts in the funds for agencies that traditionally support public broadcasting programming.

   It would seem that a concerted effort is under way to deprive the American people of anything that remotely resembles cultural intelligence.

   People who think that the public deserves more than the insulting rubbish put out by the commercial networks should notify their congressional representatives of the fact. The alternative is all wasteland.

3. Science is attacked these days from both the left and the right. The disputes described here represent a reaction from a highly conservative population; but interest in local control, increased participation, objection to the dominance of scientific values and the role of expertise will be familiar to people located anywhere along the spectrum of political ideologies. The textbook disputes provide a means of exploring the relationships between science and its public, and of examining how the growing criticism of science can bear on public policy.

   —Dorothy Melkin, *The Creation Controversy*.

4. As Americans, we don't worry much about terror. Not the way people do in the Middle East or Northern Ireland, where it has been an instrument of government policy for decades. In the United States, we have seen relatively little organized violence for political ends. But terror is not really about violence. It's about fear, and the climate of intimidation, repression and chaos such fear creates.

   In the eighties, America has given birth to a new form of terror, a campaign of fear and intimidation aimed at the hearts of millions. It is in two great American arenas—religion and politics—that this new terror has raised its head. In the past few years, a small group of preachers and political strategists have begun to use religion and all that Americans hold sacred to seize power across a broad spectrum of our lives. They are exploiting this cherished and protected institution—our love of country—in a concerted effort to transform our culture into one altogether different from the one we have known. It is an adventurous thrust: with cross and flag to pierce the heart of Americans without bloodshed. And it is already well under way.

—Flo Conway and Jim Siegelman, *Holy Terror.*

5. This is nuts! Your tax dollars help foreigners compete with U.S. Farmers!

While American peanut farmers are battling to survive a crop surplus and low prices, lunatic bureaucrats are squandering nearly $10 million of your tax money to help foreigners compete with them.

Peanut-brains at the Federal Agency for International Development (AID) are shelling out $9.85 million in an insane scheme to aid peanut growers in Senegal, Trinidad, Cameroon, Nigeria, Sudan, Thailand and the Philippines.

* 6. We have concluded from studying lonely people that some of these reactions are much more helpful than others. Solitary television viewing—the most common diversion—seems to be almost as destructive as solitary drinking or pill taking, for example. Other, more active forms of solitude, such as reading and letter or journal writing, contribute to personal strength, self-awareness, and creativity. Establishing intimate ties with others is even more helpful. Since loneliness reflects a need for intimacy, friendship, or community, remedies that don't include these provisions won't work.

—Carin Rubenstein and Phillip Shaver, *In Search of Intimacy.*

7. I have just been visually assaulted for a second time today by Channel 11. The latest of these insults, aired in the name of the Humane Society, showed, in slow motion, parts of certain rodeo events where bovine creatures were wrestled to the ground by their horns, or were thrown by means of a lasso. Both scenes focused on the twisting of the animals' necks in these events. The PSA then went on to bluntly question the "humanity" of the people partaking in the rodeo.

This position, when taken to its logical end, calls for the elimination of the "beastly" humans (and conversely, the worship of the "divine" cows).

Are we in India? Should we all become Hindus?

Perhaps the Humane Society should also condemn any dog owner—as every dog born will strain any chain beyond belief. And all who eat chicken must surely be savages, having wrung their necks as the cowboys do to the poor "divine" cows.

Personally, I am appalled at this blatant attempt to denigrate America and the sport which was derived from the success of her pioneers in their battle to subdue the wilderness.

8. I was asked by the Medical Committee for Human Rights to speak on sex. All right, I really will.

The oppression of women by men is the source of all the corrupt values throughout the world. Between men and women, we brag about domination, surrender, inequality, conquest, trickery, exploitation. Men have robbed women of their lives.

A human being is not born from the womb; it must create itself. It must be free, self-generative. A human being must feel that it can grow in a world where injustice, inequity, hatred, sadism are not directed at it. No person can grow into a life within these conditions; it is enough of a miracle to survive as a functioning organism.

Now let's talk about function. Women have been murdered by their so-called function of childbearing exactly as the black people were murdered by their function of color. The truth is that childbearing isn't the function of women. The function of childbearing is the function of men oppressing women.

It is the function of men to oppress. It is the function of men to exploit. It is the function of men to lie, and to betray, and to humiliate, to crush, to ignore, and the final insult: It is the function of men to tell women that man's iniquities are woman's function!

I'm telling it to you as straight as I can. Marriage and the family are as corrupt institutions as slavery ever was. They must be abolished as slavery was. By definition they necessarily oppress and exploit their subject groups. If women were free, free to grow as people, free to be self-creative, free to go where they like, free to be where they like, free to choose their lives, there would be no such institutions as marriage or family. If slaves had had those freedoms, there wouldn't have been slavery.

—Ti-Grace Atkinson, *Amazon Odyssey*.

9. When definitions are attempted of such concepts as Freedom or Duty, they usually consist of other abstract terms that need defining. The more careful and elaborate the definition, the harder it is for the arguing parties to remember and observe it. In the end, human agreement depends on good speechways, that is, on controlling one's language by visualizing—and not switching—the same concrete embodiments of the concepts being tossed about in discussion.

For it is easy to think and talk without the image of anything at all present to the mind; words—the bare familiar sounds ending in -tion or -ity or -ness—will keep thought moving and make the thinker feel rational though he literally does not know what he is talking about. If the user of concepts—which is to say everybody—habitually fails to think of persons, things, events as they happen in the world of particulars, steady abstraction will land him in sheer nonsense or dangerous folly. For abstractions form a ladder which takes the climber into the clouds where diagnostic differences disappear.

—Jacques Barzun, *A Stroll with William James*.

* 10. Featherbrained bureaucrats are wasting a staggering $130,000 of your hard-earned tax money to study the mating habits of song sparrows.

The outrageous $130,000 grant by the National Science Foundation was awarded to a researcher at a New York university.

He'll tiptoe through field and forest spending your money to explore "the influences of a mate and territory on the fine temporal adjustment of reproductive functions in the sparrow."

Rep. Ron Paul (R-Texas) charges: "It's incredible that our fine-feathered friends at the NSF would find this unbelievable way to squander our hard-earned tax money.

"It's unthinkable to spend $130,000 to find whether the presence of a mate, territory, nest sites, nesting materials and a sufficient supply of food for feeding young affects the mating of the song sparrow.

"With many taxpayers worrying how to provide food to feed their young, wasting $130,000 of their money is not birdseed."

11. "You are dictatorial." My dear sirs, just as you say. That is just what we are. All the experiences of the Chinese people, accumulated in the course of several decades, tell us to put into effect a people's democratic dictatorship. This means that the reactionaries must be deprived of the right to voice their opinions; only the people have that right.

Who are the "people"? At the present stage in China, they are the working class, the peasantry, the petty bourgeoisie, and the national bourgeoisie.

Under the leadership of the working class and the Communist Party these classes unite to create their own state and elect their own government so as to enforce their dictatorship over the henchmen of imperialism—the landlord class and bureaucratic capitalist class, as well as the reactionary clique of the Kuomintang, which represents these classes, and their accomplices. The people's government will suppress such individuals. It will only tolerate them if they prove tractable in speech or action. If they are intractable, they will be instantly curbed and punished. Within the ranks of the people, the democratic system is carried out by giving freedom of speech, assembly, and association. The right to vote is given only to the people, not to the reactionaries.

These two aspects, democracy for the people and dictatorship for the reactionaries, when combined, constitute the people's democratic dictatorship.

—Mao Ze dong, *On People's Democratic Dictatorship.*

12. Individuals form classes according to the similarity of their interests, they form syndicates according to differentiated economic activities within these interests; but they form first, and above all, the State, which is not to be thought of numerically as the sum-total of individuals forming the majority of a nation. And consequently, Fascism is opposed to Democracy, which equates the nation to the majority, lowering it to the level of that majority; nevertheless it is the purest form of democracy if the nation is conceived, as it should be, qualitatively and not quantitatively, as the most powerful idea (most powerful because most moral, most coherent, most true) which acts within the nation as the conscience and the will of a few, even of One, which ideal tends to become active within the conscience and the will of all—that

is to say, of all those who rightly constitute a nation by reason of nature, history or race, and have set out upon the same line of development and spiritual formation as one conscience and one sole will. Not a race, nor a geographically determined region, but as a community historically perpetuating itself, a multitude unified by a single idea, which is the will to existence and to power; consciousness of itself, personality.

This higher personality is truly the nation in so far as it is the State. It is not the nation that generates the State, as according to the old naturalistic concept which served as the basis of the political theories of the national States of the nineteenth century. Rather the nation is created by the State, which gives to the people, conscious of its own moral unity, a will and therefore an effective existence. The right of a nation to independence derives not from a literary and ideal consciousness of its own being, still less from a more or less unconscious and inert acceptance of a *de facto* situation, but from an active consciousness, from a political will in action and ready to demonstrate its own rights: that is to say, from a state already coming into being. The State, in fact, as the universal ethical will, is the creator of right.

—Benito Mussolini, *The Doctrine of Fascism*.

13. The bourgeoisie, wherever it has got the upperhand, has put an end to all feudal, patriarchal, idyllic relations. It has pitilessly torn asunder the motley feudal ties that bound man to his "natural superiors," and has left no other bond between man and man than naked self-interest, than callous "cash payment." It has drowned the most heavenly ecstasies of religious fervor, of chivalrous enthusiasm, of philistine sentimentalism, in the icy water of egotistical calculation. It has resolved personal worth into exchange value, and in place of the numberless indefeasible chartered freedoms, has set up that single, unconscionable freedom—Free Trade. In one word, for exploitation, veiled by religious and political illusions, it has substituted naked, shameless, direct, brutal exploitation.

The bourgeoisie has stripped of its halo every occupation hitherto honored and looked up to with reverent awe. It has converted the physician, the lawyer, the priest, the poet, the man of science, into its paid wage-laborers.

The bourgeoisie has torn away from the family its sentimental veil, and has reduced the family relation to a mere money relation.

—Karl Marx and Friedrich Engels, *The Communist Manifesto*.

6.4 Understatement and the Fallacy of Hedging

In Sections 6.2 and 6.3, we discussed two pervasive features of language, vagueness and ladenness, which are not necessarily bad but can be exploited to commit fallacies. We now turn to a third such feature of language: understatement.

6.4 / Understatement and the Fallacy of Hedging

To **understate** a claim is to use words that diminish the content or force of the claim. For example, the claim "No pigs can fly" may be understated as "Almost no pigs can fly." We need to clarify the ways that claims can be understated, and then the uses to which understatement is put.

A common way to understate a claim is to insert a detensifier. A **detensifier** is an adverb of degree used to downtone (to diminish the power of) a predicate. The predicate in a sentence is the word or words that express what is affirmed or denied of the subject. For example, the word 'nearly' is a common detensifier and accordingly can be used to understate claims:

1a. I am exhausted.
2a. He is finished.

1b. I am nearly exhausted.
2b. He is nearly finished.

English is rich in detensifiers, the most common of which are listed in Table 6.1. Detensifiers are often derogatorily called 'weasel words,' since, as we shall see, they may be used to weasel out of a rhetorical corner into which we have painted ourselves.

Detensification is one method of understatement. *Qualification* is another. Consider the following sequence of claims.

1. I will tell the truth.
2. I will tell you the truth.
3. I will tell you the truth about this matter.
4. I will tell you the truth about this matter today.

Table 6.1 *Common English Detensifiers*

Sort of	More or less	Scarcely
Pretty much	A little	Barely
Rather	Quite	Almost
In many/some respects	Partially	Technically
In most respects	Enough	Virtually
Relatively	Mainly	Basically
Slightly	To some extent	Nearly/very nearly
Kind of	In part	Essentially
Somewhat	Moderately	Practically
Typically	Mildly	A bit
A fair amount	Possibly	

Notice that the claims get weaker: (1) commits the speaker to tell the truth without qualification; (2) commits the speaker only to telling the truth to the listener; (3) commits the speaker only to telling the truth to the listener about the area under discussion; (4) further qualifies the promise to apply only to the day of the discussion. A *qualifier* is a phrase that limits the application of the predicate.

A third technique of understatement is *substitution of contradictories for contraries*. For instance, you can understate the claim "Maria is bad" by saying "Maria is not good." The contradictory of 'good' is 'not good,' while the contrary is 'bad.' Calling a person 'not good' leaves open whether the person is positively bad (as opposed to merely neutral), and hence is a weaker claim than calling the person 'bad.'

So far we have been dealing with unmodalized assertions, that is, statements that do not contain modalities. (You should review Section 3.4 at this point.) The following modal words indicate that no doubts are had about the claim: *be sure, surely; be obvious, obviously; be evident, evidently; be certain, certainly; be clear, clearly;* and *must, has to.*

The following modal words indicate confidence, but some doubt: *think, suppose, believe, probably, presumably, supposedly,* and *should.*

The following modal words indicate less confidence, more doubt: *guess, conjecture, seem, can, could, may, maybe, might, possibly, perhaps,* and *conceivably.*

Modalized assertions are open to a fourth sort of understatement, namely, where the *modality is weakened*, as in the following examples.

    1a. I am sure that he is a thief.
    1b. I believe that he is a thief.
    1c. I suspect that he is a thief.

    2a. It is obviously a plane.
    2b. It is presumably a plane.
    2c. It is conceivably a plane.

    3a. The rabbit must have gone down this hole.
    3b. The rabbit should have gone down this hole.
    3c. The rabbit may have gone down this hole.

As a special case of the previous example, unmodalized statements can be understated by adding modalities, typically parenthetically, as in the examples that follow.

    1a. The president is honest.
    1b. The president is, I believe, honest.

6.4 / Understatement and the Fallacy of Hedging

2a. Pigs cannot dance.
2b. Pigs, to the best of my knowledge, cannot dance.

The point is that when a speaker asserts a statement without a modality (say, pigs cannot dance), the listener assumes that the speaker has no particular doubts about the statement being made. To parenthetically add a modality is a way to signal doubt, hence to understate.

These four devices for understatement (namely, the use of detensifiers, the use of qualifiers, the substitution of contradictories for contraries, and the substitution of a weaker for a stronger modality, including the addition of a modality) can be employed together, as in the cases that follow:

1a. The president is honest.
1b. The president is probably honest about this matter.
2a. Hitler was evil.
2b. In my opinion, Hitler was not good.
3a. Fred is dead.
3b. Fred, I suspect, is sort of dead.

In (1b), we inserted a detensifier and a qualifier. In (2b), we inserted a parenthetical and substituted a contradictory for a contrary. In (3b), we used a parenthetical and a detensifier.

Understatement has many uses. We can understate to make a joke. Examples:

Scene: Michelle and Tracy are watching Zach, a 3-year-old boy, trying to pour milk. Zach spills the milk all over the kitchen.
Michelle: Zach is not exactly the most coordinated kid I have ever seen.
Scene: Two police officers watch a thoroughly inebriated man grope his way out of a bar.
First Officer: I could be wrong, but I really do suspect that man may not be entirely sober.

People understate for other reasons as well. Sometimes we wish to be courteous; for example, "I'm not sure you are entirely correct in what you say" is a very polite way of calling somebody wrong. Sometimes we understate to avoid a dispute.

But we can commit a fallacy by understatement. **Hedging** is the fallacy committed when an arguer changes the claim by understating it during the course of an argument. Typically, the fallacy of hedging occurs when one participant tries to weaken the person's claim (so it is easier to defend) without the other participant noticing. For example:

> *A:* Why do you say that the poor don't need government assistance?
>
> *B:* It just seems to me pretty clear that a lot of them could hold jobs if they really wanted to.

Hedging is the converse of strawman. In *strawman*, an arguer tries to overstate the opponent's position so as to make it harder to defend. In *hedging*, an arguer tries to understate the arguer's own position so as to make it easier to defend.

EXERCISES 6.4

Understate each of the following modalized claims in all four ways discussed in the text. (An asterisk indicates problems that are answered in the back of the book.)

Example: I am sure that Sara hates jazz.
  a. Detensifier: I am sure that Sara probably hates jazz.
  b. Qualifier: I am sure that Sara hates Jazz of this sort.
  c. Contradictory substituted: I am sure that Sara does not like Jazz.
  d. Weaker modality substituted: I believe that Sara hates Jazz.

1. Dogs certainly bark.
* 2. I am positive that pizza tastes good.
3. Certainly, cats can climb trees.
4. Clearly, not everyone can dance.
5. People obviously go where the work is.
* 6. Clearly, Darwin devised modern evolutionary theory.
7. Beavers surely are smart animals.
8. Clearly, education is useful.
9. I am sure that photographs can lie.
* 10. It is a fact that people hate the truth.

6.5 Ambiguity and Fallacies of Ambiguity

We say a word is **ambiguous** if it has more than one meaning. For example, the word 'code' is ambiguous: it means both "a system of laws" and "a system of symbols used to transmit messages."

Many words are ambiguous. But ambiguity, though a widespread feature of language, is not normally a problem because context usually makes clear the intended meaning. For instance, the sentence:

Let's cut the turkey.

employs an ambiguous word, 'turkey,' which can mean a type of bird eaten at Thanksgiving, or a fool. If that sentence is uttered sitting at the table at Thanksgiving dinner, it would be obvious to the listeners that the speaker intended the first meaning. If a gang member uttered the word to the rest of the gang (in the presence of a member of a rival gang), the second meaning would be intended. Yet another meaning would be intended if a baseball manager in a clubhouse meeting uttered the sentence.

Thus ambiguity, like vagueness and ladenness, is a common feature of words, which is not usually troublesome. Words can be ambiguous though not used ambiguously. When a person uses an ambiguous word, phrase, or sentence one way in one premise of an argument, and a different way in another premise or in the conclusion, that person is committing the fallacy of **equivocation**, and we say this is 'equivocating.'

As an example of the fallacy of equivocation, consider this case. A person once came to my door representing a religion that believes in sending its members door-to-door in search of converts. One of his arguments was that while his religion was completely poor, the Catholic Church was wealthy. (He used this claim to support another claim to the effect that members of his faith were more devout.) His argument went:

1. The Catholic Church owns many churches.
2. Religion X does not own any churches.
3. Owning churches makes a religion wealthy.

∴ Religion X is less wealthy than the Catholic Church.

Whatever other defects his argument had, it involved equivocation, because upon questioning, it turned out that religion X does own property, but members do not call their houses of worship 'churches.'

The point is that the word 'church' is ambiguous. It can mean "mainstream Christian house of worship," and in that sense neither Jews nor Moslems go to church. But 'church' more often means "any house of worship," and in that sense, Jews and Moslems do go to church.

So far we have focused upon ambiguous words. But sentences, too, can be ambiguous. We call any sentence that is ambiguous due to poor grammatical construction an **amphiboly**, as for instance:

The tuna are biting off Washington Coast.

This can mean that off the coast of Washington the tuna are biting (the baited hooks of anglers), or else that the fish are chewing away at the coastline itself! Note that in amphiboly the individual terms are not ambiguous. Instead, the double meanings occur because the reference of one of the terms is unclear, or else because of some other grammatical problem.

Amphibolies typically occur in three situations. These situations are not arguments, though important in their own right.

First, unwary editors occasionally let amphibolous headlines slip by them, as in these examples taken from newspapers.

Death Prompts Coach to Quit
Paramedics Help Dog Bite Victim
Firefighters Threaten to Sue if Killed on Job
Alcoholic Rats May Aid Humans
Toronto Police Shoot Dead Man Robbing Gas Station

Some newspapers will put sneakily amphibolous headlines above trivial, uninteresting stories that make those stories seem more interesting to trick the reader into buying the paper. Examples:

MY HUSBAND USED TO PUSH ME IN A BABY CARRIAGE.

You buy the paper, expecting to read a disgusting story about some new kind of sexual perversion, but the story merely says that when she was a baby, her future husband—then 7 years old—had to push her baby carriage.

NUN WALKS HUNDREDS OF MILES TO LOSE WEIGHT.

The story merely says that this nun walks a few blocks every day, which adds up over the years to hundreds of miles.

The third place we encounter amphibolies is in fortune-telling and prognostication. Fortune-tellers will often use sentences that can be interpreted several ways, so their predictions will always come true. For example, King Croesus is alleged to have asked an oracle whether he (the king) should attack a neighboring king, Cyrus. The oracle replied, "When Croesus shall o'er Halys River go, He will a mighty kingdom overthrow." Croesus decided to attack and was defeated. When he later expressed some disappointment with the oracle's powers, the oracle pointed out that the prediction did come true—for in going to war against Cyrus, Croesus had destroyed his own kingdom!

The fallacy of accent, like the fallacy of equivocation, involves change of meaning. But whereas in equivocation the word or phrase is already ambiguous as it stands, **accent** is the fallacy of changing the meaning of an unambiguous sentence, either by (orally or typographically) stressing part of it, or else by omitting words. As examples of the first, consider Figures 6.1A–6.1D.

Figure 6.1A

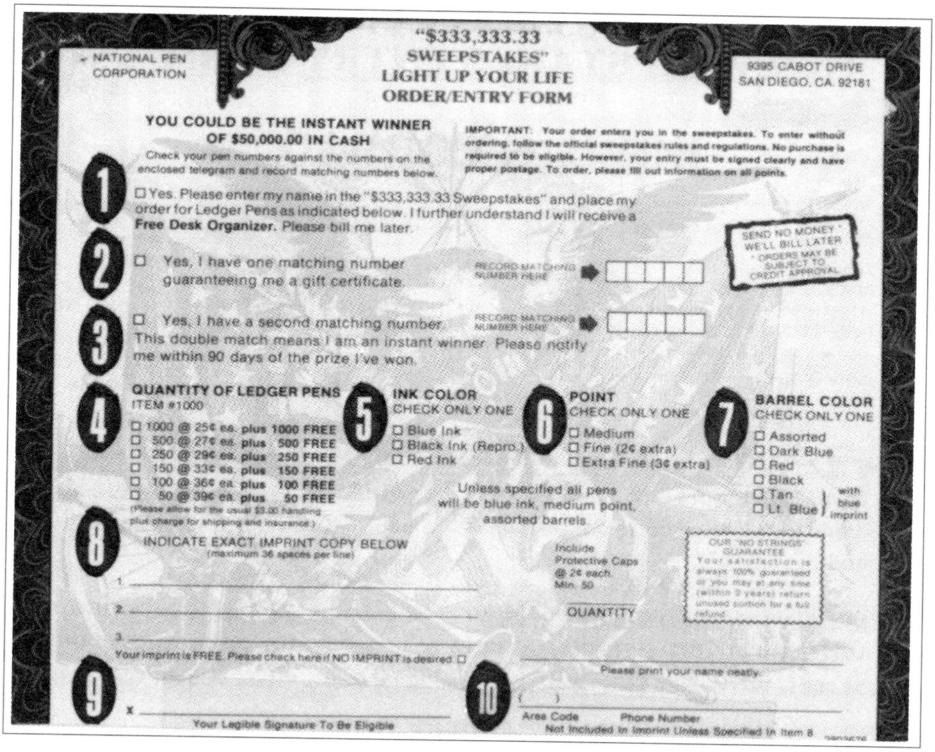
NATIONAL PEN CORPORATION

"$333,333.33 SWEEPSTAKES"
LIGHT UP YOUR LIFE
ORDER/ENTRY FORM

9395 CABOT DRIVE
SAN DIEGO, CA 92181

YOU COULD BE THE INSTANT WINNER OF $50,000.00 IN CASH

Check your pen numbers against the numbers on the enclosed telegram and record matching numbers below.

IMPORTANT: Your order enters you in the sweepstakes. To enter without ordering, follow the official sweepstakes rules and regulations. No purchase is required to be eligible. However, your entry must be signed clearly and have proper postage. To order, please fill out information on all points.

1 ☐ Yes. Please enter my name in the "$333,333.33 Sweepstakes" and place my order for Ledger Pens as indicated below. I further understand I will receive a **Free Desk Organizer.** Please bill me later.

2 ☐ Yes, I have one matching number guaranteeing me a gift certificate. RECORD MATCHING NUMBER HERE

SEND NO MONEY * WE'LL BILL LATER * ORDERS MAY BE SUBJECT TO CREDIT APPROVAL

3 ☐ Yes, I have a second matching number. This double match means I am an instant winner. Please notify me within 90 days of the prize I've won. RECORD MATCHING NUMBER HERE

4 QUANTITY OF LEDGER PENS
ITEM #1000
☐ 1000 @ 25¢ ea. plus 1000 FREE
☐ 500 @ 27¢ ea. plus 500 FREE
☐ 250 @ 29¢ ea. plus 250 FREE
☐ 150 @ 33¢ ea. plus 150 FREE
☐ 100 @ 36¢ ea. plus 100 FREE
☐ 50 @ 39¢ ea. plus 50 FREE
(Please allow for the usual $3.00 handling plus charge for shipping and insurance.)

5 INK COLOR
CHECK ONLY ONE
☐ Blue Ink
☐ Black Ink (Repro.)
☐ Red Ink

6 POINT
CHECK ONLY ONE
☐ Medium
☐ Fine (2¢ extra)
☐ Extra Fine (3¢ extra)

7 BARREL COLOR
CHECK ONLY ONE
☐ Assorted
☐ Dark Blue
☐ Red
☐ Black
☐ Tan
☐ Lt. Blue } with blue imprint

Unless specified all pens will be blue ink, medium point, assorted barrels.

8 INDICATE EXACT IMPRINT COPY BELOW
(maximum 36 spaces per line)
1. ______
2. ______
3. ______
Your imprint is FREE. Please check here if NO IMPRINT is desired ☐

Include Protective Caps @ 2¢ each. Min. 50
QUANTITY

OUR "NO STRINGS" GUARANTEE
Your satisfaction is always 100% guaranteed or you may at any time (within 2 years) return unused portion for a full refund.

9 x ______
Your Legible Signature To Be Eligible

______
Please print your name neatly.

10 ( ) ______
Area Code Phone Number
Not Included In Imprint Unless Specified In Item 8

Reprinted with permission of National Pen Corporation (San Diego, CA).

6 / *Fallacies of Language*

Figure 6.1B

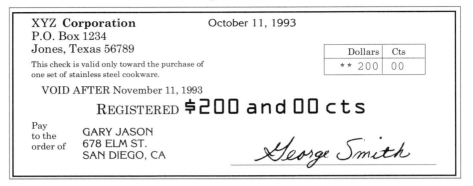

Figure 6.1C

```
METRO BANK                              No. 02345
    2/25/93              REQUEST CERTIFICATE

Fill in any amount from $4,000 to $20,000      AMOUNT

      Gary Jason
      P.O. Box 789
      Generic, CA                    VOID AFTER May 15, 1993
                                     OXL 72
Please complete the form attached to this
voucher and mail in the enclosed prepaid    James Rich
envelope.                                   Authorized Signature
```

In Figure 6.1A, a sweepstakes shrinks all the qualifying words in small print. Specifically, the fact that one can enter the sweepstakes without ordering something (required by law) is stated in small print *after* the word 'important'! In Figures 6.1B and 6.1C, advertisements are cited in the form of checks with "Pay to" or "Pay to the order of" appearing in the window of the envelope to get the recipient to read them. In Figure 6.1D, an advertisement for a fitness center has key qualifications, such as "facilities may vary" and "+$10 Registration fee" in small print.

The second way to commit the fallacy of accent is by omitting words. This is usually signaled by three dots, which stands for an ellipsis. Thus, an advertiser might change this sentence from a government report:

> Very few tests have been done on the additive ABBA, but those that have been done indicate that while there is no great risk in its use, there is some hazard.

6.5 / *Ambiguity and Fallacies of Ambiguity*

Figure 6.1D

GRAND OPENING

FREE GIFTS!!!
Limited Supply

BOOM BOX STEREO STORE
1234 Smith Street
Bartlett, CA 555-1029

CDs/Tapes $5.00
With $100 purchase. Limit three.

WHOLESALE PRICING*

*To club members only.
Three-year memberships

---

to:

> ... tests ... done on the additive ABBA ... indicate ... that there is no great risk in its use ....

which changes the meaning of the original into something more favorable to ABBA.

MAD Magazine has quite often parodied the fallacy of accent by advertisers. Figures 6.2A–D are courtesy of E.C. Publications.

Figure 6.2A

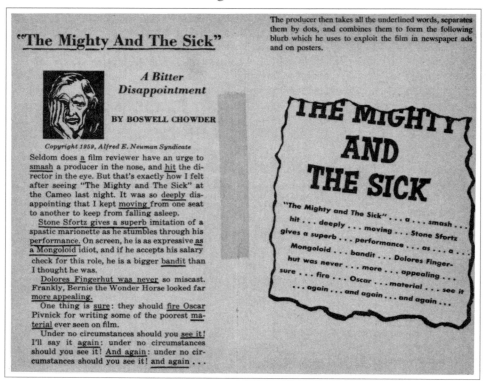

© 1959, 1963, and 1976 by E.C. Publications.

The next two fallacies of ambiguity are composition and division. These are converses of each other, and they come in two varieties. We begin with **composition**. The first form of this fallacy is arguing that what is true of the parts of a thing must be true of the whole. It is ridiculous to argue that since each part of a grandfather clock is light, the whole clock must be light. Equally illogical would be to argue that since Gloria can run the half-mile in two minutes, she can run a mile in four minutes.

The second form of the fallacy of composition can be defined only after we introduce some technical terms. By **general term** we mean a word that refers to a group of things—people, nails, slugs, chairs. Any general term is inherently ambiguous, in that it can refer to the group as a collective whole, or it can refer to the members of the group as individuals. The **collective use** of a general term is the use of it to refer to the group as a whole; the **distributive use** of a general term is the use of that term to talk about the members of the group as individuals.

6.5 / Ambiguity and Fallacies of Ambiguity

Figure 6.2B

© 1972, 1979 by Dick Debartolo, Bob Clarke, and E.C. Publications.

This ambiguity is brought out by the old school riddle: "Why do white sheep eat more than black ones? Because there are more of them!" The point of the riddle is that as individuals, all sheep eat about the same, but (since most sheep are white) white sheep collectively eat more.

We can now define the second form of composition. It is the fallacy of using a general term distributively in the premises, but collectively in the conclusion. Examples:

1. All people die at some time.
   ∴ There will come a time when all people are dead.

1. Assistant professors earn less than full professors.
   ∴ Assistant professors take up a smaller percentage of the college budget than do full professors.

Figure 6.2C

© 1972, 1979 by Dick Debartolo, Bob Clarke, and E.C. Publications.

There are similarly two forms of **division**. The first form is the fallacy of arguing that what is true of the whole must be true of the parts. It would be silly to conclude that a dorm room must be large because the whole dorm is, or that a given division of a corporation must be doing well since the corporation as a whole is.

The second form of the fallacy of division involves using a general term collectively in the premises, but distributively in the conclusion. Examples:

> Prostitutes have been practicing their profession since biblical times. How can they keep going so long?

They have been practicing their profession since biblical times collectively, not as individuals.

Figure 6.2D

© 1972, 1979 by Dick Debartolo, Bob Clarke, and E.C. Publications.

Philosophy professors average $20,000 per year; so if I become a philosophy professor, I will earn that much.

Panda bears are nearly extinct, so this panda must be nearly dead.

The words 'average' and 'extinct' indicate that a collective whole is being referred to.

People occasionally confound hasty generalization with composition, because in both fallacies the premises deal with particular cases, while the conclusions deal with whole groups. But the resemblance ends there. For in hasty generalization, the inference is from a claim about a *few* atypical cases to a claim about the whole group *as individuals*. On the other hand, in composition, the inference is from a claim about *all* individuals to a claim about the group *as a whole*. To see the difference, consider the following two arguments.

Hasty generalization:

1. These two apples I picked off this truckload of apples weigh 1 oz. a piece.

∴ All the apples on the truck weigh 1 oz. a piece.

Composition:

1. All the apples on the truck weigh 1 oz. a piece.

∴ The whole truckload of apples weighs 1 oz.

In the first, the general terms are used distributively throughout. In the second, 'apples' is used distributively in the premise (indicated by the term 'a piece'), but used collectively in the conclusion.

Some people confound accident with division, since each goes from a claim about a group to a claim about particular cases. But in accident, we go from a general statement about a group as individuals to a claim about some atypical case. On the other hand, in division, we go from a statement about a group taken collectively to a claim about all members of the group as individuals. To see the difference again compare two sample arguments.

Accident:

1. People are rational.

∴ This person will be rational, even though I ran into her car.

Division:

1. The human race will live past 2090.

∴ This human will live past 2090.

In the first, the general terms are used again distributively throughout. In the second, the word 'human' is used collectively in the premise and distributively in the conclusion.

EXERCISES 6.5A

The following words are ambiguous. Give the different meanings of each.

1. Dumb
2. Loaf
3. Sound
4. Carp
5. Dough
6. Punch
7. Kick
8. Sick
9. Unctuous
10. Action

6.5 / *Ambiguity and Fallacies of Ambiguity*

EXERCISES 6.5B

Each of the following passages contains one or more of the fallacies of ambiguity listed as follows:

Equivocation: Shifting from one meaning of an ambiguous word or phrase (or, in the case of amphibolies, a sentence or clause) to another during the course of the argument.

Accent: Changing the meaning of an unambiguous word, phrase, clause, or sentence by accenting (stressing) part of it, or by omitting part of it.

Composition: Either assuming that what is true of the parts must be true of the whole, or else using a general term distributively in the premises and collectively in the conclusion.

Division: Either assuming that what is true of the whole is true of the parts, or else using a general term collectively in the premises and distributively in the conclusion.

Identify the fallacies in each. (If an amphiboly, instead of a complete argument, is present, state the two meanings the sentence can express.) (An asterisk indicates problems that are answered in the back of the book.)

1. The ladies of the church have cast off garments of every kind, and they can be seen in the church basement as of Friday afternoon.

* 2. Headline: Forget the small car myth—big cars save more gas.
   Story: Americans are getting far better gas mileage these days—not because they're driving smaller cars, but because almost all cars are more fuel-efficient, says a government economist.
   American-made automobiles—no matter what their size—are built to get more miles to the gallon.
   In fact, the overall fuel economy of all new cars has risen by more than 75 percent since 1973, said Dr. Philip Patterson of the Department of Energy.

3. The way I figure it is this. If the ingredients are harmless, the whole mixture is harmless, too.

4. Unwanted hair removed permanently in College area at Kansas Medical Center.

5. Sign outside of a dry cleaners: "For best results, drop your pants here!"

* 6. Headline: Reds Go To War.
   Story: Candy addicts are seeing red. Manufacturers took the red M&M's off the market in 1976 in the midst of a red-dye cancer scare. Now the 35-member Society for the Restoration and Preservation of Red M&M's wants them back.
   Founder, Paul Hethmon, 19, of Knoxville, Tennessee, said the society is petitioning President Reagan and the M&M-Mars company to get their favorite snack back.

A spokesman for M&M-Mars said that, although the red dye was quite safe, there are no plans to use it again.

7. The ship was christened by Mrs. Reagan. The lines of her bottom were admired by the enthusiastic crowd.

8. Jaime: My gosh, Sue, you look scared!
Sue: I am. I just looked at the life insurance tables and my life expectancy is only ten years. I don't want to die that soon.

9. Under the New Deal, many men got jobs, and women also.

* 10. Each person's happiness is a good to that person, and the general happiness, therefore, a good to everybody.

11. Motorist wounded by sniper on freeway.

12. Should we not assume that just as the eye, the hand, the foot, and in general each part of the body clearly has its own proper function, so man too has some function over and above the function of his parts?

13.

* 14. It is predicted that the cost-of-living index will rise again next month. Consequently you can expect to pay more for butter and eggs next month.

15. MANIAC ATTACKS SDSU CAMPUS! Hordes of innocent San Diego State University students are being frightened, assaulted, and forced to submit to the needs of the "Maniac," a campus "Humor" Magazine!

16. AUSTIN, Texas (UPI)—His rhetoric is indisputable and his record is spotless. Face it, Nobody's perfect.

The Nobody for President campaign charmed a crowd of 500 at the University of Texas Monday, offering an alternative to the somebodies on the presidential ballot.

Working the crowd for Nobody in particular was a character called Wavy Gravy, Hugh Romney of the Hog Farm commune in the San Francisco Bay area. Gravy is Nobody's chief aide, a position he calls Nobody's fool.

"Who was president before George Washington?" Gravy, dressed in a clown suit, asked the crowd.

"Nobody!" responded the crowd.

"Who honored the treaties with the Indians?"

"Nobody!"

"Who do you want to run your life?"

"Nobody!"

The address was so stirring—it was obvious Nobody cares—that one man stepped forward and said, "I've never voted for anybody but this year I'm going to vote for Nobody."

Curtis Spangler, Nobody's campaign manager, went on to explain Nobody knows how to dispose of nuclear waste, Nobody has brought peace, Nobody fed the hungry and the destitute and Nobody keeps all his campaign promises.

In fact, Spangler argues, Nobody actually won in 1976, although Jimmy Carter was permitted to take office. Spangler says only 40 percent of eligible Americans voted, leaving 60 percent voting for Nobody.

6.6 Types of Definition

In the foregoing sections, we have seen how vagueness, ambiguity, understatement, and ladenness of words can impede any dialogue. The best way to help a dialogue overcome such stumbling blocks is for the participants to define key terms at the outset.

The word 'definition' refers to an activity; it is something we do with language. We usually do this defining by giving a statement of definition (often called 'a definition'—note the ambiguity). Typically, we express our definition using a declarative sentence of the form:

A is defined as B

or more compactly:

$A = (\text{def}) B$

We will use that format in what follows. A will be called the **definiendum** (the thing being defined); B will be called the **definiens** (the thing being used to do the defining).

Definition has many uses. We have already talked about four uses: to reduce vagueness, to reduce understatement, to eliminate ambiguity, and to make terms more emotionally or theoretically neutral in order to focus a discussion on the facts. But definition can also be used to increase people's vocabularies. An obvious example of this use is the student who memorizes vocabulary-builder definitions, in order to prepare for the Scholastic Aptitude Test. A lot of new terms are learned any time a person changes jobs or college majors. Indeed, one of the major tasks facing a new employee is to master the technical terminology of the field.

Definitions have other uses as well. Scientists define key terms in their fields of research (such as 'elementary particle,' 'force,' 'learning,' 'compound') in order to help explain theoretically those concepts. In such a definition, the definiens B serves to attach to the definiendum A not the meaning commonly associated with A, but those features or characteristics of A that the scientist's theory indicates are the most important. Thus a scientist might define *momentum* as the product of mass times velocity. This definition is not intended to represent what people ordinarily mean by that term, but to attach those characteristics that are most useful in explaining and predicting the behavior of physical objects.

Another purpose or use of definition is to persuade or influence attitudes. For instance, a traditional Marxist critical of the People's Republic of China might define *Chinese socialism* as state capitalism. Such a Marxist would be trying to influence other Marxists, to create in them negative feelings about the People's Republic of China by classing it with capitalist countries (countries of which any Marxist can be presumed to disapprove). Naturally, by influencing attitudes, the speaker hopes to influence action as well. However, the use of definition to persuade is outside the focus of this book, which is concerned with methods of proof (as opposed to arts of persuasion).

Let us turn from the uses of definition to the types of definition. Two main kinds of definition are reportive and suggestive.

A **reportive definition** attempts to report how people use a given word. Reportive definitions are also called 'dictionary definitions' and 'lexical definitions.' Reportive definitions are empirical claims, factual statements about actual usage, and, as such, are either true or false. For instance:

> Elephant = (def) A very large Asian or African mammal with a long, flexible trunk and long tusks.

is a true report of the meaning of the word 'elephant'; the statement:

> Elephant = (def) A left-handed person who directs small-town ballet performances.

is false.

Word usage changes, and groups of people give special meanings to many words. The people who compile dictionaries (lexicographers) try to be more specific in their reportive definitions by qualifying them with words such as 'archaic' (meaning that the word is no longer used that way), and 'colloquial' or 'slang' (meaning that the word is used that way only in informal social contexts). But it is still true that reportive definitions are intended to represent actual usage.

A **suggestive definition** suggests or proposes that a given term be used in a specific way. Suggestive definitions come in several varieties. A **coining definition** is the assignment of meaning to a new word or term (a 'neologism'). For instance, a mathematician might coin the word 'poset' by saying:

> poset = (def) a set P together with a partial ordering of the elements in it.

Since scientists and other scholars often invent neologisms, coining definitions are not rarities. We can have good reasons for introducing new terminology. For one thing, a judicious use of new words can increase the readability of the writing, by shortening many of the sentences involved. Thus, we express $2 \times 2 \times 2 \times 2 \times 2 \times 2 \times 2 \times 2 \times 2 \times 2$ more economically as 2^{10}. In mathematics we often see chains of coining definitions.

Besides increasing readability through economy of expression, coining new words serves another purpose: it can allow the scientist to substitute a less emotionally charged term in a given context. So a psychologist may prefer to use 'exceptional' instead of 'gifted' or 'retarded.'

In a **reformative definition** the proposer seeks to establish a new meaning for a term already in general use. One common source of reformative definitions is science, since improved theoretical understanding often leads scientists to want to reform the usage of terms. So for example, the definition of the term 'electricity' physicists give has changed over the centuries as theories have come and gone.

A **precising definition** seeks to reduce the vagueness of a term, that is, to make the definiens more precise. For a precising definition to succeed in clarifying a term, it must agree with the existing application of the term in the clear cases, before it goes on to assign new meaning in the unclear borderline cases.

Precising definitions are often met in the law. For example, in *Roe v. Wade*, dealing with abortion, the Supreme Court, in effect, ruled that a fetus of less than three months is not a "person" in a full legal sense of an entity bearing the right to life. This controversial ruling amounted to making the term 'person' more precise: drawing a line at the end of the first trimester of pregnancy.

The previous example brings up the point that the suggestive extension beyond the clear cases should not be purely arbitrary or subjective. In the case of legal reasoning, jurists use precedent and analogies to justify their rulings. Even where precedent is lacking, considerations about social goals and social realities are brought in. Figure 6.3 summarizes our discussion so far.

All definitions should conform to several common-sensical rules.

1. A definition should state the important properties of what is being defined. In traditional terms, this is called framing a definition by indicating essential rather than accidental properties of what is defined. For instance, the ancient Greek Diogenes, upon hearing that some philosophers of Plato's Academy had defined 'human being' "a featherless biped," plucked a live chicken and introduced it to them. The point he made (rather cruelly) was that their definition focused on accidental rather than essential features of human beings.

2. The definition should not be circular. A definition should not contain in the definiens either the definiendum or a synonym of it. Thus the definition:

Sun = (def) star that shines by day

is not good, because 'day' is itself defined in terms of the sun's shining.

3. A definition should be put in positive rather than negative terms. Saying what a thing is not usually does not separate it sufficiently from other things. For example, the definition of 'dog' as "an

Figure 6.3 Types of Definition

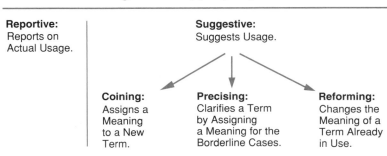

animal that is not a cat," does not distinguish a dog from a frog, and adding that "it is not a frog" still does not tell us what a dog is. Of course, we cannot always avoid negative definitions because some terms are intrinsically negative since they refer to the lack of some property. For instance, 'bald' meaning "lacking hair."

4. The definition should not include figurative (or vague, emotionally toned, or ambiguous) language. Using such language may load a term with unintended meaning. For example, the definition of businessperson = (def) a parasite who exploits workers, is both biased and figurative.

5. The definition should not be too broad. It should not make the term cover more than intended. For example, the definition of 'whale' as "a large aquatic mammal" is too broad, because that definition would include seals as whales.

6. The definition should not be too narrow. It should not make the term cover less than intended. For example, the definition of 'whale' as "a black and white mammal of the biological order Cetacea" would be too narrow, because while some whales are black and white (such as killer whales), not all are.

These rules are heuristic in nature. We cannot always avoid negative definitions (as for words with the prefix *in-* as in 'insensitive,' 'insincere,' 'inadequate,' and so on).

EXERCISES 6.6

1. For each of the following terms, give your own reportive definitions, then look up the word in a few dictionaries.
   a. Faith
   b. Hunger
   c. Love
   d. Argument
   e. Cummerbund
   f. Bolt
   g. Harpy
   h. Forte
   i. Heresy
   j. Mite

2. Give an example of a reforming definition. Give an example of a precising definition. (Do not use the examples given in the book.)

6.7 Fallacies of Technical Language

In addition to the kinds of problems that we have so far discussed, which are problems inherent in all language, technical language faces special problems. Especially common are fallacies of data presentation, by which we mean ways in which data can be presented misleadingly or deceptively.

Numerical data are often misleadingly stated. Raw numbers can seriously mislead, as when a person says, "20,000 people have resubscribed to my newsletter showing its popular success," while neglecting to mention that this was out of 1,000,000 initial subscribers. But rates and percentages can be equally misleading, as when we talk about the average income among members of some group going up 100% in 2 years, when in reality the jump was from $100 per capita to $200 per capita. A classic example was furnished by the USSR in the 1920s. One year the government decided the growth rate was not sufficiently impressive, so currency was devalued 50% and the following year the government claimed a growth rate of 100%![1]

The use of loaded language and accenting devices makes things worse. Table 6.2 gives some examples of how numbers can be covertly interpreted by accentuation and loading.

Graphs also can mislead—this is especially worrisome given their striking visual effect. For one thing, the designer of the graph chooses the scale upon which the graph is drawn. Look at the graph given in Figure 6.4A.

It looks as though Farkle's sales are soaring. But suppose we fill in the amounts on a different scale (Figure 6.4B).

Not so impressive! Expand the scale and even the smallest rise can seem wonderful. Again, scales using bar graphs can serve to understate or overstate differences (Figure 6.5).

Figures 6.5a, 6.5b, and 6.5c may be used to convey the same numbers with different impressions.

Pictographs, since by definition they employ pictures, can subtly misrepresent data. Suppose a group wanted to compare alcohol consumption in two states, and did so by the graph given in Figure 6.6.

This graph would be visually misleading in that while the Illinois bottle is twice as tall as the Kansas bottle—reflecting the double per capita consumption—the bottle area has quadrupled, giving the impression of an even greater disparity. (All illustrations used here are purely fictitious.)

---

[1] My thanks to Professor Greg Brunk of the University of Oklahoma for pointing that out to me.

6.7 / *Fallacies of Technical Language*

Table 6.2

Neutral	Positively Slanted	Negatively Slanted
Evita has two cars.	Evita only has two cars.	Evita has no less than TWO cars!
Evita's income went from $18,000 to $22,000 in 1986.	Evita's income rose a mere $4,000 in 1986.	Evita's income absolutely SOARED a stunning 22% in only one short year.
The birthrate in Bambia is 2% in a population of 10,000,000.	Bambia's population is only increasing at a 2% rate.	Bambia is flooding the planet with TWO HUNDRED THOUSAND new mouths per year!

Figure 6.4   **Sales Chart for Farkle Shoe Company**

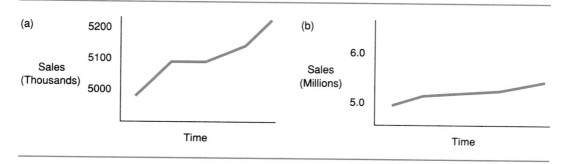

Figure 6.5

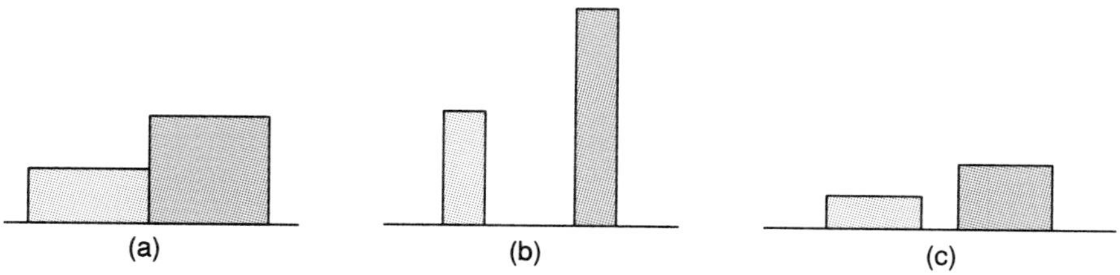

186 6 / *Fallacies of Language*

Figure 6.6

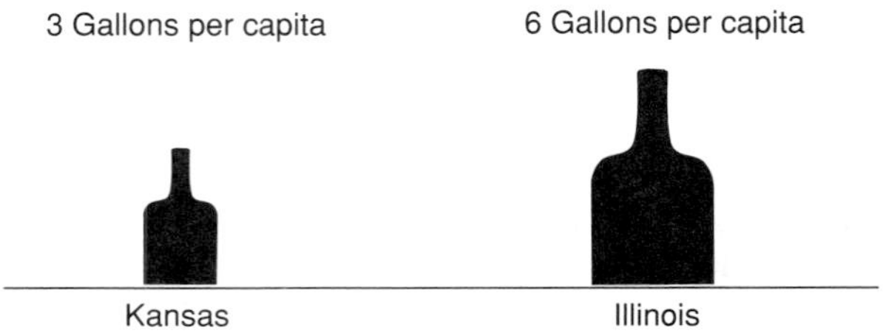

3 Gallons per capita   6 Gallons per capita

Kansas   Illinois

The pictures selected in the pictograph may arouse emotions while presenting data. In Figure 6.7, two different pictographs present capital punishment figures. Each symbol represents one hundred executions. (All illustrations used here are again fictitious.)

We are exaggerating a bit: nobody is going to be so obvious in attempting to slant negatively as to use figures of dead persons. But we often see figures on defense spending, for instance, represented by ominous men-at-arms.

Fallacies of data presentation can be viewed as varieties of special pleading, in that numerical evidence is being slanted in presentation.

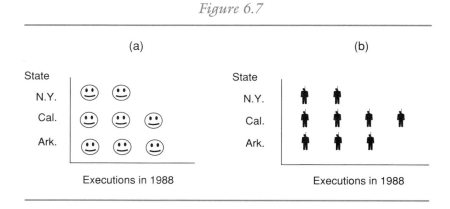

Figure 6.7

EXERCISES 6.7

Find three instances of fallacious data presentation in newspapers, magazines, or books. Analyze their use of fallacy. Specify how they should be corrected.

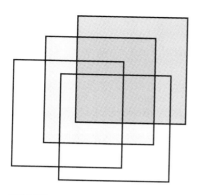

7

An Extended Example

7.1 A Review of the Fallacies

Tables 7.1–7.4 summarize the fallacies investigated in Chapters 4, 5, and 6.

With these terms as guides, we can now turn to the review exercises, which discuss the fallacies in Chapters 1–6. Asking the following sequence of questions about the passage should help dig out the fallacies.

1. Who is speaking, and whom is the speaker addressing?
2. What is the topic?
3. What is being said—what is the overall claim being made; what is the issue at hand?
4. What 'evidence' is being offered?
5. Supposing for the sake of argument that the 'evidence' is true, does it support the conclusion, and if not, why not?
6. What label fits the fallacy if the evidence doesn't support the conclusion?

Table 7.1 *Fallacies*

Fallacies of no evidence	Fallacies of little evidence	Fallacies of language

Table 7.2 *Fallacies of No Evidence (These Involve an Evasion of Your Responsibility to Prove)*

Complete evasion	Evasion by emotional appeals	Evasion by talking about irrelevancies	Evasion by presupposition	Evasion by repetition
1. Pooh-poohing 2. Shifting the burden of proof	1. Attacking the person 2. Appeal to fear 3. Appeal to pity 4. Appeal to the crowd	1. Ignoring the issue	1. Loaded question	1. Begging the question

Table 7.3 *Fallacies of Little Evidence (Inductive Reasoning Gone Awry Due to the Violation of Constraints)*

Ad vericundiam	False dilemma	False analogy		False cause
(Bad appeal to authority) (Testimony gone awry)	(Reasoning about alternatives gone awry)	(Bad comparison)	1. Hasty generalization 2. Accident (Bad transition from case to rule, or vice versa)	(Bad causal inference)

Table 7.4 *Fallacies of Language*

Fallacy of vagueness	Fallacy of ladenness	Fallacy of understatement	Fallacy of ambiguity	Fallacies of data presentation
(Slippery slope reconsidered)	(Loaded language)	(Hedging)	(Equivocation, amphiboly accent, composition, division)	(Special pleading)

EXERCISES 7.1A

Identify the fallacies in the following passages. All the fallacies we have discussed in Chapters 1–6 are represented. (An asterisk indicates problems that are answered in the back of the book.)

1. Now, Doctors at leading universities first in Europe and now worldwide have found substances where mere aroma can be used to make you appear more attractive, more impressive, and even more desirable. Scientists first described the incredibly powerful sexual attractants in insects as Pheromones. For years, many Musk fragrances have used Pheromone from animals. Now, Pheromones have been found in humans, too! American scientist and researcher, William Borneo, has captured the secret in two new formulas utilizing male and female Pheromones to create the ultimate perfumes.

* 2. Nutshell: How do you square voluntary school prayer with no change in the Bill of Rights?

Falwell: The same way we squared it for 180 years of American history. I feel that the atheist or any objecting child should have the right to leave the classroom during a time of voluntary prayer.

Nutshell: Do you know how traumatic that can be?

Falwell: It would be no problem for me.

3. There's an epidemic sweeping through medical schools: alcoholism.

And the only way it's going to be stopped is if med school staff members are trained to detect it early, say Drs. Anderson Spickard and Tremaine Billings.

All too often excessive alcohol use is overlooked by faculty members and colleagues or is written off to the stress and rigors of the profession, they note.

But all the excuses in the world won't change the fact: Alcoholism is a disease that needs urgent attention.

Over the last ten years, Spickard and Billings have successfully helped seven doctors at Vanderbilt University Medical School who were alcoholics.

"They represent only some of those on the faculty who are impaired by alcoholism and other forms of chemical dependency," they explained.

4. There will be no Christmas for "Red." "Red" has become our symbol in the fight to stop suffering and cruelty. This poor Irish Setter could barely stand when our investigators reached him. Our vets tried to save him, but we were too late. Red's owner just let him starve to death. We were alerted to this tragedy by a neighbor. This winter and its cold weather will bring us more strays, more abandoned, starving, and sick animals. We cannot help them all, but we try to help as many as we can. We're doing our best, but we cannot continue without your help. A dollar goes a long way with us. This Christmas more than any before, we need your help in trying to eliminate some of the suffering. Please! The animals need your help.

5. How to stop feeling guilty about worldwide poverty—Every caring person knows there is a great deal of poverty in the world. Just the mention of places like Uganda, Thailand, India, and Guatemala brings to mind heartbreaking pictures we've all seen.

And here we are, a people with so much material wealth. But what can one person do that will make any real difference?

You can do the single most important and loving thing of all: you can change one child's life for the better. That's the whole purpose of our fund.

* 6. Why do I know more than other people? Why, in general, am I so clever?

7. It is difficult to understand why college athletes and fans cannot support their own activities. Professional sports teams pay their players millions of dollars a year and manage to get by without a tax on the citizens of their home towns. Certainly collegiate teams, which pay their players only through scholarships, should be able to survive without squeezing money out of already hard-pressed students.

8. Nutshell: At Baylor University, Fundamentalists burned records by the Beatles, Rolling Stones, Bee Gees, Eagles . . . Do you support this kind of activity?

Falwell: We have never done that.

Nutshell: I asked if you supported it.

9. Three real old buffalo nickels, minted before 1939, only $1.90.

* 10. This is in reference to the article (Feb. 21) on animal rights. We were shocked and amazed at the attitude of these so-called animal rights activists. How can they care so much about comfortable living conditions for chickens when there are humans (read "people") living in equally inhumane conditions?

Isn't it rather contradictory to worry about the comfort of animals that are raised for slaughter? Why can't these people work up enough feeling for their fellow man to try to do something about the unjust and degrading lives so many are forced to lead?

11. Headline: City life is the most attractive to Americans.

Story: Seventy-five percent of the American population—more than 160 million people—lives in 318 metropolitan areas that make up only 16 percent of the country's land mass, the Census Bureau says.

Contrary to popular belief, the most densely populated city in the country is not New York but Jersey City, N.J., where there are 23,000 people for every square mile of its 46-mile area.

New York is next with 6,600 people per square mile, while the place for lonely hearts is Alaska, with less than one person per square mile—counting only the populated area of the icy state. In the metropolitan areas, 30 percent of the people live in central cities, and 45 percent live in the suburbs.

12. Jimmy Carter's 1976 presidential campaign slogan: For America's third century, why not the best?

13. You Americans criticise our country for so-called human rights violations. But what moral right do you have to act as preachers of freedom and democracy, given your own history of racial and social problems?

* 14. Interior Secretary James G. Watt's appearance on NBC's "Meet the Press" (Feb. 21) leads me to the following observations:

Watt's appearance was a sorry spectacle of deceit as he attempted to demean the reporters with his transparent answers to the searching questions on his former positions and policy statements on wilderness areas, environmental pollution, and national parks. It made one wonder about his professed commitment to Christ and God. To hear a man utter such calculated misstatements and lies on the Sabbath leads a person to believe he is a devil's disciple, rather than a lover of the Lord.

15. People who follow the news are more successful, friendly, and responsible than those with no interest in current events, a university study has discovered.

But if you haven't been keeping informed, you can easily change, says an expert.

"Those who don't keep up with current events tend to be more hostile, more defensive, less sociable and less likely to accept blame," says psychology professor Dr. Henry H. Schmo.

"They are the foot-stompers, the people who shake their fists at stoplights and scream at sales ladies."

However, people in the know about the news tend to be "all-around high achievers," said Dr. Schmo who conducted a study of 63 people at Long Island University (L.I.U.).

"They are easier to be around because they are more secure."

"If you don't keep up with the news, you can change," said Dr. Sherman Blatz, L.I.U. professor of psychology.

"You can too be more successful, more sociable, more willing to accept responsibility and generally more fun to be around simply by making yourself more aware of current events."

How? Read at least one newspaper a day, watch TV news at least twice a week, watch a TV talk show at least once a month and discuss current events with other people as often as possible, he suggested.

16. Pam Poco's whole life has suddenly changed, since she lost an amazing 16 lbs. in seven days and then quickly lost 99 lbs. more. As she describes it: "I was always overweight and tried everything to lose, but the Super-Loss Capsule Program is the only thing that ever worked. It's truly the #1 weapon against fat."

What is the Super-Loss Diet Program? It's a remarkable weight loss plan with an amazing Clinically Proven Pill. This incredible program can actually make your body burn off the maximum amount of fat . . . in record time. It can help you quickly lose all the weight you need to—without torturous starvation and without battling hunger pangs. Weight loss is fast and dramatic.

17. Reporter: Why does O.J. Simpson carry the ball so often?
Coach Mackay: Why not? It isn't very heavy!

* 18. Blap's Beer—America's brew!

19. Enlarge your bosom three full cup sizes in only seven days! What is the secret of larger breasts? The lush, big, round, firm bosoms that get all the love, admiration, and attention of men everywhere! The only permanently easy and safe answer is a "New" active energizer enlarger "Big Bosom Pill" called Aero-plus 3! Developed after 20 years of extensive scientific and medical research and a quarter century of interplanetary and computer experience by a famous lab. If we helped put a man on the moon, we surely can put a big beautiful bosom on a deserving lady. Know the pleasure and fulfillment of going from an A to a D cup. Only $20.00 complete.

20. The first step is registration of handguns, the second step is total ban on handguns, the third step is registration of rifles and shotguns, and the final step is total gun control—confiscation of all guns, just as in Nazi Germany.

We who oppose gun control will do all in our power to see that our Second Amendment and other constitutional rights are not violated.

21. Maybe it's time for you to do what over one million French women have done for years. Rely on the famous French Suave Beauty Treatment for your body. Its sole purpose is to make your derriere, upper arms, and legs look more beautiful. The Suave Beauty Treatment for your body.

* 22. Can the universe think about itself? We know that at least one part of it can: we ourselves. Is it not reasonable to conclude the whole can?

23. Announcing the amazing doctor-tested formula that is three times stronger than other diets. This incredible breakthrough that attacks overweight three ways is your key to total weight loss.

This doctor-tested, medically proven formula is based on published reports from professors at leading medical schools. This three-way punch is the most advanced weight loss method known to medical science.

With this doctor-tested formula, you can lose even more—like 20, 30, even 50 lbs., till you reach your ideal weight level.

24. Q: Mr. President, how certain are you that the economy is going to begin to turn around by the end of spring?

A: Well, I'm not going to pick any particular month or anything and then find myself having to be held to that . . . I'm going to tell you that I believe, in these months ahead and the coming year, I think we're going to see the recession bottomed out and we're going to see interest rates begin to fall.

25. I find it necessary as one who does not wallow in Donald Smith's pompous world of pseudointellectualism to answer his absurd article, "Roll over, Beethoven! The Disco Beat is Bach."

Smith must not be aware that it takes real people who understand and love the classics to appreciate the "Hooked on Classics" record. Obviously, he is also insensitive to talent and the innovative ability to make money for a worthy cause.

Perhaps Smith can use his enormous ego, limited insights, and questionable knowledge to implement a program as successful as "Hooked on Classics" to save our San Diego Symphony instead of wasting his time criticizing a gutsy and talented individual like Louis Clarke.

* 26. Under the ERA, its foes contended, women would be drafted and sent into combat, separate lavatories for men and women would vanish from the public scene, homosexual marriages would be legalized, and legal protection for divorced women would be undermined.

"Every argument made by Phyllis Schlafy is wrong, and I can answer them," says Elizabeth Griffith, former vice chairman of the National Women's Political Caucus. "But you have to be very specific and very technical, and it takes a longer time to answer a charge than to make one."

27. One of the "intellectual" arguments advanced by those who oppose the death penalty is that it is not a deterrent. Then how in the world can we account for the monstrous increase in armed robbery and murder since our 10-year moratorium on that penalty? Or is that too simplistic for "intellectuals" to understand?

28. Dodge—America's Driving Machine.

29. Who's really covering up the truth about Laetrile?

* 30. The patronizing tone of your Feb. 5 editorial makes it clear just where your self-righteous mentality lives: in the dreary, drab world of uncreative sameness. To assert that student fees should not be used to pay for art is indicative of your ill-informed small-mindedness. This is exactly the way student fees should be spent.

31. Ban is more effective than Right Guard, Secret, Arrid, and Sure. In fact, Ban Roll-on is more effective than all leading aerosols.

32. Q: Should we develop solar power?

A: The problem lies in converting sunshine into usable, affordable energy to heat and light our homes and operate our appliances. At today's rates, if we tried to convert all our oil, gas, and nuclear power systems to solar, it would cost hundreds of billions of dollars.

33. Assault, murder, and other crimes against another person are both immoral and illegal in our society. Why, then, should someone with a bad habit, like smoking, be allowed to slowly murder those around him?

* 34. The idea that a gun is a good method of self-defense is a fallacy. With all the firepower the Secret Service possessed, it didn't save the most protected man in the country, Ronald Reagan, from being shot.

35. Charles Smith, who is so concerned about guns killing people (Letters, Oct. 4), failed to mention a worse killer. Thousands of adults and children are killed or injured for life by autos.

Let's start a "Ban-the-Car" movement. The right to life is more important than the right to drive a car.

We can all go back to riding horses.

36. We who support the ERA have never dismissed the draft issue as rubbish. What we continue to say is that Congress has always had this power to draft anyone. By raising the question of the Pentagon and Congress contemplating a selective service registration for men and women, Buchanan makes our point. Namely—if and when we are needed, we will be called, with or without the ERA.

37. Lose 4–8 pounds in one hour! With the Space Suit slenderizing system as documented by a U.S. government expert.

* 38. I'm thoroughly disgusted with all the slobbering going on about the cut in school lunches! Will someone please explain to me why the taxpayers should be providing food for children? Where are their parents? Don't parents accept any responsibility for their offspring at all?

Why don't we go all the way and just let the government take over altogether, just as they do in the communist countries? I've come to the conclusion that what the people in this country really want is communism. Why else are our citizens always clamoring for the government to provide them with food, clothing and shelter? Oh, for the good old days when individuals took responsibility for their own actions instead of whining away at the government's doorstep.

39. David Morehouse became the first UCLA undergraduate president to be recalled Wednesday when election results revealed that 73.9 percent of 3,827 students voted to remove him from office.

40. Those who love honey can follow the advice of Thelma McElhiney of Moberly, Mo., who will receive $5.00 for her ingenuity.

She stretches a 40-oz. jar of honey by mixing it with a pint of light corn syrup—which is less expensive than honey.

41. Every human group is born, grows, declines, and dies, as it must if it is an aggregation of individual living beings.

* 42. Remember when life was so simple.

43. Omoro—The only one you need! Millions of people have counted on Omoro for their vitamins and mineral protection. Omoro is the only one you need. With 31 vitamins, minerals and other important nutrients, it's the multi-vitamin that looks out for your well being. Try this exclusive master formula and get the healthful vitamin and mineral insurance you deserve.

44. Don't take chances! Take a vitamin supplement!

45. Great American Car Sale!! Uncle Sam says "lower interest rates mean low payments!!"

* 46. People who kill their mates were almost all driven to murder by humiliating remarks or other ego-destroying blows at the hands of their spouses.

That's the finding of sociologist Dr. Fred Smith, who spoke with 34 convicted spouse-killers to find out what caused them to resort to bloodshed.

EXERCISE 7.1B

Find the fallacies in the following speech.

THE "CHECKERS" SPEECH
Richard M. Nixon

My fellow Americans: I come before you tonight as a candidate for the vice presidency and as a man whose honesty and integrity has been questioned. Now, the usual political thing to do when charges are made against you is to either ignore them or to deny them without giving details. I believe we've had enough of that in the United States, particularly with the present Administration in Washington, D.C. To me the office of the Vice Presidency of the United States is a great office, and I feel that the people have got to have confidence in the integrity of the men who run for that office and who might attain it. I have a theory, too, that the best and only answer to a smear or to an honest misunderstanding of the facts is to tell the truth. And that's why I am here tonight. I want to tell you my side of the case.

I'm sure you have read the charge, and you have heard it, that I, Senator Nixon, took $18,000.00 from a group of my supporters. Now, was that wrong? And let me say that it was wrong—I am saying, incidentally, that it was wrong [sic, asking, incidentally, if it was wrong], not just illegal, because it isn't a question of whether it was legal or illegal, that isn't enough. The question is "Was it morally wrong?" I say that it was morally wrong if any of the $18,000.00 went to Senator Nixon for my personal use. I say that it was morally wrong if it was secretly given and secretly handled. And I say that it was morally wrong if any of the contributors got special favors for the contributions that they made.

And now to answer those questions let me say this: Not one cent of the $18,000.00 or any other money of that type ever went to me for my personal use. Every penny of it was used to pay for political expenses that I did not think should be charged to the taxpayers of the United States. It was not a secret fund. As a matter of fact, when I was on "Meet the Press"—some of you may have seen it last Sunday—Peter Edson came up to me after the program and he said, "Dick, what about this fund we hear about?" And I said, "Well, there's no secret about it. Go out and see Dana Smith," who was the administrator of the fund. And I gave him his address, and I said, "You will find that the purpose of the fund simply was to defray political expenses that I did not feel should be charged to the Government." And third, let me point out, and I want to make this particularly clear, that no contributor to this fund, no contributor to any of my campaign, has ever received any consideration that he would not have received as an ordinary constituent. I just don't believe in that and I can say that never, while I have been in the Senate of the United States, as far as the people that contributed to this fund are concerned, have I made a telephone call for them to an agency, or have I gone down to an agency in their behalf. And the records will show that, the records which are in the hands of the Administration.

Well, then, some of you will say, and rightly, "Well, what did you use the fund for, Senator? Why did you have to have it?" Let me tell you in just a word how a Senate office operates. First of all, a senator gets $15,000.00 a year in salary. He gets enough money to pay for one trip a year, a round trip, that is, for himself and his family between his home and Washington, D.C. And then he gets an allowance to handle the people that work in his office, to handle his mail. And the allowance for my state of California is enough to hire thirteen people. And let me say, incidentally, that that allowance is not paid to the senator—it's paid directly to the individuals that the senator puts on his payroll—but all of these people and all of these allowances are for strictly official business. Business, for example, when a constituent writes in and wants you to go down to the Veterans' Administration and get some information about his GI policy. Items of that type, for example.

But there are other expenses which are not covered by the Government. And I think I can best discuss those expenses by asking you some questions. Do you think that when I or any other senator makes a political speech, has it printed, should charge the printing of that speech and the mailing of that speech to the taxpayers? Do you think, for example, when I or any other senator makes a trip to his home state to make a purely political speech that the cost of that trip should be charged to the taxpayers? Do you think when a senator makes political broadcasts or political television broadcasts, radio or television, that the expense of those broadcasts should be charged to the taxpayers? Why, I know what your answer is. It's the same answer that audiences give me whenever I discuss this particular problem. The answer is, "No, the taxpayers shouldn't be required to finance items which are not official business but which are primarily political business."

Well then the question arises: You say, "Well, how do you pay for these and how can you do it legally?" And there are several ways that it can be done, incidentally, and that it is done legally in the United States Senate and in the Congress. The first way is to be a rich man. I don't happen to be a rich man so I couldn't use that way. Another way that is used is to put your wife on the payroll. Let me say, incidentally, that my opponent, my opposite number for the Vice Presidency on the Democratic ticket, does have his wife on the payroll. And has had it, her on his payroll for the ten years—for the past ten years. Now, just let me say this. That's his business and I'm not critical of him for doing that. You will have to pass judgment on that particular point. But I

7.1 / A Review of the Fallacies

have never done that for this reason: I have found that there are so many deserving stenographers and secretaries in Washington that needed the work that I just didn't feel it was right to put my wife on the payroll. My wife's sitting over here. She's a wonderful stenographer. She used to teach stenography and she used to teach shorthand in high school. That was when I met her. And I can tell you folks that she's worked many hours at night and many hours on Saturdays and Sundays in my office and she's done a fine job. And I'm proud to say tonight that in the six years I've been in the House and the Senate of the United States, Pat Nixon has never been on the government payroll. What are other ways that these finances can be taken care of? Some who are lawyers, and I happen to be a lawyer, continue to practice law. But I haven't been able to do that. I'm so far away from California and I've been so busy with my senatorial work that I have not engaged in any legal practice. And also as far as law practice is concerned, it seemed to me that the relationship between an attorney and the client was so personal that you couldn't possibly represent a man as an attorney and then have an unbiased view when he presented his case to you in the event that he had one before the Government.

And so I felt that the best way to handle these necessary political expenses, of getting my message to the American people, and the speeches I made, the speeches that I had printed, for the most part, concerned this one message—of exposing this Administration, the communism in it, the corruption in it—the only way that I could do that was to accept the aid which people in my home state of California who contributed to my campaign and who continued to make these contributions after I was elected, were glad to make. And let me say I'm proud of the fact that not one of them has ever asked me for a special favor. I'm proud of the fact that not one of them has ever asked me to vote on a bill other than as my own conscience would dictate. And I'm proud of the fact that the taxpayers by subterfuge or otherwise have never paid one dime for expenses which I thought were political and shouldn't be charged to the taxpayers.

Let me say, incidentally, that some of you may say, "Well, that's all right, Senator; that's your explanation, but have you got any proof?" And I'd like to tell you this evening that just an hour ago we received an independent audit of this entire fund. I suggested to Governor Sherman Adams, who is the chief of staff of the Dwight Eisenhower campaign, that an independent audit and legal report be obtained. And I have that audit here in my hand. It's an audit made by the Price, Waterhouse and Company firm, and the legal opinion by Gibson, Dunn and Crutcher, lawyers in Los Angeles. I'm proud to be able to report to you tonight that this audit and this legal opinion is being forwarded to General Eisenhower. And I'd like to read to you the opinion that was prepared by Gibson, Dunn and Crutcher and based on all the pertinent laws and statues, together with the audit report prepared by the certified public accountants, quote;

> It is our conclusion that Senator Nixon did not obtain any financial gain from the collection and disbursement of the fund by Dana

Smith; that Senator Nixon did not violate any Federal or state law by reason of the operation of the fund, and that neither their portion of the fund paid by Dana Smith directly to third persons or the portions paid the Senator Nixon to reimburse him for designated office expenses constituted income to the Senator which was either reportable or taxable as income under applicable tax laws.

(Signed) Gibson, Dunn and Crutcher by Elmo H. Conway

Now that, my friends, is not Nixon speaking, but that's an independent audit which was requested because I want the American people to know all the facts and I'm not afraid of having independent people go in and check the facts, and that is exactly what they did.

But then I realize that there are still some who may say, and rightly so—and let me say that I recognize that some will continue to smear regardless of what the truth may be, but that there has been understandably some honest misunderstanding on this matter—and there's some that will say, "Well, maybe you were able, Senator, to fake this thing. How can we believe what you say? After all, is there a possibility that maybe you got some sums in cash? Is there a possibility that you may have feathered your own nest?" And so now what I am going to do—and incidentally this in unprecedented in the history of American politics—I am going at this time to give to this television and radio audience, a complete financial history; everything I've earned; everything I've spent; everything I owe. And I want you to know the facts.

I'll have to start early. I was born in 1913. Our family was one of modest circumstances and most of my early life was spent in a store out in East Whittier. It was a grocery store—one of those family enterprises. The only reason we were able to make it go was because my mother and dad had five boys and we all worked in the store. I worked my way through college and to a great extent through law school. And then, in 1940, probably the best thing that ever happened to me happened. I married Pat—she's sitting over here. We had a rather difficult time after we were married, like so many of the young couples who may be listening to us. I practiced law; she continued to teach school. Then in 1942, I went into the service. Let me say that my service record was not a particularly unusual one. I went to the South Pacific. I guess I'm entitled to a couple of battle stars. I got a couple of letters of commendation, but I was just there when the bombs were falling and then I returned. I returned to the United States and in 1946 I ran for the Congress. When we came out of the war, Pat and I—Pat during the war had worked as a stenographer and in a bank and as an economist for a Government agency—and when we came out, the total of our savings from both my law practice, her teaching, and all the time that I was in the war—the total for that entire period was just a little less than $10,000.00. Every cent of that, incidentally, was in Government bonds.

Well, that's where we start when I go into politics. Now what have I earned since I went into politics? Well, here it is—I jotted it down—let me read the notes. First of all, I've had my salary as a congressman and as a senator; second, I have received a total in this past six years of $1,600.00 from estates which were in my law firm at the time that I severed my connection with it. And, incidentally, as I said before, I have not engaged in any legal practice and have not accepted any fees from business that came into the firm after I went into politics. I have made an average of approximately $1,500.00 a year from nonpolitical speaking engagements and lectures. And then, fortunately, we've inherited a little money. Pat sold her interest in her father's estate for $3,000.00 and I inherited $1,500.00 from my grandfather. We lived rather modestly. For four years we lived in an apartment in Park Fairfax in Alexandria, Virginia. The rent was $80.00 a month. And we saved for the time that we could buy a house.

Now, that was what we took in. What did we do with this money? What do we have today to show for it? This will surprise you, because it is so little, I suppose, as standards generally go, of people in public life. First of all, we've got a house in Washington which cost $41,000.00 and on which we owe $20,000.00. We have a house in Whittier, California, which cost $13,000.00 and on which we owe $3,000.00. My folks are living there at the present time. I have just $4,000.00 in life insurance, plus my GI policy which I've never been able to convert and which will run out in two years. I have no life insurance whatever on Pat. I have no life insurance on our two youngsters, Patricia and Julie. I own a 1950 Oldsmobile car. We have our furniture. We have no stocks and bonds of any type. We have no interest of any kind, direct or indirect, in any business.

Now, that's what we have. What do we owe? Well, in addition to the mortgage, the $20,000.00 mortgage on the house in Washington, the $10,000.00 [Nixon actually owed $3,000.00] one on the house in Whittier, I owe $4,500.00 to the Riggs Bank in Washington D.C., with interest at $4\frac{1}{5}$ percent. I owe $3,500.00 to my parents and the interest on that loan—which I pay regularly, because it's the part of the savings they made through the years they were working so hard—I pay regularly 4 per cent interest. And then I have a $500.00 loan which I have on my life insurance. Well, that's about it. That's what we have and that's what we owe. It isn't very much, but Pat and I have the satisfaction that every dime that we've got is honestly ours. I should say this—that Pat doesn't have a mink coat. And I always tell her that she'd look good in anything.

One other thing I probably should tell you because if I don't they'll probably be saying this about me too. We did get something—a gift—after the election [nomination]. A man down in Texas heard Pat on the radio mention the fact that our two youngsters would like to have a dog. And believe it or not, the day before we left on this campaign trip, we got a message from the Union Station in Baltimore saying they had a package for us. We went down to get it. You know what it was? It was a little cocker spaniel dog, in a crate that he had sent all the way from Texas, black and white, spotted, and our little girl, Tricia, the six-year-old, named it Checkers. And, you know, the kids, like all kids, love the dog, and I just want to say this, right now, that regardless of what they say about it, we're going to keep it.

It isn't easy to come before a nation-wide audience and bare your life, as I have done. But I want to say some things before I conclude, that I think most of you will agree on. Mr. Mitchell, the Chairman of the Democratic National Committee, made the statement that if a man couldn't afford to be in the United States Senate, he shouldn't run for the Senate. And I just want to make my position clear. I don't agree with Mr. Mitchell when he says that only a rich man should serve his Government in the United States Senate or in the Congress. I don't believe that represents the thinking of the Democratic Party, and I know that it doesn't represent the thinking of the Republican Party. I believe that it's fine that a man like Governor Stevenson, who inherited a fortune from his father, can run for President. But I also feel that it is essential in this country of ours that a man of modest means can also run for President, because, you know—remember Abraham Lincoln—you remember what he said—"God must have loved the common people, he made so many of them."

And now I'm going to suggest some courses of conduct. First of all, you have read in the papers about other funds, now. Mr. Stevenson apparently had a couple. One of them in which a group of business people paid and helped to supplement the salaries of state employees. Here is where the money went directly into their pockets, and I think that what Mr. Stevenson should do should be to come before the American people, as I have, give the names of the people who contributed to that fund, give the names of the people who put this money into their pockets at the same time that they were receiving money from their state government and see what favors, if any, they gave out for that. I don't condemn Mr. Stevenson for what he did, but until the facts are in there is a doubt that will be raised. And as far as Mr. Sparkman is concerned, I would suggest the same thing. He's had his wife on the payroll. I don't condemn him for that, but I think that he should come before the American people and indicate what outside sources of income he has had. I would suggest that under the circumstances both Mr. Sparkman and Mr. Stevenson should come before the American people, as I have, and make a complete financial statement as to their financial history, and if they don't it will be an admission that they have something to hide. And I think you will agree with me—because, folks, remember, a man that's to be President of the United States, a man that's to be Vice President of the United States, must have the confidence of all the people. And that's why I'm doing what I'm doing, and that's why I suggest that Mr. Stevenson and Mr. Sparkman, since they are under attack, should do what they're doing.

Now, let me say this: I know that this is not the last of the smears. In spite of my explanation tonight, other smears will be made. Others have been made in the past. And the purpose of the smears, I know, is this, to silence me, to make me let up. Well, they just don't know who they are dealing with. I'm going to tell you this: I remember in the dark days of the Hiss case, some of the same radio commentators who are attacking me now and misrepresenting my position, were violently opposing me at the time I was after Alger Hiss.

But I continued to fight, because I knew I was right, and I can say to this great television and radio audience that I have no apologies to the American people for my part in putting Alger Hiss where he is today. And as far as this is concerned, I intend to continue to fight.

Why do I feel so deeply? Why do I feel that in spite of the smears, the misunderstanding, the necessity for a man to come up here and bare his soul, as I have, why is it necessary for me to continue this fight? And I want to tell you why. Because, you see, I love my country. And I think my country is in danger. And I think the only man that can save America at this time is the man that's running for President on my ticket, Dwight Eisenhower. You say, "Why do I think it's in danger?" and I say, "Look at the record." Seven years of the Truman-Acheson administration and what's happened? Six hundred million people lost to the Communists. And a war in Korea, in which we have lost 117,000 American casualties. And I say to all of you that a policy that results in a loss of six hundred million people to the Communists and a war which costs us 117,000 American casualties isn't good enough for America. And I say that those in the State Department that make the mistakes which caused that war and which resulted in those losses should be kicked out of the State Department just as fast as we get them out of there. And let me say that I know Mr. Stevenson won't do that because he defends the Truman policies. And I know that Dwight Eisenhower will do that and that he will give America the leadership that it needs. Take the problem of corruption. You've read about the mess in Washington. Mr. Stevenson can't clean it up because he was picked by the man, Truman, under whose administration the mess was made. You wouldn't trust the man who made the mess to clean it up—that's Truman—and by the same token, you can't trust the man who was picked by the man that made the mess to clean it up—and that's Stevenson. And so I say, Eisenhower, who owes nothing to Truman, nothing to the big city bosses. He is the man that can clean up the mess in Washington.

Take Communism. I'd say, as far as that subject is concerned, the danger is great to America. In the Hiss case, they got the secrets which enabled them to break the American secret State Department code. They got secrets in the Atomic Bomb case which enabled them to get the secret of the atomic bomb five years before they would have gotten it by their own devices. And I say that any man who calls the Alger Hiss case a red herring, isn't fit to be President of the United States. I say that a man, who like Mr. Stevenson, has pooh-poohed and ridiculed the Communist threat in the United States—he said that they are phantoms among ourselves; he has accused us that have attempted to expose the Communists of looking for Communists in the Bureau of Fisheries and Wild Life—I say that a man who says that isn't qualified to be President of the United States. And I say that the only man who can lead us in this fight to rid the government of both those who are Communists and those who have corrupted this government is Eisenhower because Eisenhower, you can be sure, recognizes the problem and he knows how to deal with it.

And let me say, finally this evening, I want to read to you, just briefly, excerpts from a letter which I received. A letter which after all this is over no one can take away from me. It reads as follows:

Dear Senator Nixon:

Since I am only nineteen years of age I can't vote in this presidential election, but believe me, if I could, you and General Eisenhower would certainly get my vote. My husband is in the Fleet Marines in Korea. He's a Corpsman on the front lines, and we have a two-month-old son he's never seen. And I feel confident that with great Americans like you and General Eisenhower in the White House, lonely Americans like myself will be united with their loved ones now in Korea. I only pray to God that you won't be too late. Enclosed is a small check to help you in your campaign. Living on eighty-five dollars a month, it is all I can afford at present, but let me know what else I can do.

Folks, it's a check for ten dollars, and it's one that I will never cash. And just let me say this; we hear a lot about prosperity these days, but I say, "Why can't we have prosperity and an honest government in Washington, D.C., at the same time?" Believe me, we can. And Eisenhower is the man that can lead this crusade to bring us that kind of prosperity.

And now, finally, I know that you wonder whether or not I am going to stay on the Republican ticket or resign. Let me say this: I don't believe that I ought to quit, because I am not a quitter. And, incidentally, Pat's not a quitter. After all, her name was Patricia Ryan, and she was born on St. Patrick's Day, and you know the Irish never quit. But the decision, my friends, is not mine. I would do nothing that would harm the possibilities of Dwight Eisenhower to become President of the United States, and for that reason, I am submitting to the Republican National Committee tonight, through this television broadcast, the decision which it is theirs to make. Let them decide whether my position on the ticket will help or hurt. And I'm going to ask you to help them decide. Wire and write the Republican National Committee whether you think I should stay on or whether I should get off. And whatever their decision is, I will abide by it.

But just let me say this last word. Regardless of what happens, I'm going to continue this fight. I'm going to campaign up and down in America until we drive the crooks and Communists and those that defend them out of Washington.

And remember, Folks, Eisenhower is a great man. Believe me, he's a great man, [Announcer in background, simultaneously: This program has been sponsored by the Republican Senatorial Committee and the Republican Congressional Committee.] and a vote for Eisenhower is a vote for what's good for America.

7.2 An Extended Example

A common complaint about textbook discussions of fallacies is that the theory of fallacies doesn't really apply to real life. Consider what one scholar has said:[1]

> What is wrong with such (i.e., textbook) accounts of fallacies? One problem concerns the paucity of actual examples. . . . It is in fact puzzling that logic shouldn't be able to come up with more examples of fallacies actually committed given that fallacies are supposed to be *common* errors in reasoning. One gets the suspicion that logically incorrect arguments are not that common in practice, that their existence may be largely restricted to logic textbooks and examples.

The examples given in this text are actual cases of bad reasoning. But to make sure that the student sees that fallacies are indeed common, we will examine an actual extended example: The second debate between President Ronald Reagan and Senator Walter Mondale, which took place on October 21, 1984. This debate was historically crucial; indeed, it may have ensured the reelection of Reagan. We will point out some of the fallacies committed. (They are numbered and explained at the end of the debate.) The reader should look for other fallacies not caught in the text.

Preliminary

The following is a transcript of the televised debate between President Reagan and Walter F. Mondale sponsored by the League of Women Voters in Kansas City, Missouri as recorded by the *New York Times*. (Reprinted with permission.)

DOROTHY S. RIDINGS: Good evening from the Municipal Auditorium in Kansas City. I am Dorothy Ridings, the president of the League of Women Voters, the sponsor of this final Presidential debate of the 1984 campaign between Republican Ronald Reagan and Democrat Walter Mondale.

Our panelists for tonight's debate on defense and foreign policy issues are Georgie Anne Geyer, syndicated columnist for Universal Press Syndicate, Marvin Kalb, chief diplomatic correspondent for *NBC News*, Morton Kondracke, executive editor of the *New Republic*

---

[1] Maurice Finocchiaro, "Fallacies and the Evaluation of Reasoning," *American Philosophical Quarterly* 18 (1) (Jan. 1981).

magazine and Henry Trewhitt, diplomatic correspondent for the *Baltimore Sun*.

Edwin Newman, formerly of *NBC News* and now a syndicated columnist for King Features is our moderator. Ed.

EDWIN NEWMAN: Dorothy Ridings, thank you. A brief word about our procedure tonight. The first question will go to Mr. Mondale. He'll have two and a half minutes to reply. Then the panel member who put the question will ask the follow-up. The answer to that will be limited to one minute. After that, the same question will be put to President Reagan. Again, there will be a follow-up and then each man will have one minute for rebuttal.

The second question will go to President Reagan first. After, the alternating will continue. At the end there will be four-minute summations with President Reagan going last.

We have asked the questioners to be brief.

Let's begin, Miss Geyer, your question to Mr. Mondale.

Debate[2]

Q: Mr. Mondale, two related questions on the crucial issue of Central America. You and the Democratic Party have said that the only policy toward the horrendous civil wars in Central America should be on the economic developments and negotiations with, perhaps a quarantine of, Marxist Nicaragua. Do you really believe that there is no need to resort to force at all? Are not these solutions to Central America's gnawing problems simply again too weak and too late?

MONDALE: I believe that the question oversimplifies the difficulties of what we must do in Central America. Our objectives ought to be to strengthen the democracy, to stop Communist and other extremist influences and stabilize the community in that area.

1. To do that, we need a three-pronged attack. One is military assistance to our friends who are being pressured.

Secondly, a strong and sophisticated economic aid program and human rights program that offers a better life and a sharper alternative to the alternative offered by the totalitarians who oppose us. And finally, a strong diplomatic effort that pursues the possibilities of peace in the area.

That's one of the big disagreements that we have with the President, that they have not pursued the diplomatic opportunities either

[2] Numbers that appear in the lefthand margin correspond to the numbered list of comments that follows the debate.

within El Salvador or as between the country and its neighbors and have lost time during which we might have been able to achieve peace.

This brings up the whole question of what Presidential leadership is all about. I think the lesson in Central America, this recent embarrassment in Nicaragua where we are giving instructions for hired assassins, hiring criminals and the rest—all of this has strengthened our opponent.

A President must not only assure that we're tough. But we must also be wise and smart in the exercise of that power. We saw the same thing in Lebanon where we spent a good deal of America's assets, but because the leadership of this government did not pursue wise policies, we have been humiliated and our opponents are stronger.

2. The bottom line of national strength is that the President must be in command. He must lead. And when a President doesn't know that submarine missiles are recallable, says that 70 percent of our strategic forces are conventional, discovers three years into his Administration that our arms control efforts have failed because he didn't know that most Soviet missiles were on land—these are things a President must know to command. A President is called the Commander-in-Chief. And he's called that because he's supposed to be in charge of facts and run our government and strengthen our nation.

3. Q: Mr. Mondale, if I could broaden the question just a little bit. Since World War II, every conflict that we as Americans have been involved with has been in nonconventional or irregular terms and yet we keep fighting in conventional or traditional military terms. The Central American wars are very much in the same pattern as China, as Lebanon, as Iran, as Cuba in the early days. Do you see any possibility that we are going to realize the change in warfare in our time or react to it in those terms?

MONDALE: We absolutely must, which is why I responded to your first question the way I did. It's much more complex. You must understand the region, you must understand the politics in the area, you must provide a strong alternative, and you must show strength—and all at the same time.

4. That's why I object to the covert action in Nicaragua. That's a classic example of a strategy that's embarrassed us, strengthened our opposition and undermined the moral authority of our people and our country in the region.

Q: Mr. President, in the last few months it has seemed more and more that your policies in Central America were beginning to work. Yet just at this moment we are confronted with the extraordinary story of the

C.I.A. guerrilla manual for the anti-Sandinista Contras, whom we are backing, which advocates not only assassinations of Sandinistas but the hiring of criminals to assassinate the guerrillas we are supporting in order to create martyrs. Is this not in effect our own state-supported terrorism?

REAGAN: No, but I'm glad you asked that question because I know it's on many people's minds. I have ordered an investigation; I know that the C.I.A. is already going forward with one. We have a gentleman down in Nicaragua who is on military tactics, the Contras. And he drew up this manual. It was turned over to the agency head of the C.I.A. in Nicaragua to be printed, and a number of pages were excised by that agency head there, the man in charge, and he sent it on up here to C.I.A., where more pages were excised before it was printed. But some way or other, there were 12 of the original copies that got out down there and were not submitted for this printing process by the C.I.A. Now those are the details as we have them, and as soon as we have an investigation and find out where any blame lies for the few that did not get excised or changed, we certainly are going to do something about that. We'll take the proper action at the proper time.

I was very interested to hear about Central America and our process down there, and I thought for a moment that instead of a debate I was going to find Mr. Mondale in complete agreement with what we're doing because the plan that he has outlined is the one we've been following for quite some time, including diplomatic processes throughout Central America and working closely with the Contadora Group. So I can only tell you, about the manual, that we're not in the habit of assigning guilt before there has been proper evidence produced in proof of that guilt; but if guilt is established, whoever is guilty, we will treat with that situation then and they will be removed.

Q: Well, Mr. President, you are implying then that the C.I.A. in Nicaragua is directing the Contras there. I'd also like to ask whether having the C.I.A. investigate its own manual in such a sensitive area is not sort of like sending the fox into the chicken coop a second time.

REAGAN: I'm afraid I misspoke when I said a C.I.A. head in Nicaragua. There's not someone there directing all of this activity. There are, as you know, C.I.A. men stationed in other countries in the world, and certainly in Central America, and so it was a man down there in that
5. area that this was delivered to. And he recognized that what was in that manual was a direct contravention of my own executive order in December of 1981, that we would have nothing to do with regard to political assassinations.

MODERATOR: Mr. Mondale, your rebuttal?

MONDALE: What is a President charged with doing when he takes his oath of office? He raises his right hand and takes an oath of, oath of office to take care, to faithfully execute the laws of the land. Presidents can't know everything but a President has to know those things that are essential to his leadership and the enforcement of our laws.

This manual, several thousands of which were produced, was distributed ordering political assassination, hiring of criminals and other forms of terrorism. Some of it was excised but the part dealing with political terrorism was continued.

How can this happen? How can something this serious occur in an Administration and have a President of the United States in a situation like this say he didn't know. A President must know these things.

I don't know which is worse—not knowing or knowing and not stopping it.

And what about the mining of the harbors in Nicaragua, which violated international law? This has hurt this country and a President's supposed to command.

MODERATOR: Mr. President, your rebuttal.

REAGAN: Yes. I have so many things there to respond to, I'm going to pick out something you said earlier.

6. You've been all over the country repeating something that I will admit the press has also been repeating—that I believe that nuclear missiles could be fired and then called back. I never conceived of such a thing. I never said any such thing.

In a discussion of our strategic arms negotiations, I said that submarines carrying missiles and airplanes carrying missiles were more conventional-type weapons, not as destabilizing as the land-based missiles and that they were sent out and there was a change, you could call them back before they had launched their missiles. But I hope that from here on, you will no longer be saying that particular thing, which is absolutely false. How anyone could think that any sane person would believe you could call back a nuclear missile I think is as ridiculous as the whole concept has been.

So, thank you for giving me a chance to straighten the record. I'm sure that you appreciate that.

MODERATOR: Mr. Kalb, your question to President Reagan.

Q: Mr. President, you have often described the Soviet Union as a powerful evil empire intent on world domination. But this year, you

7. have said, and I quote: "If they want to keep their Mickey Mouse system, that's O.K. with me." Which is it, Mr. President—Do you want to contain them within their present borders and perhaps try to reestablish détente or what goes for détente or do you really want to roll back their empire?

REAGAN: I have said, on a number of occasions, exactly what I believe about the Soviet Union. I retract nothing that I have said. I believe that many of the things they have done are evil in any concept of morality that we have. But I also recognize that as the two great superpowers in the world, we have to live with each other. And I told Mr. Gromyko we don't like their system. They don't like ours. And
8. we're not gonna change their system and they sure better not try to change ours. But, between us, we can either destroy the world or we can save it. And I suggested that certainly it was to their common interest, along with ours, to avoid a conflict and to attempt to save the world and remove the nuclear weapons. And I think that perhaps we established a little better understanding.

I think that in dealing with the Soviet Union, one has to be realistic. I know that Mr. Mondale, in the past, has made statements as if they were just people like ourselves and if we were kind and good and did something nice, they would respond accordingly. And the result was unilateral disarmament. We canceled the B-1 under the previous Administration. What did we get for it? Nothing.

The Soviet Union has been engaged in the biggest military build-up in the history of man at the same time that we tried the policy of unilateral disarmament, of weakness, if you will. And now, we are putting up a defense of our own. And I've made it very plain to them. We seek no superiority. We simply are going to provide a deterrent so that it will be too costly for them if they are nursing any ideas of aggression against us.

Now they claim they're not. And I made it plain to them that we're not. But, this, there's been no change in my attitude at all. I just thought when I came into office it was time that there was some realistic talk to and about the Soviet Union. And we did get their attention.

Q: Mr. President, on perhaps the other side of the coin, a related question, sir. Since World War II, the vital interests of the United States have always been defined by treaty commitments and by presidential proclamations. Aside from what is obvious, such as NATO, for example, which countries, which regions in the world do you regard as vital national interests of this country, meaning that you would send American troops to fight there if they were in danger?

REAGAN: Ah, well now you've added a hypothetical there at the end, Mr. Kalb, about that where we would send troops in to fight. I am not going to make the decision as to what the tactics could be, but obviously there are a number of areas in the world that are of importance to us.

One is the Middle East. And that is of interest to the whole Western world and the industrialized nations, because of the great supply of energy upon which so many depend there.

The—our neighbors here in America are vital to us. We're working right now in trying to be of help in southern Africa with regard to the independence of Namibia and the removal of the Cuban surrogates, the thousands of them, from Angola.

So, I can say there are a great many interests. I believe that we have a great interest in the Pacific basin. That is where I think the future of the world lies.

But I am not going to pick out one and in advance and hypothetically say, oh, yes, we should send troops there. I don't...

MODERATOR: Sorry, Mr. President. Sorry, your time was up.

Q: Mr. Mondale, you have described the Soviet leaders as, and I'm quoting, cynical, ruthless and dangerous, suggesting an almost total lack of trust in them. In that case, what makes you think that the annual summit meetings with them that you've proposed will result in agreements that would satisfy the interests of this country?

9. MONDALE: Because the only type of agreements to reach with the Soviet Union are the types that are specifically defined, so we know exactly what they must do, subject to full verification. Which means we know every day whether they're living up to it, and follow-ups wherever we find suggestions that they're violating it, and the strongest possible terms.

I have no illusions about the Soviet Union leadership or the nature of that state. They are a tough and a ruthless adversary, and we must be prepared to meet that challenge. And I would.

Where I part with the President is that despite all of those differences, we must, as past Presidents before this one have done, meet on the common ground of survival.

And that's where the President has opposed practically every arms control agreement, by every President of both political parties, since the bomb went off.

And he now completes this term with no progress toward arms control at all, but with a very dangerous arms race underway instead.

10. There are now over 2,000 more warheads pointed at us today than there were when he was sworn in, and that does not strengthen us.

We must be very, very realistic in the nature of that leadership, but we must grind away and talk to find ways to reducing these differences, particularly where arms races are concerned and other dangerous exercises of Soviet power.

There will be no unilateral disarmament under my Administration. I will keep this nation strong. I understand exactly what the Soviets are up to. But that, too, is a part of national strength.

To do that, a President must know what is essential to command and to leadership and to strength. And that's where the President's failure to master, in my opinion, the essential elements of arms control has cost us dearly.

These four years—three years into this Administration he said he just discovered that most Soviet missiles are on land and that's why his proposal didn't work.

11. I invite the American people tomorrow, [to follow up] because I will issue the statement quoting President Reagan. He said exactly what I said he said. He said that these missiles were less dangerous than ballistic missiles because you could fire them and you could recall them if you decided there'd been a miscalculation. A President must know those things.

MODERATOR: I'm sorry.

Q: A related question, Mr. Mondale, on Eastern Europe: Do you accept the conventional diplomatic wisdom that Eastern Europe is a Soviet sphere of influence, and if you do, what could a Mondale Administration realistically do to help the people of Eastern Europe achieve the human rights that were guaranteed to them as a result of the Helsinki accords.

MONDALE: I think the essential strategy of the United States ought not accept any Soviet control over Eastern Europe. We ought to deal with each of these countries separately, we ought to pursue strategies with each of them—economic and the rest—that help them pull away from their dependence upon the Soviet Union.

Where the Soviet Union has acted irresponsibly, as they have in many of those countries—especially in Poland—I believe we ought to insist that Western credits extended to the Soviet Union bear the market rate, make the Soviets pay for their irresponsibility. That is a very important objective to make certain that we continue to look forward to progress toward greater independence by these nations and work with each of them separately.

7.2 / An Extended Example

MODERATOR: Mr. President, your rebuttal.

REAGAN: Yes, I'm not going to continue trying to respond to these repetitions of the falsehoods that have already been stated here, but with regard to whether Mr. Mondale would be strong, as he said he would be, I know that he has a commercial out where he is appearing on the deck of the Nimitz and watching the F-14's take off, and that's an image of strength—except that if he had had his way when the Nimitz was being planned he would have been deep in the water out there because there wouldn't have been any Nimitz to stand on. He was against it.

12. He was against the F-14 fighter, he was against the M-1 tank, he was against the B-1 bomber, he wanted to cut the salary of all of the military, he wanted to bring home half of the American forces in Europe, and he has a record of weakness with regard to our national defense that is second to none. Indeed, he was on that side virtually throughout all his years in the Senate and he opposed even President Carter when toward the end of his term President Carter wanted to increase the defense budget.

MODERATOR: Mr. Mondale, your rebuttal.

MONDALE: Mr. President, I accept your commitment to peace, but I want you to accept my commitment to a strong national defense. I propose a budget, I have proposed a budget, which would increase our nation's strength by, in real terms, by double that of the Soviet Union. I tell you where we disagree. It is true, over 10 years ago I voted to delay production of the F-14 and I'll tell you why. The plane wasn't flying the way it was supposed to be, it was a waste of money.

13. Your definition of national strength is to throw money at the Defense Department. My definition of national strength is to make certain that a dollar spent buys us a dollar's worth of defense. There's a big difference between the two of us. A President must manage that budget. I will keep us strong, but you'll not do that unless you command that budget and make certain we get the strength that we need. When you pay $500.00 for a $5.00 hammer, you're not buying strength.

MODERATOR: I would ask the audience not to applaud. All it does is take up time that we would like to devote to the debate. Mr. Kondracke, your question to Mr. Mondale.

Q: Mr. Mondale, in an address earlier this year you said that before this country resorts to military force, and I'm quoting, American interests should be sharply defined, publicly supported, Congressionally sanctioned, militarily feasible, internationally defensible, open to independent scrutiny and alert to regional history. Now aren't you setting up such a gauntlet of tests here that adversaries could easily suspect that as President you would never use force to protect American interests?

14. MONDALE: No; as a matter fact, I believe every one of those standards is essential to the exercise of power by this country. And we can see that in both Lebanon and in Central America. In Lebanon this President exercised American power all right, but the management of it was such that our marines were killed, we had to leave in humiliation, the Soviet Union became stronger, terrorists became emboldened, and it was because they did not think through how power should be exercised, did not have the American public with them on a plan that worked, that we ended up the way we did.

Similarly, in Central America, what we're doing in Nicaragua with this covert war which the Congress, including many Republicans, have tried to stop is finally end up with the public definition of American power that hurts us, where we get associated with political assassins and the rest. We have to decline for the first time in modern history jurisdiction of the World Court because they'll find us guilty of illegal actions, and our enemies are strengthened from all of this.

We need to be strong. We need to be prepared to use that strength, but we must understand that we are a democracy; we are a government by the people, and when we move, it should be for very severe and extreme reasons that serve our national interest and end up with a stronger country behind us. It is only in that way that we can persevere.

Q: You've been quoted as saying that you might quarantine Nicaragua. I'd like to know what that means. Would you stop Soviet ships as President Kennedy did in 1962 and wouldn't that be more dangerous than President Reagan's covert war?

15. MONDALE: What I'm referring to there is the mutual self-defense provisions that exist in the inter-American treaty, the so-called Rio Pact, that permits the nations, our friends in that region, to combine to take some steps, diplomatic and otherwise, to prevent Nicaragua when she acts irresponsibly in asserting power in other parts outside of her border, to take those steps, whatever they might be, to stop it.

The Nicaraguans must know that it is the policy of our Government that those people, that leadership must stay behind the boundaries of their nations, not interfere in other nations. And by working with all of the nations in the region, unlike what the President said the [the Administration] have not supported negotiations in that region, we will be much stronger because we'll have the moral authority that goes with those efforts.

Q: President Reagan, you introduced U.S. forces into Lebanon as neutral peacekeepers but then you made them combatants on the side of the Lebanese Government. Eventually you were forced to withdraw them under fire and now Syria, a Soviet ally, is dominant in the country. Doesn't Lebanon represent a major failure on the part of your Administration and raise serious questions about your capacity as a foreign policy strategist and as Commander-in-Chief?

REAGAN: No, Morton, I don't agree to all of those things. First of all, when we and our allies, the Italians, the French and the United Kingdom, went into Lebanon, we went in there at the request of what was left of the Lebanese Government, to be a stabilizing force while they tried to establish a government. But first, pardon me, the first time we went in at their request because the war was going on right in Beirut between Israel and the P.L.O. terrorists. Israel could not be blamed for that. Those terrorists had been violating their northern border consistently and Israel chased them all the way to there.

16. Then, we went in, with the multinational force, to help remove and did remove more than 13,000 of those terrorists from Lebanon. We departed and then the Government of Lebanon asked us back in as a stabilizing force while they established a government and sought to get the foreign forces all the way out of Lebanon and that they could then take care of their own borders. And were succeeding. We were there for the better part of a year. Our position happened to be at the airport or there were occasional snipings and sometimes some artillery fire, but we did not engage in conflict that was out of line with our mission.

I will never send troops anywhere on a mission of that kind without telling them that if somebody shoots at them they can darn well shoot back. And this is what we did. We never initiated any kind of action, we directed ourselves there. But, we were succeeding to the point that the Lebanese Government had been organized, if you will remember there were the meetings in Geneva in which they began to meet with the hostile factional forces and try to put together some kind of a peace plan. We were succeeding and that was why the terrorist acts began. There are forces there—and that includes Syria, in my mind—who don't want us to succeed, who don't want that kind of a peace with a dominant Lebanon, dominant over its own territory. And so the terrorist acts began and led to the one great tragedy when they were killed in that suicide bombing of the building. Then the multilateral force withdrew for only one reason. We withdrew because we were no longer able to carry out the mission for which we had been sent in. But we went in, in the interest of peace, and to keep Israel and Syria from getting into the sixth war between them. And I have no apologies for our going on a peace mission.

Q: Mr. President, four years ago you criticized President Carter for ignoring ample warning that our diplomats in Iran might be taken hostage. Haven't you done exactly the same thing in Lebanon, not once, but three times, with 300 Americans, not hostages, but dead? And you vowed swift retaliation against terrorists but doesn't our lack of response suggest that you're just bluffing?

REAGAN: Morton, no. I think there's a great difference between the Government of Iran threatening our diplomatic personnel and there is a Government that you can see and can put your hand on. In the terrorist situation there are terrorist factions all over—in a recent 30 day period 37 terrorist actions in 20 countries have been committed. The most recent has been the one in Brighton. In dealing with terrorists, yes, we want to retaliate, but only if we can put our finger on the people responsible and not endanger the lives of innocent civilians there in the various communities and in the city of Beirut where these terrorists are operating. I have just signed legislation to add to our ability to deal, along with our allies, with this terrorist problem, and it's going to take all the nations together, just as when we banded together we pretty much resolved the problem of skyjackings some time ago. Well, the red light went on—I could have gone on forever.

MODERATOR: Mr. Mondale, your rebuttal?

MONDALE: Groucho Marx said, "Who do you believe, me or your own eyes?" And what we have in Lebanon is something that the American people have seen. The Joint Chiefs urged the President not to put our troops in that barracks because they were undefensible. They urged—they went to five days before they were killed and said please take them out of there. The Secretary of State admitted that this morning. He did not do so. The report following the explosion in the barracks disclosed that we had not taken any of the steps that we should have taken. That was the second time. Then the embassy was blown up a few weeks ago and once again none of the steps that should have been taken were taken and we were warned five days before that explosives were on their way and they weren't taken. The terrorists have won each time. The President told the terrorists he was going to retaliate. He didn't. They called his bluff. And the bottom line is the United States left in humiliation and our enemies are stronger.

7.2 / *An Extended Example*

MODERATOR: Mr. President, your rebuttal?

REAGAN: Yes, first of all, Mr. Mondale should know that the President of the United States did not order the Marines into that barracks. That was a command decision made by the commanders on the spot and based with what they thought was best for the men there. That is one. One of the other things that you've just said about the terrorists—I'm tempted to ask you what you would do. Those are unidentified people, and after the bomb goes off they're blown to bits because they are suicidal individuals who think that they're going to go to paradise if they perpetrate such an act and lose their life in doing it. We are going to, as I say—we are busy trying to find the centers where these

17. operations stem from and retaliation will be taken, but we are not going to simply kill some people to say, oh look, we got even. We want to know when we retaliate that we're retaliating with those who are responsible for the terrorist acts. And terrorist acts are such that our own United States Capitol in Washington has been bombed twice.

MODERATOR: Mr. Trewhitt, your question to President Reagan?

Q: Mr. President, I want to raise an issue that I think has been lurking out there for two or three weeks, and cast it specifically in national security terms. You already are the oldest President in history, and some of your staff say you were tired after your most recent encounter with Mr. Mondale. I recall, yes, that President Kennedy, who had to go for days on end with very little sleep during the Cuba missile crisis. Is there any doubt in your mind that you would be able to function in such circumstances?

REAGAN: Not at all, Mr. Trewhitt and I want you to know that also I will not make age an issue of this campaign. I am not going to exploit for political purposes my opponent's youth and inexperience.

    If I still have time, I might add, Mr. Trewhitt, I might add that it was Seneca or it was Cicero, I don't know which, that said if it was not for the elders correcting the mistakes of the young, there would be no

18. state.

Q: Mr. President, I'd like to head for the fence and try to catch that one before it goes over but—without going to another question. The—you and Mr. Mondale have already disagreed about what you had to say about recalling submarine-launched missiles. There's another similar issue out there that relates to your—you said at least that you were unaware that the Soviet retaliatory power was based on land-missiles. First, is that correct? Secondly, if it is correct, have you informed yourself in the meantime and, third, is it even necessary for the President to be so intimately involved in strategic details?

REAGAN: Yes. This had to do with our disarmament talks and the whole controversy about land missiles came up because we thought that the strategic nuclear weapons—the most destabilizing are the land-based. You put your thumb on a button and somebody blows up 20 minutes later.

So we thought that it would be simpler to negotiate first with those, and then we made it plain, a second phase, take up the submarine-launched—the airborne missiles. The Soviet Union, to our surprise and not just mine—made it plain when we brought this up that they placed, they thought, a greater reliance on the land-based missiles and therefore they wanted to take up all three and we agreed. We said all right, if that's what you want to do.

19.

But, it was a surprise to us because they outnumbered us 64 to 36 in submarines and 20 percent more bombers capable of carrying nuclear missiles than we had. So, why should we believe that they had placed that much more reliance on land-based? But even after we gave in and said all right, let's discuss it all, they walked away from the table. We didn't.

Q: Mr. Mondale, I'm going to hang in there. Should the President's age and stamina be an issue in the political campaign?

MONDALE: No. And I have not made it an issue nor should it be. What's an issue here is the President's application of his authority to understand what a President must know to lead this nation, secure our defense and make the decisions and judgments that are necessary.

A minute ago, the President quoted Cicero, I believe. I want to quote somebody a little closer to home, Harry Truman. He said the buck stops here. We just heard the President's answer for the problems at the barracks in Lebanon where 241 Marines were killed. What happened?

First, the Joint Chiefs of Staff, with the President, said don't put those troops there. They did it. And then five days before the troops were killed, they went back to the President through the Secretary of Defense, and said please, Mr. President, take those troops out of there because we can't defend them. They didn't do it. And we know what's—what happened.

After that, once again our embassy was exploded. This is the fourth time this happened—an identical attack in the same region, despite warnings even public warnings from the terrorists. Who's in charge? Who's handling this matter? That's my main point.

7.2 / An Extended Example

Now on arms control—we're completing four years—this is the first Administration since the bomb went off that made no progress. We have an arms race under way instead. A President has to lead his Government or it won't be done. Different people with different views fight with each. For three and a half years, this Administration avoided arms control, resisted tabling arms control proposals that had any hope of agreeing, rebuked their negotiator in 1981 when he came close to an agreement, at least in principle, on medium-range weapons and we have this arms race under way. And a recent book that just came out by, perhaps, the nation's most respected author in this field,

20. Strobe Talbott, called *Deadly Gambit*, concludes that this President has failed to master the essential details needed to command and lead us both in terms of security and terms of arms control. That's why they call the President the Commander-in-Chief. Good intentions, I grant, but it takes more than that. He must be tough and smart.

Q: This question of leadership keeps arising in different forms in this discussion already. And the President, Mr. Mondale, has called you whining and vacillating, among the more charitable phrases—weak, I believe. It is a question of leadership. And he has made the point that you have not repudiated some of the semi-diplomatic activity of the Rev. Jackson, particularly in Central America. Do you, did you approve of his diplomatic activity? And are you prepared to repudiate him now?

MONDALE: I, I read his statement the other day. I don't admire Fidel Castro at all. And I have said that. Che Guevara was a contemptible figure in civilization's history. I know the Cuban state as a police state. And all my life, I've worked in a way that demonstrates that.

21. But Jesse Jackson is an independent person. I don't control him. And, let's talk about people we do control. In the last debate, the Vice President of the United States said that I said the marines had died
22. shamefully and died in shame in Lebanon. I demanded an apology from Vice President Bush because I had instead honored these young men, grieved for their families and think they were wonderful Americans that honored us all. What does the President have to say about taking
23. responsibility for a Vice President who won't apologize for something like that?

MODERATOR: Mr. President, your rebuttal.

REAGAN: Yes, I know it'll come as a surprise to Mr. Mondale, but I am in charge. And as a matter of fact we haven't avoided arms control talks with the Soviet Union. Very early in my Administration, I proposed—and I think something that had never been proposed by any previous Administration—I proposed a total elimination of intermediate range missiles where the Soviets had better than a, and still have better than a, ten-to-one advantage over the allies in Europe. When they protested that and suggested a smaller number, perhaps, I went along with that. The so-called negotiation that you said I walked out on was the so-called "walk in the woods" between one of our representatives and one of the Soviet Union and it wasn't me that turned it down. The Soviet Union disavowed it.

MODERATOR: Mr. Mondale, your rebuttal.

24. MONDALE: There are two distinguished authors of arms control in this country. There are many others, but two that I want to cite tonight. One is Strobe Talbott in his classic book *Deadly Gambit*. The other is John Newhouse, who's one of the most distinguished arms control specialist in our country. Both said that this Administration turned down the "walk in the woods" agreement first and that would have been a perfect agreement from the standpoint of the United States and Europe and our security. When Mr. Nitze, a good negotiator returned, he was rebuked and his boss was fired. This is the kind of leadership that we've had in this Administration in the most deadly issue of our time. Now we have a runaway arms race. All they've got to show for four years in U.S.-Soviet relations is one meeting in the last weeks of an Administration and nothing before. They're tough negotiators, but all previous Presidents have made progress. This one has not.

MODERATOR: Miss Geyer, your question to Mr. Mondale.

Q: Mr. Mondale, many analysts are now saying that actually our No. 1 foreign policy problem today is one that remains almost totally unrecognized. Massive illegal immigration from economically collapsing countries. They are saying that it is the only real territorial threat to the American nation-state. You yourself said in the 1970's that we had a "hemorrhage on our borders" yet today you have backed off on immigration reform such as the balanced and highly-crafted Simpson-Mazzoli bill. Why? What would you do instead today, if anything?

MONDALE: Ah, this is a very serious problem in our country and it has to be dealt with. I object to that part of the Simpson-Mazzoli bill which I think is very unfair and would prove to be so. That is the part that requires employers to determine the citizenship of an employee before they're hired. I am convinced that the result of this would be that people who are Hispanic, people who have different languages or speak with an accent would find it difficult to be employed.

I think that's wrong. We've never had citizenship tests in our country before. And I don't think we should have a citizenship card today. That is counterproductive. I do support the other aspects of the Simpson-Mazzoli bill that strengthens enforcement at the border, strengthens other ways of dealing with undocumented workers in this difficult area and dealing with the problem of setting people who have lived here for many many years and do not have an established status.

I further strongly recommend that this Administration do something it has not done. And that is to strengthen enforcement at the border, strengthen the officials in this Government that deal with undocumented workers and to do so in a way that's responsible and
25. within the Constitution of the United States.

We need an answer to this problem. But it must be an American answer that is consistent with justice and due process. Everyone in this room, practically, here tonight, is an immigrant. We came here loving this nation, serving it and it has served all of our most bountiful dreams. And one of those dreams is justice. And we need a measure, and I will support a measure that brings about those objectives, but avoids that one aspect that I think is very serious.

The second part is to maintain and improve relations with our
26. friends to the south. We cannot solve this problem all on our own. And that's why the failure of this administration to deal in an effective and good-faith way with Mexico, with Costa Rica, with the other nations in trying to find a peaceful settlement to the dispute in Central America has undermined our capacity to effectively to deal diplomatic in this, diplomatically in this area as well.

Q: Sir, people as well-balanced and just as Father Theodore Hesburgh at Notre Dame, who headed the Select Commission on Immigration, have pointed out repeatedly that there will be no immigration reform without employer sanctions because it would be an unbalanced bill and there would be simply no way to enforce it. However, putting that aside for the moment, your critics have also said repeatedly that you have not gone along with the bill, or with any immigration reform, because of the Hispanic groups—or Hispanic leadership groups, who actually do not represent what the Hispanic Americans want because polls show that they overwhelmingly want some kind of immigration reform. Can you say, or how can you justify your position on this, and how do you respond to the criticism that this is another, or that this is an example of the American nation?

MONDALE: I think you're right that the polls show that the majority of Hispanics want that bill, so I'm not doing it for political reasons. I'm doing it because all my life I've fought for a system of justice in this country, a system in which every American has a chance to achieve the fullness of life without discrimination. This bill imposes upon employers the responsibility of determining whether somebody who applies for a job is an American or not, and just inevitably they're going to be reluctant to hire Hispanics or people with a different accent.

If I were dealing with politics here the polls show the American people want this. I am for reform in this area for tough enforcement at the borders and for many other aspects of the Simpson-Mazzoli bill, but all my life I've fought for a fair nation and, despite the politics of it, I stand where I stand, and I think I'm right. And before this fight is over, we're going to come up with a better bill, a more effective bill, that does not undermine the liberties of our people.

Q: Mr. President, you too have said that our borders are out of control. Yet this fall, you allowed the Simpson-Mazzoli bill, which would at least have minimally protected our borders and the rights of citizenship, to die because of a relatively unimportant issue of reimbursement to the states for legalizing aliens. Given that, may I ask what priority can we expect you to give this forgotten national security element; how sincere are you in your efforts to control, in effect, the nation's states, that is, the United States.

27. REAGAN: Georgie, and we, believe me, supported that Simpson-Mazzoli bill strongly, and the bill that came out of the Senate. However, there were things added in the House side that we felt made it less of a good bill; as a matter of fact, made it a bad bill. And in conference, we stayed with them in conference all the way to where even Senator Simpson did not want the bill in the manner in which it would come out of the conference committee. There were a number of things in there that weakened the bill—I can't go into detail about them here. But it is true our borders are out of control, it is also true that this has been a situation on our borders back through a number of Administrations.

And I supported this bill, I believe in the idea of amnesty for those who have put down roots and who have lived here, even though some time back they may have entered illegally. With regard to the employer sanctions, we must have that—not only to ensure that we can identify the illegal aliens but also, while some keep protesting about what it would do to employers, there is another employer that we shouldn't be so concerned about, and those are employers down through the years who have encouraged the illegal entry into this country because they then hire those individuals and hire them at starvation wages and with none of the benefits that we think are normal and natural for workers in our country. And the individuals can't complain because of their illegal status. We don't think that those people should be allowed to continue operating free, and this was why the provisions that we had in with regard to sanctions and so forth.

And I'm going to do everything I can, and all of us in the administration are, to join in again when Congress is back at it to get an immigration bill that will give us once again control of our borders. And with regard to friendship below the border with the countries down there, yes, no Administration that I know has established the relationship that we have with our Latin friends. But as long as they have an economy that leaves so many people in dire poverty and unemployment, they are going to seek that employment across our borders. And we work with those other countries.

Q: Mr. President, the experts also say that the situation today is terribly different from what it has been in the past because of the gigantic population growth. For instance, Mexico's population will go from about 60 million today to 120 million at the turn of the century. Many of these people will be coming into the United States not as citizens but as illegal workers. You have repeatedly said recently that you believe that Armageddon, the destruction of the world, may be imminent in our times. Do you ever feel that we are in for an Armageddon or a situation, a time of anarchy, regarding the population explosion in the world?

REAGAN: No, as a matter of fact the population explosion, if you look at the population explosion, if you look at the actual figures, has been vastly exaggerated—over-exaggerated. As a matter of fact, there are

28. some pretty scientific and solid figures about how much space there still is in the world and how many more people can have. It's almost like going back to the Malthusian theory, when even then they were saying

29. that everyone would starve with the limited population they had then.

But the problem of population growth is one here with regard to our immigration. And we have been the safety valve, whether we wanted to or not, with the illegal entry here; in Mexico, where their population is increasing and they don't have an economy that can absorb them and provide the jobs. And this is what we're trying to work out, not only to protect our own borders but to have some kind of fairness and recognition of that problem.

MODERATOR: Mr. Mondale, your rebuttal.

MONDALE: One of the biggest problems today is that the countries to our south are so desperately poor that these people who will almost lose their lives if they don't come north, come north despite all the risks. And if we're going to find a permanent, fundamental answer to this, it goes to American economic and trade policies that permit these nations to have a chance to get on their own two feet and to get prosperity so that they can have jobs for themselves and their people.

And that's why this enormous national debt, engineered by this Administration, is harming these countries and fueling this immigration.

30.   These high interest rates, real rates, that have doubled under this Administration, have had the same effect on Mexico and so on, and the cost of repaying those debts is so enormous that it results in massive unemployment, hardship and heartache. And that drives our friends to the north—to the south—up into our region, and the need to end those deficits as well.

MODERATOR: Mr. President, your rebuttal.

REAGAN: Well, my rebuttal is I've heard the national debt blamed for a lot of things, but not for illegal immigration across our border, and it has nothing to do with it.

31.   But with regard to these high interest rates, too, at least give us the recognition of the fact that when you left office, Mr. Mondale, they were 22½, the prime rate; it's now 12¼, and I predict it'll be coming down a little more shortly. So we're trying to undo some of the things that your Administration did.

MODERATOR: Mr. Kalb. No applause, please. Mr. Kalb, your question to President Reagan.

Q: Mr. President, I'd like to pick up this Armageddon theme. You've been quoted as saying that you do believe deep down that we are heading for some kind of biblical Armageddon. Your Pentagon and your Secretary of Defense have plans for the United States to fight and prevail in a nuclear war. Do you feel that we are now heading, perhaps, for some kind of nuclear Armageddon? And do you feel that this country and the world could survive that kind of calamity?

REAGAN: Mr. Kalb, I think what has been hailed as something I'm supposedly, as President, discussing as principle is the result of just some philosophical discussions with people who are interested in the same things. And that is the prophecies down through the years, the biblical prophecies of what would portend the coming of Armageddon and so forth. And the fact that a number of theologians for the last decade or more have believed that this was true, that the prophecies are coming together that portend that.

But no one knows whether Armageddon—those prophecies—mean that Armageddon is a thousand years away or day after tomorrow. So I have never seriously warned and said we must plan according to Armageddon.

32. Now, with regard to having to say whether we would try to survive in the event of a nuclear war—of course we would.

But let me also point out that to several parliaments around the world, in Europe and in Asia, I have made a statement to each one of them, and I'll repeat it here: A nuclear war cannot be won and must never be fought.

And that is why we are maintaining a deterrent and trying to achieve a deterrent capacity to where no one would believe that they could start such a war and escape with limited damage. But the deterrent—and that's what it is for—is also what led me to propose what is now being called the Star Wars concept, but propose that we research to see if there isn't a defensive weapon that could defend against incoming missiles. And if such a defense could be found, wouldn't it be far more humanitarian to say that now we can defend against a nuclear war by destroying missiles instead of slaughtering millions of people?

Q: Mr. President, when you made that proposal, the so-called Star Wars proposal, you said, if I'm not mistaken, that you would share this very super-sophisticated technology with the Soviet Union. After all of the distrust over the years, sir, that you have expressed towards the Soviet Union, do you really expect anyone to take seriously that offer—that you would share the best of America's technology in this weapons area with our principal adversary?

33. REAGAN: Why not? What if we did and I hope we can, we're still researching. What if we come up with a weapon that renders those missiles obsolete? There has never been a weapon invented in the history of man that has not led to a defensive, a counter-weapon, but suppose we came up with that. Now, some people have said, "Ah, that would make a war imminent" because they would think that we could launch a first strike because we could defend against that enemy. But why not do what I have offered to do and asked the Soviet Union to do? Say look, here's what we can do, we'll even give it to you, now will

you sit down with us and once and for all get rid—all of us—of these nuclear weapons and free mankind from that threat. I think that would be the greatest use of a defensive weapon.

Q: Mr. Mondale you've been very sharply critical of the President's strategic defense initiative and yet what is wrong with a major effort by this country to try to use its best technology to knock out as many incoming nuclear warheads as possible?

MONDALE: First of all, let me sharply disagree with the President on sharing the most advanced, the most dangerous, that most important technology in America with the Soviet Union. We have had, for many years, understandably, a system of restraints on high technology because the Soviets are behind us and any research or development along the Star Wars schemes would inevitably involve our most advanced engineering and the thought that we would not let the Soviet Union is, in my opinion, a total non-starter. I would not let the Soviet Union get their hands on it at all.

34. Now, what's wrong with Star Wars? There's nothing wrong with the theory of it. If we could develop a principle that would say both sides could fire all their missiles and no one would get hurt, I suppose it's a good idea. But the fact of it is, we're so far away from research that even comes close to that that the director of engineering research in the Defense Department said to get there we would have to solve eight problems, each of which are more difficult than the atomic bomb and the Manhattan Project. It would cost something like a trillion dollars to test and deploy weapons. The second thing is this all assumes that the Soviets wouldn't respond in kind, and they always do. We don't get behind and that's been the tragic story of the arms race. We have more at stake in space satellites than they do. If we could stop right now the testing and the deployment of these space weapons and the President's proposals go clear beyond research. If it was just research, we wouldn't have any argument, because maybe some day somebody will think of something. But to commit this nation to a buildup of anti-satellite and space weapons at this time in their crude state would bring about an arms race that's very dangerous indeed.

35. One final point: The most dangerous aspect of this proposal is for the first time we would delegate to computers the decision as to whether to start a war. That's dead wrong.

There wouldn't be time for a President to decide. It would be decided by these remote computers. It might be an oil fire, it might be a jet exhaust, the computer might decide it's a missile and off we go.

Why don't we stop this madness now and draw a line and keep the heavens free from war?

Q: Mr. Mondale, in this general area, sir, or arms control, President Carter's national security adviser Zbigniew Brzezinski said, "A nuclear freeze is a hoax," yet the basis of your arms proposals as I understand them is a mutual and verifiable freeze on existing weapons systems. In your view, which specific weapons systems could be subject to a mutual and verifiable freeze and which could not?

MONDALE: Every system that is verifiable should be placed on the table for negotiations for an agreement. I would not agree to any negotiations or any agreement that involved conduct on the part of the Soviet Union that we couldn't verify every day. I would not agree to any agreement in which the United States' security interest was not fully recognized and supported. That's why we say mutual and verifiable freezes.

Now, why do I support the freeze? Because this ever-rising arms race madness makes both nations less secure, it's more difficult to defend this nation, it is putting a hair trigger on nuclear war. This Administration, by going into the Star Wars system, is going to add a dangerous new escalation. We have to be tough on the Soviet Union, but I think the American people and the people of the Soviet Union want it to stop.

MODERATOR: Time is up, Mr. Mondale. President Reagan, your rebuttal.

REAGAN: Yes, my rebuttal once again is that this invention that has just been created here of how I would go about rolling over to the Soviet Union—No, Mr. Mondale, my idea would be with that defensive weapon, that we would sit down with them and then say, now, are you willing to join us? Here's what we can—give them a demonstration, and then say, here's what we can do. Now, if you're willing to join us in getting rid of all the nuclear weapons in the world, then, we'll give you this one so that we would both know that no one can cheat—that we've both got something that if anyone tries to cheat—but when you keep star-warring it—I never suggested where the weapons should be or what kind. I'm not a scientist. I said, and the Joint Chiefs of Staff agreed with me, that it was time for us to turn our research ability to seeing if we could not find this kind of defensive weapon. And suddenly somebody says, "Oh, it's got to be up there"—star wars—and so forth. I don't know what it would be, but if we can come up with one, I think the world will be better off.

MODERATOR: Mr. Mondale, your rebuttal?

36. MONDALE: Well, that's what a President's supposed to know—where those weapons are going to be. If they're space weapons, I assume they'll be in space. If they're anti-satellite weapons, I assume they're going to be armed against any satellite. Now, this is the most dangerous technology that we possess. The Soviets try to spy on us—steal this stuff—and to give them technology of this kind, I disagree with. You haven't just accepted research, Mr. President, you've set up a strategic defense initiative and agency. You're asking for a budget of some $30 billion for this purpose. This is an arms escalation, and we will be better off—far better off—if we stop right now, because we have more to lose in space than they do. If someday somebody comes along with an answer, that's something else. But that there would be an answer in our lifetime is unimaginable. Why do we start things that we know
37. the Soviets will match and make us all less secure? That's what a President is for.

MODERATOR: Mr. Kondracke, your question to Mr. Mondale?

Q: Mr. Mondale, you say that with respect to the Soviet Union, you want to negotiate a mutual nuclear freeze. Yet you would unilaterally give up the MX missile and the B-1 bomber before the talks have even begun; and you have announced in advance that reaching an agreement with the Soviets is the most important thing in the world to you. Now aren't you giving away half the store before you even sit down to talk?

MONDALE: As a matter of fact we have a vast range of technology and weaponry right now that provides all the bargaining chips that we need, and I support the air launch cruise missile, ground launch cruise missile, Pershing missile, the Trident submarine, the D-5 submarine, the Stealth technology, the Midgetman—we have a whole range of technology. Why I disagree with the MX is that it's a sitting duck. It'll draw an attack. It puts a hair trigger, and it is a dangerous destabilizing weapon. And the B-1 is similarly to be opposed because for 15 years the Soviet Union has been preparing to meet the B-1, the Secretary of Defense himself said it would be a suicide mission, if it were built. Instead, I want to build the Midgetman which is mobile and thus less vulnerable, contributing to stability, and a weapon that will give us security and contribute to an incentive for arms control. That's why I'm for Stealth technology to build the Stealth bomber, which I supported for years, that can penetrate the Soviet air defense system without any hope that they can perceive where it is because their radar system is frustrated. In other words, a President has to make choices. This makes us stronger.

The final point is that we can use this money that we save on these weapons to spend on things that we really need. Our conventional strength in Europe is under strength. We need to strengthen that in order to assure our Western allies of our presence there, a strong defense, but also to diminish and reduce the likelihood of a commencement of a war and the use of nuclear weapons. It's by this way by making wise choices that we're stronger, we enhance the chances of arms control. Every President until this one has been able to do it, and this nation, the world is more dangerous as a result.

Q: I want to follow up on Mr. Kalb's question. It seems to me that on the question of verifiability that you do have some problems with the extent of the freeze. It seems to me, for example, that testing would be very difficult to verify because the Soviets encode their telemetry. Research would be impossible to verify. Now a view of that, what is going to be frozen?

38. MONDALE: I will not agree to any arms control agreement, including a freeze, that's not verifiable. Let's take your warhead principle. The warhead principle, they've been counting rules for years. Whenever a weapon is tested, we counted the number of warheads on it, and whenever that warhead is used we count that number of warheads, whether they have that number or less on it or not. These are standard rules. I will not agree to any production restrictions or agreement unless we have the ability to verify those agreements. I don't trust the Russians. I believe that every agreement we reach must be verifiable, and I will not agree to anything that we cannot tell every day. In other words, we've got to be tough, but in order to stop this arms madness we've got to push ahead with tough negotiations that are verifiable so that we know the Soviets are agreeing and living up to their agreements.

Q: Mr. President, I want to ask you about negotiating with friends. You severely criticized President Carter for helping to undermine two friendly dictators who got into trouble with their own people, the Shah of Iran and President Somoza of Nicaragua. Now there are other such leaders heading for trouble, including President Pinochet of Chile and President Marcos of the Philippines. What should you do and what can you do to prevent the Philippines from becoming another Nicaragua?

39. REAGAN: Morton, I did criticize the President because of our undercutting of what was a stalwart ally, the Shah of Iran. And I am not at

all convinced that he was that far out of line with his people or that they wanted that to happen.

The Shah had done our bidding and carried our load in the Middle East for quite some time and I did think that it was a blot on our record that we let him down. Had things gotten better, the Shah, whatever he might have done, was building low-cost housing, had taken land away from the mullahs and was distributing it to the peasants so they could be landowners, things of that kind. But we turned it over to a maniacal fanatic who has slaughtered thousands and thousands of people calling it executions.

The matter of Somoza, no, I never defended Somoza. And as a matter of fact, the previous Administration stood by and so did I—not that I could have done anything in my position at that time. But for this revolution to take place and the promise of the revolution was democracy, human rights, free labor unions, free press. And then just as Castro had done in Cuba, the Sandinistas ousted the other parties to the revolution. Many of them are now the Contras. They exiled some, they jailed some, they murdered some. And they installed a Marxist-Leninist totalitarian Government.

And what I have to say about this is, many times—and this has to do with the Philippines also—I know there are things there in the Philippines that do not look good to us from the standpoint right now of democratic rights. But what is the alternative? It is a large Communist movement to take over the Philippines.

They have been our friends for—since their inception as a nation. And I think that we've had enough of a record of letting, under the guise of revolution, someone that we thought was a little more right than we would be, letting that person go and then winding up with totalitarianism pure and simple as the alternative and I think that we're better off, for example, with the Philippines of trying to retain our friendship and help them right the wrongs we see rather than throwing them to the wolves and then facing a Communist power in the Pacific.

Q: Mr. President, since the United States has two strategic bases in the Philippines, would the overthrow of President Marcos constitute a threat to vital American interests, and, if so, what would you do about it?

40. REAGAN: Well, as I say we have to look at what an overthrow there would mean and what the government would be that would follow. And there is every evidence, every indication that that government would be hostile to the United States and that would be a severe blow to the—to our abilities there in the Pacific.

Q: And what would you do about it?

MODERATOR: Sorry, sorry, you've asked the follow-up question. Mr. Mondale your rebuttal.

MONDALE: Perhaps in no area do we disagree more than this Administration's policies on human rights. I went to the Philippines as Vice President, pressed for human rights, called for the release of Aquino and made progress that had been stalled on both the Subic and the Clark airfield bases. What explains this Administration cozying up to the Argentine dictators after they took over? Fortunately a democracy took over but this nation was embarrassed by this current Administration's adoption of their policies. What happens in South Africa, where, for example, the Nobel Prize winner two days ago said this Administration is seen as working with the oppressive government of that region, of South Africa.

That hurts this nation. We need to stand for human rights. We need to make it clear we're for human liberty. National security and human rights must go together, but this Administration time and time again has lost its way in this field.

MODERATOR: President Reagan, your rebuttal.

41. REAGAN: Well, the invasion of Afghanistan didn't take place on our watch. I have described what has happened in Iran and we weren't here then either. I don't think that our record of human rights can be assailed. I think that we have observed ourselves and have done our best to see that human rights are extended throughout the world.

Mr. Mondale has recently announced a plan of his to get the democracies together and to work with the whole world to turn to democracy. And I was glad to hear him say that because that's what we've been doing ever since I announced to the British Parliament that I thought we should do this.

And human rights are not advanced when at the same time you then stand back and say, "Whoops, we didn't know the gun was loaded," and you have another totalitarian power on your hands.

MODERATOR: In this, in this segment, because of the pressure of time, there will be no rebuttals and there will be no follow-up questions. Mr. Trewhitt, your question to President Reagan.

Q: One question to each candidate?

MODERATOR: One question to each candidate.

Q: Mr. President, could I take you back to something you said earlier? And if I'm misquoting you please correct me. But I understood you to say that if the development of space military technology was successful, you might give the Soviets a demonstration and say, "Here it is," which sounds to me as if you might be trying to gain the sort of advantage that would enable you to dictate terms, and which I would then suggest to you might mean scrapping a generation of nuclear strategy called mutual deterrence, in which we in effect hold each other hostage. Is that your intention?

42. REAGAN: Well, I can't say that I have roundtabled that and sat down with the Chiefs of Staff, but I have said that it seems to me that this could be a logical step in what is my ultimate dream. And that the elimination of nuclear weapons in the world. And it seems to me that this could be an adjunct, or certainly a great assisting agent, in getting that done. I am not going to roll over, as Mr. Mondale suggests, and give them something that could turn around and be used against us. But I think it's a very interesting proposal to see if we can find first of all something that renders those weapons obsolete, incapable of their mission.

But Mr. Mondale seems to approve MAD—MAD is Mutual Assured Destruction, meaning if you use nuclear weapons on us, the only thing we have to keep you from doing it is that we'll kill as many people of yours as you will kill of ours. I think that to do everything we can to find, as I say, something that would destroy weapons and not humans is a great step forward in human rights.

Q: Mr. Mondale, could I ask you to address the question of nuclear strategy. Formal doctrine is very arcane, but I'm going to ask you to deal with it anyway. Do you believe in MAD, Mutual Assured Destruction, mutual deterrence, as it has been practiced for the last generation?

43. MONDALE: I believe in a sensible arms control approach that brings down these weapons to manageable levels. I would like to see their elimination. And in the meantime, we have to be strong enough to make certain that the Soviet Union never attempts this.

44. Now here we have to decide between generalized objectives and reality. The President says he wants to eliminate or reduce the number of nuclear weapons, but in fact these last four years have seen more weapons built, a wider and more vigorous arms race than in human history. He says he wants a system that will make nuclear arms wars safe, so nobody's going to get hurt. Well, maybe someday somebody can dream of that. But why start an arms race now? Why destabilize our relationship? Why threaten our space satellites, upon which we depend? Why pursue a strategy that would delegate to computers the question of starting a war?

A President, to defend this country and to get arms control, must master what's going on. I accept his objective and his dreams, we all do. But the hard reality is that we must know what we're doing and pursue those objectives that are possible in our time. He's opposed every effort of every President to do so; in the four years of his Administration he's failed to do so. And if you want a tough President who uses that strength to get arms control, and draws the line in the heavens, vote for Walter Mondale.

45.

MODERATOR: Please, I must again ask the audience not to applaud, not to cheer, not demonstrate its feelings in any way. We've arrived at the point in the debate now where we call for closing statements. You have the full four minutes, each of you. Mr. Mondale, will you go first.

MONDALE: I want to thank the League of Women Voters, the good citizens of Kansas City and President Reagan for agreeing to debate this evening.

This evening we talked about national strength. I believe we need to be strong, and I will keep us strong. But I think strength must also require wisdom and smarts in its exercise—that's key to the strength of our nation.

A President must know the essential facts, essential to command. But a President must also have a vision of where this nation should go.

Tonight, as Americans you have a choice. And you're entitled to know where we would take this country if you decide to elect us.

As President, I would press for long-term vigorous economic growth. That's why I want to get these debts down and these interest rates down, restore America's exports, help rural America's exports, help rural America which is suffering so much, and bring the jobs back here for our children.

I want this next generation to be the best educated in American history; to invest in the human mind and science again, so we're out front.

I want this nation to protect its air, its water, its land and its public health. America is not temporary. We're forever. And as Americans, our generation should protect this wonderful land for our children.

I want a nation of fairness, where no one is denied the fullness of life or discriminated against, and we deal compassionately with those in our midst who are in trouble.

And above all, I want a nation that's strong. Since we debated two weeks ago, the United States and the Soviet Union have built 100 more warheads, enough to kill millions of Americans and millions of Soviet citizens.

This doesn't strengthen us, this weakens the chances of civilization to survive.

I remember the night before I became Vice President. I was given the briefing and told that any time, night or day, I might be called upon to make the most fateful decision on earth—whether to fire these atomic weapons that could destroy the human species.

That lesson tells us two things. One, pick a President that you know will know, if that tragic moment ever comes, what he must know. Because there'll be no time for staffing committees or advisers; a President must know right then.

But above all, pick a President who will fight to avoid the day when that God-awful decision ever needs to be made. And that's why this election is so terribly important.

America and Americans decide not just what's happening in this country; we are the strongest and most powerful free society on earth. When you make that judgment, you are deciding not only the future of our nation; in a very profound respect, you're providing the future—deciding the future of the world.

We need to move on. It's time for America to find new leadership. Please join me in this cause to move confidently and with a sense of assurance and command to build the blessed future of the nation.

MODERATOR: President Reagan, your summation, please.

REAGAN: Yes, my thanks to the League of Women Voters, to the panelists, to the moderator, and to the people of Kansas City for their warm hospitality and greeting.

I think the American people tonight have much to be grateful for: an economic recovery that has become expansion, freedom, and most of all, we are at peace. I am grateful for the chance to reaffirm my commitment to reduce nuclear weapons and one day to eliminate them entirely.

The question before comes down to this: do you want to see America return to the policies of weakness of the last four years, or do we want to go forward marching together as a nation of strength and that's going to continue to be strong?

We shouldn't be dwelling on the past or even the present. The meaning of this election is the future, and whether we're going to grow and provide the jobs and the opportunities for all Americans that they need. Several years ago I was given an assignment to write a letter. It was to go into a time capsule and would be read in 100 years when that time capsule was opened. I remember driving down the California coast one day. My mind was full of what I was going to put in that letter about the problems and the issues that confront us in our time

7.2 / An Extended Example

and what we did about them, but I couldn't completely neglect the beauty around me—the Pacific out there on one side of the highway shining in the sunlight, the mountains of the coast range rising on the other side, and I found myself wondering what it would be like for someone, wondering if someone 100 years from now would be driving down that highway and if they would see the same thing.

And with that thought I realized what a job I had with that letter. I would be writing a letter to people who know everything there is to know about us. We know nothing about them. They would know how we solved them and whether our solution was beneficial to them down throughout the years or whether it hurt them. They would also know that we lived a world with terrible weapons, nuclear weapons of terrible destructive power aimed at each other, capable of crossing the ocean in a matter of minutes and destroying civilization as we know it.

You know, I am grateful for all of you for giving the opportunity to serve you for these four years and I seek reelection because I want more than anything else to try to complete the new beginning that we charted four years ago.

George Bush, who I think is one of the finest Vice Presidents this country has ever had, George Bush and I have crisscrossed the country and we've had in these last few months a wonderful experience. We have met young America. We have met your sons and daughters.

MODERATOR: Mr. President, I'm obliged to cut you off there under the rules of the debate. I'm sorry.

REAGAN: All right, I was just going to . . .

MODERATOR: Perhaps I should point out that the rules under which we did that were agreed upon by the two campaigns.

REAGAN: I know, yes.

MODERATOR: Thank you, Mr. President. Thank you, Mr. Mondale. Our thanks also to the panel, finally to our audience. We thank you and the League of Women Voters asks me to say to you: Don't forget to vote on Nov. 6.

Comments on the Debate

1. The question about whether military force is needed is never specifically addressed. Instead, Mondale talks in generalities. Ignoring the issue.

2. Presupposes that arms control failed because Reagan did not know that Soviet missiles are primarily based on land. Complex question.

3. The question is phrased in vague language: what does *conventional* mean here? In what way was the war in the Pacific in World War II conventional, while the Korean War was not?

4. Mondale answers the question (how should the U.S. handle unconventional wars?) by speaking in generalities. Ignoring the issue.

5. Reagan hedges—he doesn't fully deny the claim that a CIA chief is in Nicaragua directing the Contras.

6. Reagan changes the subject from the issue of whether he is truly in command to whether he said that missiles could be recalled after launch.

7. The questioner poses two false dilemmas. First, either Reagan can say that the Soviet Union is an evil empire or else he can say they are welcome to their system (couldn't Reagan say both, that if the Soviets want to keep their system, that is fine with him, but they should not impose it on others?). Second, either we can engage in détente or try to roll back their empire (again, can't Reagan want to do both?).

8. Strawman.

9. Mondale's answer (that any treaty we sign with the Soviets should be verifiable) doesn't answer the question whether we can expect the Soviets to agree to such treaties.

10. There were no agreements and more bombs under Reagan, so Reagan was the cause. False cause.

11. Now Mondale changes the issue back to whether Reagan said ballistic missiles could be recalled.

12. Reagan presupposes that those programs were vital to American defense. (Mondale goes on to contest that presupposition.)

13. Strawman.

14. Mondale ignores the question whether his restrictions on the use of American military force would lead enemies to doubt his resolve, and talks instead about how ineptly Reagan has handled the military. Ignoring the issue.

15. Mondale does not say what specifically he means by quarantining (he only says "take some steps, diplomatic and otherwise . . .").

16. Reagan hedges here so as to lessen his responsibility—the marines "happened" to be positioned at the airport (which Reagan's critics claimed to have been a vulnerable location).

17. Reagan shifts the burden of proof here.

18. Perhaps the most famous moment in the debate—Reagan turns aside Trewhitt's question with a quip that even made Mondale laugh.

19. Reagan never explicitly addresses the question whether he was initially unaware that the Soviet retaliating power is primarily land-based. Ignoring the issue.

7.2 | An Extended Example

20. Strobe Talbott's qualifications aren't given. Ad verecundiam.

21. Mondale does not answer the question whether he approves of Jackson's diplomatic activity, or whether he (Mondale) will repudiate Jackson. Ignoring the issue.

22. Mondale changes the subject abruptly. Ignoring the issue.

23. Shifting the burden, or perhaps tu quoque.

24. Again, an authority is cited without being fully identified. Ad verecundiam.

25. Glittering generalities.

26. Glittering generalities throughout.

27. Pooh-poohing, or perhaps ignoring the issue—"there are lots of flaws, but I can't go into them right now."

28. What scientific figures? Ad verecundiam.

29. Malthus did not say "everyone would starve." Strawman.

30. The national debt doubled under this Administration so this Administration engineered (caused) it; there is high debt here and high debt in Mexico so our high debt causes their high debt (and thus their emigration). (Reagan catches both causal claims.) Double false cause.

31. Reagan (in return) argues that since the prime rate was high under the last Administration, their policies caused it. False cause.

32. To say that we would try to survive a nuclear war and could not win one does not directly address the question whether we could survive one. Ignoring the issue.

33. The President never answers the question how the Russians can believe that he would indeed turn over Star Wars technology to them. Ignoring the issue.

34. The Star Wars proposal did not envision that "both sides could fire all their missiles and no one would get hurt," only that a significant number of missiles could be stopped. Strawman.

35. Does the Star Wars proposal really include delegation of war powers to computers? Possible strawman.

36. Mondale presupposes that some determination has been reached about the nature of the Star Wars systems, and concludes that the President is remiss in not knowing about it. Loaded question.

37. Presupposes that the technology under discussion, if generally available, would make us all less secure. Loaded question.

38. Mondale never spells out what specific weapon systems he would freeze. He goes on to speak in generalities. Ignoring the issue.

39. Reagan never specifically says what he would do to stop the Philippines from turning into another Nicaragua. Ignoring the issue.

40. Reagan again does not answer the question (what he would do about

an attempt to overthrow Marcos). The moderator stops the questioner from reiterating the question.

41. False cause if Reagan is concluding that the previous administration caused the invasion of Afghanistan.

42. Reagan does not answer the question whether he really wants Star Wars in order to be able to dictate terms to the Soviets. Ignoring the issue.

43. Mondale does not specifically say that he supports MAD; instead, he speaks in generalities. Ignoring the issue.

44. More weapons have been built in the last 4 years, so the Administration is at fault. False cause.

45. A bit of appeal to the crowd here.

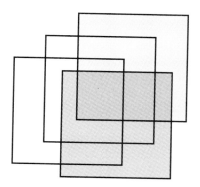

Analogies

8.1 The Uses of Analogy

In this chapter, we will look at a common mode of nondeductive reasoning: analogy. An **analogy** is a comparison of two things. Consider these two examples:

> Where does all of this leave us? Economics is in a state of turmoil. The economics of the textbooks and of the graduate schools not only still teaches price-auction model but is moving toward narrower and narrower interpretations. The mathematical sophistication intensifies as an understanding of the real world diminishes.
>
> Nevertheless, one can see signs of countercurrents beginning to develop. Economic models are being constructed that are designed to better reflect the world as we can see and measure it and also enhance possibilities of exercising economic control. . . .
>
> The transition from one mode of thought to another is difficult, since it involves abandoning a beautiful sailing ship—the equilibrium price-auction model—that happens to be torn apart and sinking in a riptide. So a raft must be built to catch whatever winds may come by. That raft won't match the beauty or mathematical elegance of the sailing ship, although it has one undeniable virtue—it floats.
>
> —Lester Thurow, *Dangerous Currents*.

None of these theorists, however, have gone on to remark what seems obvious: that if clothing is a language, it must have a vocabulary and a grammar like other languages. Of course, as with human speech, there is not a single language of dress, but many: some (like Dutch and German) closely related and others (like Basque) almost unique. And within every language of clothes there are many different dialects and accents, some almost unintelligible to members of the mainstream culture. Moreover, as with speech, each individual has his own stock of words and employs personal variations of tone and meaning. . . . The vocabulary of dress includes not only items of clothing, but also hair styles, accessories, jewelry, make-up and body decoration.

—Alison Lurie, *The Language of Clothes.*

In the first example, we do not have to follow the technical details to see the analogy being made, or why it is being made: a specific economic model is compared to a beautiful ship breaking up, and the author is urging us to adopt a less elegant, but more workable, model much as we would jump on an inelegant, but functional, raft. In the second example, the author compares clothing fashions to language—with items of clothing and jewelry likened to words.

An analogy compares one thing called 'the subject' to one or more other things called 'the analogues.' Thus in the first example, the subject is an economic theory and the analogue is a beautiful ship. In the second passage, the subject is clothing and the analogue is language. Any analogy can be put in the standard form:

<Subject> is like <analogue(s)>

(Don't confound the standard form of an analogical statement with the standard form of an argument.)

In evaluating reasoning by analogy, we must explicitly state the analogy, that is, put it in the standard form. This helps us focus on the key issues, namely, the similarities and dissimilarities of the two things compared.

Analogies have four basic uses: (1) the **descriptive** use of an analogy is the use of that analogy to describe a situation or state of affairs, (2) the **definitional** use is to define a word or phrase, (3) the **argumentative** use is as a premise in an argument, and (4) the **heuristic** use of an analogy is to aid in research. We will explore each of these uses in turn.

EXERCISES 8.1

For each of the following passages, determine whether or not it contains an analogy, and if so, put the analogical statement in standard form. (An asterisk indicates problems that are answered in the back of the book.)

1. Science can make a bulldozer. But it cannot train the emotions of the driver so that he will stop it before a crawling baby and not run over it. This point has been brought out succinctly by Bertrand Russell, when he says: "Science can tell him how certain ends might be reached. What it cannot tell him is that he should pursue one end rather than the other."

—Sampooran Singh, *The Dynamic Interplay between Science and Religion.*

* 2. It was my ambition to make a difficult subject accessible to the general reader, but students familiar with it will, I hope, nevertheless find some new information in these pages. This refers mainly to Johannes Kepler, whose works, diaries and correspondence have so far not been accessible to the English reader; nor does a serious English biography exist. Yet Kepler is one of the few geniuses who enables one to follow, step by step, the tortuous path that led him to his discoveries, and to get a really intimate glimpse, as in a slow-motion film, of the creative act. He accordingly occupies a key-position in the narrative.

—Arthur Koestler, *The Sleepwalkers.*

3. How do political reformers get away with continuing to use the whole population as guinea pigs, despite a trail of failures and broken promises that would embarrass a used-car dealer?

Whether it is the environmental protection racket today or the Prohibition amendment sixty years ago, it has been one disastrous "noble experiment" after another. It took us more than a decade to face the fact that Prohibition was not prohibiting. It was just corrupting. There are already studies showing that "environmental protection" doesn't protect the environment, that criminal "rehabilitation" programs don't rehabilitate, and that school "integration" programs don't integrate.

Like Prohibition, these other experiments did not simply fail. They have created massive new problems of their own. Prohibition was the greatest boost ever given to organized crime. We are still paying for that today. There are so many bureaucrats protecting the environment that it is literally a federal case to try to drill for oil, dig some coal, or build a hydroelectric dam. All the while, we keep wringing our hands over not having enough energy.

—Thomas Sowell, *Pink and Brown People.*

4. . . . it is essential for teachers to make clear what they expect of children. This is like giving a vine a pole on which to grow.

Benjamin Spock, in *Today's Education* 64(1), (1975).

5. Canada is not, as has been often stressed, an entirely precise mosaic of nations; in fact, it is, like the U.S.A., gradually becoming a melting pot of nationalities, the difference between the two countries being that whereas the United States and the individual provinces of Canada allow their various

national groups to exist and develop their national cultures, only Canada has recently begun meagerly to assist them materially in their specific endeavours.

—J.R.C. Perkin, *The Undoing of Babel.*

* 6. Now let's consider the universe. The universe resembles the skin of the balloon and the surface of the earth, except that the universe is three-dimensional, while the skin and the surface are essentially two-dimensional objects curved through a third dimension. The universe is a three-dimensional object curved through a fourth dimension.

—Isaac Asimov, *The Roving Mind.*

7. Glaser and Pellegrino, also working with inductive reasoning tasks, are concerned with both the strategic and factual knowledge necessary to solve a variety of analogy problems. Their approach differs somewhat from that of Sternberg *et al.* in that they use more difficult problems, thereby increasing the likelihood of errors. As a result, they must be concerned not only with the sequencing of component executions and the efficiency with which they can be carried out, but also with the factors that may affect the subjects' likelihood or ability to carry out the components correctly. Glaser and Pellegrino share with Sternberg *et al.* the approach of carefully delineating various potential methods of dealing with the items. Whereas Sternberg *et al.* begin with a set of analytic models, their approach is somewhat more data-driven. Glaser and Pellegrino are guided by the various solutions attempted by their subjects. These observations are then used to construct a theoretical framework within which to view individual differences. Armed with such a theoretical framework, it is possible to design instruction based on areas of particular difficulty for problem learners. Thus, the framework not only provides a description of individual differences, it also suggests the potential form and focus of remediation.

—Douglas Detterman and Robert Sternberg (eds.), *How and How Much Can Intelligence Be Increased?*

8. Just as once Turgenev in his *Hunter's Sketches* made readers acquainted with his native Orel region, in the same way Belov has introduced them to the region roundabout Vologda where he was born and still lives.

—V. Belov, *Morning Rendezvous.*

9. France was the leader in noncommercial avant-garde film. Directors like René Claré, Marcel L'Herbier, Jacques Feyder and Abel Gance all made films of individual importance.

—John Martin, *The Golden Age of French Cinema.*

* 10. Hence Descartes framed his famous theory of vortices in a primary matter or aether, invisible but filling all space. As a straw floating on water is caught in an eddy and whirled to the center of motion, so a falling stone is drawn to the Earth and a satellite towards its planet, while the Earth and the planet, with their attendant and surrounding vortices, are whirled in a greater vortex round the Sun.

—Sir William Dampier, *A History of Science.*

8.2 Descriptive and Definitional Analogies

Often we use analogies in description. For instance, someone might say that Keisha "ran like a roach across a griddle" to describe how quickly she ran. Analogies are valuable in description for two reasons. First, a well-chosen analogy can condense several literal descriptions into a single, short phrase. For example, the description:

> Jerzy is short, squat, hirsute, with arms longer than are normal for a person his size, and he has very close-set eyes.

can be economically expressed, though pejoratively, by the analogy:

> Jerzy looks like a baboon.

A second reason that analogies are useful in description is that they can increase the aesthetic quality of the description. An analogy can be moving, amusing, disturbing, or vivid. Compare the pairs of the following descriptions. The first of each pair is literal, the second analogical.

1a. Juan is pleased with his current situation.
1b. Juan is as happy as a clam!

2a. Helen, you look good.
2b. Helen, thy beauty is to me like those Nicaean barks of yore, that gently o'er a perfumed sea, the weary, way-worn wander bore, to his own native shore.

In (1a), Juan is described literally as happy; in (1b), the description is made by a humorous analogy. In (2a), Helen is described literally; in (2b), the description is made by Edgar Allan Poe's evocative and beautiful analogy.

A second use of analogy is in definition. For instance, we might define 'wolf' as "an animal like a dog but larger." Often, an analogical definition defines a word by setting up a proportion of meaning: "a is to b as c is to d." We might have defined 'wolf' in this manner: "wolf is to dog as tiger is to cat." We often see such definitions put as proportional equations:

$$\frac{a}{b} = \frac{c}{d}$$

Such an equation asserts that the relation between the two terms on the left-hand side of the equal sign corresponds to the relation between the terms on the right-hand side. A wolf is an animal of the same genus as the dog, but bigger. A tiger is an animal of the same genus as the house cat, but bigger.

The terms above and below the line can be related in a variety of ways:

They may be synonymous: $\dfrac{\text{Kitten}}{\text{Infant cat}}$

They may be related as cause and effect: $\dfrac{\text{Heart failure}}{\text{Death}}$

One may be the goal or purpose of the other: $\dfrac{\text{Win}}{\text{Play}}$

One may be a part of the other: $\dfrac{\text{Wheel}}{\text{Car}}$

One may be related to the other as type or special case: $\dfrac{\text{Chevrolet}}{\text{Car}}$

One may be the antonym of the other (i.e., they may be opposites): $\dfrac{\text{Hot}}{\text{Cold}}$

Any of these relations can be exploited to define a term analogically.

Students taking standardized exams often are called upon to recognize or complete analogical definitions. It would be profitable to work through a couple of examples to see what strategy can be used to solve such problems.

Example: Select the term which, if substituted for x, is correctly defined analogically.

$$\dfrac{\text{Mechanic}}{\text{Car}} = \dfrac{x}{\text{Human body}}$$ (a) officer, (b) repair specialist, (c) doctor, (d) dog

Answer: The best strategy to follow in solving analogical definition problems is to find the most obvious relation that holds between the terms on the left-hand side of the equal sign, and then substitute each of the possible answers for x to see if the comparable relation holds between the terms on the right-hand side. If none of the answers fit well, go back to the left-hand side to find another relation that holds between the terms, and repeat the process again. In our example, 'mechanic' is related to 'car' in the obvious sense that mechanics fix cars. Substitute 'officer' for x, and we do not have the same relation: officers do not fix bodies. Try 'repair specialist.' Repair specialists do repair, but not human bodies. Doctors do fix bodies, so perhaps the answer is (c). Plug in the last possibility, just to make sure—do dogs fix human bodies?—far from it!

Example:

$$\frac{\text{Scalpel}}{\text{Surgeon}} = \frac{x}{\text{Electronics technician}} \quad \text{(a) car, (b) dog, (c) knife, (d) oscilloscope}$$

Begin by asking what relations hold between a scalpel and a surgeon. Well, a scalpel is a knife that a surgeon uses. 'Car' and 'dog' bear no noticable relation to electronics technicians, so our interest focuses on (c) 'knife.' Electronics technicians do occasionally use knives, but no more than do truck drivers or ballet dancers. An oscilloscope is not any kind of knife. So the relation of being a commonly used knife is not what is utilized in the analogical definition.

Now work backwards. An oscilloscope is a tool, a tool electronics technicians use. Go back to the original pair:

$$\frac{\text{Scalpel}}{\text{Surgeon}}$$

Sure enough, a scalpel is a tool which specifically surgeons use. Thus (d) is the most adequate answer.

In the previous example, had we immediately quit when we found a plausible answer ('knife'), we would have overlooked a more adequate answer (d). On any exam containing analogical definition problems, you are well advised to examine all the possible answers.

EXERCISES 8.2A

For each of the following literal descriptions, find a shorter analogical description. (An asterisk indicates problems that are answered in the back of the book.)

1. Fred is vicious, aggressive, and enjoys hurting people.

*   2. Sue is quiet, intense, focused on her work, and has intense concentration.

3. That summer was exceptionally hot, with no breeze.

4. The people were running around in all directions, murmuring angrily and moving quickly.

5. The fire was intense, burning many houses quickly, with the wind gusting and sounding a loud roar.

EXERCISES 8.2B

For each of the following, choose the word that is best analogically defined by the proportion equation.

1. $\dfrac{\text{House}}{\text{Apartment building}} = \dfrac{x}{\text{Bus}}$ (a) truck, (b) car, (c) towel, (d) flashlight

* 2. $\dfrac{\text{Cat}}{\text{Dog}} = \dfrac{x}{\text{Cat}}$ (a) pig, (b) car, (c) mouse, (d) cheese

3. $\dfrac{\text{Cowardice}}{\text{Soldier}} = \dfrac{x}{\text{Scholar}}$ (a) friendliness, (b) patience, (c) hostility, (d) ignorance

4. $\dfrac{\text{Dark}}{\text{Light}} = \dfrac{x}{\text{Happy}}$ (a) fearful, (b) sad, (c) eager, (d) lonely

5. $\dfrac{\text{Profligacy}}{\text{Poverty}} = \dfrac{x}{\text{Obesity}}$ (a) hostility, (b) overeating, (c) ignorance, (d) slyness

* 6. $\dfrac{\text{Fast}}{\text{Slow}} = \dfrac{x}{\text{Short}}$ (a) tall, (b) blonde, (c) terrific, (d) quick

7. $\dfrac{\text{Wrench}}{\text{Mechanic}} = \dfrac{x}{\text{Writer}}$ (a) pen, (b) park, (c) book, (d) cab

8. $\dfrac{\text{Actor}}{\text{Play}} = \dfrac{x}{\text{Novel}}$ (a) writer, (b) artist, (c) character, (d) book

9. $\dfrac{\text{General}}{\text{Army}} = \dfrac{x}{\text{Corporation}}$ (a) soldier, (b) sailor, (c) president, (d) friend

* 10. $\dfrac{\text{Tiger}}{\text{Mammal}} = \dfrac{x}{\text{Bird}}$ (a) chicken, (b) dog, (c) cat, (d) eagle

8.3 Analogical Arguments

Often an analogy is employed as a premise in an argument. For example:

> Pudding is very much like cheese, but cheese is nutritious, so pudding is probably nutritious.

When an analogical premise is used in an argument, we speak of the argument as being an **argument by analogy**. An argument by analogy has the form:

1. A is like $B_1, B_2, B_3 \ldots$ in sharing $P_1, P_2, P_3 \ldots$
2. $B_1, B_2, B_3 \ldots$ all have property Q.

∴ A has Q.

Arguments by analogy are inductively strong or weak, depending upon whether criteria are met. First, all things being equal, the more numerous the B cases (the analogues), the stronger the influence.

For example, the argument:

1. The Toyota I own now is like the ten Toyotas I owned previously.
2. Those others lasted 100,000 miles.

∴ This one will last 100,000 miles.

is stronger than

1. The Toyota I own now is like the two Toyotas I owned previously.
2. Those others each lasted 100,000 miles.

∴ This one will last 100,000 miles.

You cannot say that since the first argument invokes five times the number of Toyotas as does the second, that it is therefore five times as good. We can say only that, all things being equal, the first presents stronger evidence.

Second, the more numerous the relevant similarities that hold between the analogues B and the subject A, the stronger the inference. For example, the argument:

1. This Toyota is like the other one I owned in that both have the same engine size, power equipment, and will be or were driven in the same environment.
2. The other I owned lasted 100,000 miles.

∴ So this Toyota will last 100,000 miles.

is stronger than

1. This Toyota is like the one I owned in that both have the same size engine.
2. The other Toyota lasted 100,000 miles.

∴ So this Toyota will last 100,000 miles.

The key word here is 'relevant': a property is *relevant* if we have reason to think that its presence or absence has something to do with the presence or absence of the property mentioned in the conclusion. For instance, the presence or absence of red body paint has nothing to do with engine life.

Third, the more numerous the relevant disanalogies, the weaker the argument. For example, the argument:

1. The Toyota I now own is like the other Toyota I owned in engine size and power equipment, but I am using this one to haul a trailer every day.
2. The last Toyota I owned lasted 100,000 miles.

∴ So this one will last 100,000 miles.

is weaker than

1. The Toyota I now own is like the other Toyota I owned in engine size and power equipment.
2. The last Toyota lasted 100,000 miles.

∴ So this one will last 100,000 miles.

You might wonder what the difference is between the last two rules, if two things do not share a relevant similarity, does that not mean they have a relevant dissimilarity? The answer is that (as we discussed in Chapter 3) the addition of extra premises can make an inductive argument stronger or weaker. And explicitly stating a disanalogy is different from omitting a statement of similarity.

Fourth, the more the property attributed to the subject precisely matches the one attributed to the analogues, the less likely the inference. For example, suppose Lori and Anne are identical twins raised in similiar conditions, and that Lori has an IQ of 125. The inference that Anne has an IQ of precisely 125 is going to be weaker than the inference that she has an IQ of more than 120, which in turn is going to be weaker than the inference that she has an IQ of more than 110. Inductive evidence typically supports conclusions that merely specify ranges much more strongly than conclusions that specify exact numbers.

Fifth, the less numerous the irrelevant similarities between the analogues, the stronger the inference. For our analogical inference to be strong, we need to be sure that the property Q, which the analogues all share, is indeed the one we want to attribute to the subject. For example, suppose we know that Suzie, Eddie, Manny, and Yoshi all went to Wilson High School, all had B+ grade averages, and all did well at Orange Coast College. Suppose we also know that Terry went to Wilson, had a B+ average, and is now attending Orange Coast. It would be reasonable to infer that Terry also will do well at Coast. But the inference would be weaker if we knew that Suzie, Eddie, Manny, and Yoshi all were physics majors (at Wilson), whereas Terry was a chemistry major, because, for all we know, physics majors at that school are more academically adept than chemistry students. Maybe they are, maybe they aren't—in other words, how do we know whether the major taken is or is not relevant in this case? A stronger inference would be if we took several Wilson students who had different majors at Wilson (who had B+ averages at Wilson, and who have done well at Coast) and, on that basis, inferred that Terry would do well.

EXERCISES 8.3

For each of the following passages:

  a. State whether it contains an analogy.

  b. If so, put the analogy in the form "*A* is like *B*."

  c. Determine whether the analogy is used to argue; if it is, put the argument in the form discussed earlier.

  d. Rate the argument as *silly, weak, plausible,* or *strong.* Justify your answer using the criteria given earlier.

(An asterisk indicates problems that are answered in the back of the book.)

1. Karla danced all night, then she went home and made an omelette. After eating the omelette, she tuned up her car and went for a drive.

* 2. The way I look at it, the liver is like a muscle. Just like you keep a muscle in shape by exercising it, so you must exercise the liver by drinking lots of booze.

3. Margo and Kelly are identical twins and grew up together. I know Margo, and she is brilliant: three earned Ph.D.'s, a multimillionaire stockbroker, and she's a whip at fixing diesel engines. So I figure that Kelly is bright, too.

4. My dog is a veritable Einstein among dogs. I mean, she is amazingly smart.

5. People can be so cruel. My girlfriend just told me that she thinks I'm stupid.

6. I've dined at that restaurant many times and always had good food, so I think we may as well eat there tonight.

7. The Cowboys have defeated the Chargers the last four times they played each other. Moreover, this Cowboys team has even better players than it did before. So I'm betting that they will win again.

8. Sue watched Actor A in four movies, and she thought he gave a poor performance in all of them. So she is reluctant to see his latest movie.

9. Both Luisa and Audrey are successful. Luisa is a major corporate attorney who has published over five hundred articles in respected law journals, and Audrey built a multibillion dollar business from scratch.

10. The way Lorenzo sucked up that pizza, you'd think he was a vacuum cleaner!

8.4 The Heuristic Use of Analogies

We have seen how analogies can be used to describe, to define, and to argue, but analogies also can be used to aid the process of research and discovery. They serve as heuristic aids. To see how, let us rethink the nature of analogical statements.

We can view any analogical statement

A is like B

as saying implicitly three things.

1. A and B share properties P_1, P_2, P_3. . . .
2. A and B do not share (they differ in) properties N_1, N_2, N_3. . . .

3. A and B may or may not share properties O_1, O_2, O_3. . . .

We will call the shared properties, $P_1, P_2, P_3, \ldots$, the *positive analogy*; the unshared properties, $N_1, N_2, N_3, \ldots$, the *negative analogy*; and the open properties (the properties that A and B may or may not share), $O_1, O_2, O_3, \ldots$, the *neutral analogy*.[1]

As an example, consider the analogy "Jason ran like a bug across a griddle." The analogy involves:

1. Positive analogy: Jason and the bug share the properties of having legs and moving quickly by means of those legs.
2. Negative analogy: Jason and the bug differ in size, shape, and intelligence.
3. Neutral analogy: Jason and the bug may or may not both be in pain and desperate.

In research of any kind (scientific, legal, philosophical, or otherwise), we can use an analogy to point out new areas for investigation. If, for example, we visualize computer software as a kind of product by viewing the programs produced by a software company as the cars produced by an automobile company, then many areas open to investigation. Should the software company be as legally liable for the damage to the consumer their defective software caused, as the automobile manufacturer is for the damage its defective cars caused? That is an area of current legal research.

Put in terms we have just discussed, when we set up an analogy on the basis of the positive analogy, we have at our disposal a whole range of possible further positive analogy, namely, the neutral analogy. A good heuristic strategy is to set up an analogy with a strong positive analogy and weak negative analogy, and then systematically explore the neutral analogy to discover any new properties that the subject and analogue might have in common.

EXERCISES 8.4

For each of the following analogies, indicate the most obvious positive analogy, negative analogy, and neutral analogy. (An asterisk indicates problems that are answered in the back of the book.)

---

[1] This terminology is from Mary B. Hesse, *Models and Analogies in Science* (London: Sheed and Ward, 1963).

  1. Cats are like tigers.
* 2. People are like animals.
  3. Her kisses were like wine.
  4. Hatred is like a corrosive poison.
  5. Football is like war.
* 6. Life is like poker.
  7. Writing is like playing tennis.
  8. Business is like war.
  9. Dancing is like singing.
* 10. College is like prison.

8.5 Models in Science

We have discussed the use of analogies to describe, to define, to argue, and to discover. As a special case, let us discuss the use of analogies in science.

Analogies are usually termed 'models' in science. We run across phrases such as the "liquid-drop model of the atomic nucleus," "the planetary model of the atom," "the equilibrium model of the economic system," and so on, in scientific literature. What is the purpose of models in science? The answer is that scientists use models for all four basic functions discussed earlier.

For example, consider the Bohr ('planetary' or 'solar') model of the hydrogen atom. In this model, the structure of the hydrogen atom is likened to that of the solar system, with the proton being at the center and the electron orbiting it. That model serves to describe the atom. Moreover, the model stipulates that the electron can travel only in certain orbits, although the electron can jump from one orbit to another. That image serves to define the concept of a 'transition' from one state of the atom to another. Moreover, this analogy serves in an argument to explain why light emitted or absorbed by hydrogen is always at a fixed wavelength (the electron can only move from one set orbit to another, with a unit of energy being given off or absorbed). Finally, the model serves to direct research (for example, research on whether we extend it to cover the helium atom, an atom more complex than the hydrogen atom). Analogies are, therefore, valuable, not just in ordinary contexts of reasoning, but in science as well.

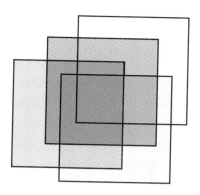

9

The Method of Truth Tables

9.1 What Is a Theory of Deduction?

In this chapter let us begin the subject variously called 'symbolic logic,' 'mathematical logic,' or 'formal logic' with a review of a few of the definitions laid down in Chapter 2. We said that an **argument** (in the strict sense, an argument as opposed to a dialogue) consists of a set of statements (called 'the premises') taken as evidence for another statement (called 'the conclusion'). Out of all possible arguments that we could dream up, only a small percentage have the property of being (deductively) valid. An argument is **valid** if it is impossible for the premises to be true while the conclusion is false.

But what is this property we call 'validity'? How can we tell if an argument has this property? Granted, we can easily see that the following argument is valid:

1. Ahmed is rich.
2. If Ahmed is rich, then he is happy.

∴ Ahmed is happy.

But what about a peculiar argument like this one:

1. If Jason is anxious, then either he is rich or smart.
2. If Jason is smart, then I'm both happy and unhappy.

∴ Either Jason is not anxious or he is smart.

Telling valid from invalid calls for a systematic scientific theory. A *theory of deduction* is a theory that does three things. First, it enables us to tell, given an argument, whether the argument is valid. That is, we call upon a theory of deduction to give us a *decision procedure*. Second, a theory of deduction should enable us to demonstrate to other people that any argument we claim to be valid is indeed valid. It should give us a *proof procedure*. Third, a theory of deduction should provide an *explanation* of what validity is.

Symbolic logic is a theory of deduction that involves symbols. The symbolic theory of deduction involves the construction of an artificial language, especially designed to reveal the logical structure of arguments.

9.2 Object Language and Metalanguage

In Section 9.1, we said that we were going to construct an artificial language. An **artificial language** is a language designed specifically to accomplish a set purpose.

Natural languages, such as English, Spanish, or Chinese, evolve naturally and by informal convention. Natural languages are spoken as well as written, and have to be flexible enough to accomplish dozens of purposes. But artificial languages are absolutely precise and designed for one purpose—although they may wind up serving other uses as well.

Hundreds of artificial languages exist. Some are *logistical systems*, artificial languages designed to clarify aspects of logical reasoning. Others are *computer programming languages*, including FORTRAN, COBOL, LOGO, BASIC, ALGOL, Pascal, and Ada, which enable computer programmers more easily to lay out instructions that a computer

Figure 9.1

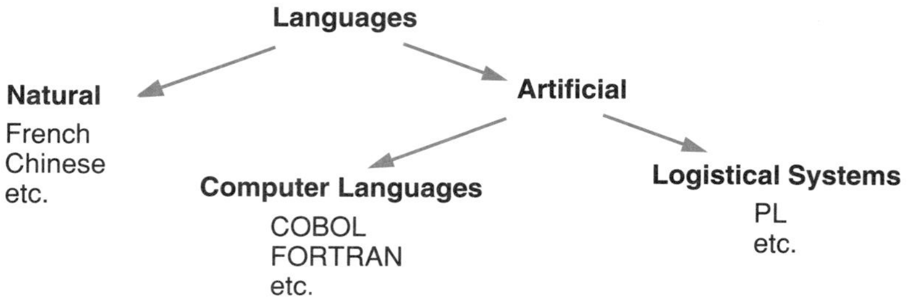

Languages

Natural
French
Chinese
etc.

Artificial

Computer Languages
COBOL
FORTRAN
etc.

Logistical Systems
PL
etc.

will follow to solve problems. Figure 9.1 summarizes the relation among languages.

Logicians customarily distinguish object language from metalanguage. The **object language** is the language being examined, that is, the object of study. The **metalanguage** is the language used to talk about the object language, that is, the language used to carry out our study.

In what follows, the object language will be symbolic logic, and our metalanguage will be English. There is nothing privileged about English—we could use Spanish or Vietnamese instead as our metalanguage. But, since you are reading this book in English, that is our most practical choice.

Any language can be an object of study. Linguists study English and all other natural languages. Indeed, the same language can be both an object language and a metalanguage. A grammarian of English may use English to talk about English, although the grammarian may also talk about English grammar in Russian.

9.3 Intended vs. Unintended Interpretation

As we embark upon our journey, consider the practical value of symbolic logic. Quite some time back, mathematicians developed a subject we now call 'Probability and Statistics' primarily to answer questions about gambling. In the twentieth century, probability theory has turned out to have a wide variety of uses: in genetics, physics, and the social sciences. A similar situation has arisen with mathematical logic. Although it was devised to exhibit validity in arguments given in ordinary language, our artificial language turns out to have other applications as well—especially in linguistics, mathematics, and computer science. This point illustrates a general feature of any symbolic system. A formal system is developed with one kind of use or 'interpretation' in mind (the 'intended interpretation'), but as the system is elaborated and developed, other interpretations are discovered as well.

9.4 The First Three Connectives

We seek to design an artificial language—let us call it *propositional logic* (PL)—that allows us to check arguments for validity. The arguments we will focus upon turn on the sentence connectives. Recalling our plan to break complicated tasks down into simple subtasks, we will construct a bit of PL, next learn to 'calculate' in it, then learn to translate from English into it, then go back and construct more of PL, and so on. We will find that PL will be an extremely useful tool in analyzing arguments and investigating their validity.

Let us distinguish simple from compound statements. A simple statement is a statement that does not contain other statements as parts. Here are some simple statements:

Whales are fat.
Pat is ugly.
Yoshi is dead.

On the other hand, a compound statement is any statement that does contain other statements as parts ('components'). Examples:

If whales are fat, then pigs are skinny.
Juanita is fat and friendly.
I believe that Bonita is a good singer.

Not all compounds are alike. A few have special nature, which we call 'truth functionality.' A **truth-functional compound** (or 'truth function') is a compound whose truth value [whether it is true (T) or false (F)] can be determined absolutely, given only the truth values of its components. A *non truth-functional compound* is not truth functional, that is, one whose truth value cannot be determined merely from determining the truth values of the components.

To get a more concrete idea of this distinction, compare these two compounds:

1. It is not true that Fred is dead.
2. Billie Mae believes that Fred is dead.

Suppose we are told that Fred is dead—or, to put it more technically, suppose that we are told that the truth value of the statement "Fred is dead" is T. Then we can see immediately that the whole compound "It is not true that Fred is dead" is F.

Now look at (2). Even if we discover that Fred is dead, can we tell whether Billie Mae believes it? Not without knowing Billie Mae personally. Thus sentence (1) is truth functional while (2) is not.

9.4 / The First Three Connectives

It is helpful to compare truth functions with numerical functions. All that is needed to determine the value of $[(7 - x)(x + 3)]$ is the value of x. Similarly with truth functions: all that is needed to determine the truth value of the compound is the truth value of the components. But we need more than the value of x to compute this odd expression:

$3x +$ my favorite number.

The *sentential connectives* of interest to us are all truth functional. This means that the connectives can be defined by the way in which they combine truth values to give other truth values. We will examine negation, conjunction, and disjunction in turn.

A **negation** states that some other statement is false. For example, the negation:

Fred is not rich.

asserts that the statement:

Fred is rich.

is F. If we adopt the prudent habit of stating denials as:

It is not the case that ○

then we can easily see that the negation or denial of A is T whenever A is F, while the negation of A is F whenever A is T. We will use a circle as a 'place-holder' for statements in what follows. This is summarized in Table 9.1, (We call any table of truth values a 'truth table.') where we have introduced the symbol—(the dash) to express negation in PL.

Note that while the English language has several words or prefixes to words that express negation (such as 'no,' 'not,' 'non-,' 'un-,' and 'dis-'), PL has only one symbol.

We must recognize that Table 9.1 completely defines the dash. If it sounds odd to speak of a table defining a symbol (or word), just

Table 9.1 *Truth Table for Negation*

○	−○
T	F
F	T

remember back to the third grade or so, when your teacher defined the multiplication symbol by means of the multiplication table.

In laying out Table 9.1, we had to employ the symbol ○. We call such symbols **variables**, since they vary in their reference. We will use the following variables: ○, □, △, ◇, and ⌂. Think of them as boxes into which we can plug statements.

The second truth-functional compound is a conjunction, which is usually expressed in English by the word 'and.' A **conjunction** is a statement that two other statements are true. Consider the compound statement:

Xavier is cold and aloof.

We prudently adopt the procedure of filling in omitted words, which allows us to see that this statement is a conjunction:

Xavier is cold and Xavier is aloof.

That conjunction is T just in case Xavier is indeed both cold and aloof. On the other hand, the conjunction is F if Xavier is not cold, or not aloof, or neither cold nor aloof.

Accordingly, Table 9.2 defines the symbol used in PL to express conjunction (the symbol & is called the 'ampersand'). What Table 9.2 for the ampersand says is that the compound:

A & B

is T if both components are T; it is F if either one or both of the components is F. Notice that the left side of Table 9.2 lists all possible combinations of truth values for two components. The right side of Table 9.2 then shows what the truth value of the compound is under each assignment of truth values to the components.

While English has many words that express a conjunction (as will be discussed in Section 9.5), PL has only the &. Moreover, while the word 'and' has other uses in English besides expressing conjunction, the & has only that use in PL.

Table 9.2 *Truth Table for Conjunction*

○	□	○ & □
T	T	T
T	F	F
F	T	F
F	F	F

9.4 / The First Three Connectives

The third type of truth function is **disjunction**, also called 'alternation.' In such a compound statement, the asserter says that either A is T or else B is T, or maybe both. In English, we usually use the word 'or' to express disjunction.

To get the truth table, consider the example:

Either Juan or Luisa will buy a car.

As usual, fill in the omitted words:

Either Juan will buy a car or Luisa will buy a car.

Such a statement is T if Juan buys a car. It is also T if Luisa buys a car. And it is T if both buy cars. If and only if both do not buy cars is the disjunction F.

Table 9.3 defines our third PL connective (the wedge, v).

While English has a number of words to express disjunction ('or' and 'unless' are the most common), PL has only the wedge.

MEMORIZE THESE TABLES. Use flash cards if necessary. The calculations you will be doing will be made much easier if you know these tables thoroughly.

So far, not much of PL has been constructed. The vocabulary consists of the three connectives $-$, &, and v, which the truth tables define. But if the connectives are used to connect components, what should we use to symbolize particular simple statements?

We will use capital letters A, B, C, ... to symbolize simple statements in PL. Thus, for example, the simple statement "John is rich" would be written in PL as J (we could have chosen any other letter as well). We call those capital letters 'sentential constants.' Look at it this way: the same statement (that John is rich) can be expressed in English by the sentence "John is rich," in Spanish by the sentence "Juan es rico," and in PL by J. J, however, does not stand for John, but for the statement.

The vocabulary of PL now consists of sentential connectives and sentential constants. What else do we need? The answer is indicated

Table 9.3 *Truth Table for Disjunction*

○	□	○ v □
T	T	T
T	F	T
F	T	T
F	F	F

by reconsidering amphiboles, which we met in Chapter 6. The sentence: "He said she was crying" is ambiguous as it stands, but punctuation can resolve the ambiguity ("He," said she, "was crying" vs. He said, "she was crying"). Similarly, the PL sentence:

A v *B* & *C*

is ambiguous. Do we mean the disjunction of *A* with *B* & *C*, or the conjunction of *A* v *B* with *C*?

PL employs parentheses and brackets as punctuation devices. Thus in the previous example, we could write either *A* v (*B* & *C*) or else (*A* v *B*) & *C*, depending upon what we wanted to say. Brackets can be used to clarify statements with many parentheses. For example, —{*A* & [*B* v (*C* & *D*)]} is easier to read than —(*A* & (*B* v (*C* & *D*))). The vocabulary of PL is summarized in Table 9.4.

9.5 Calculations in Propositional Logic

We have constructed the core part of PL. Let us turn to the task of calculating in it, that is, calculating the truth values of compound statements from the truth values of the simple components.

Recall that to calculate the numerical value of an expression like [$3x+7$], we must first plug in the value of *x*, then multiply by three, then add seven. The trick to calculating a long numerical expression is to do it one step at a time (one addition, one subtraction, one multiplication, or one division), and in the right order.

Calculating truth-functional compounds is done the same way. First, we are given the truth values of the basic parts (we are given an 'assignment of truth values to variables'). We then do the calculations one at a time. Since we have only three connectives, only three types of calculations need to be done, performed using Tables 9.1 to 9.3.

Table 9.4 *Vocabulary of PL*

Symbols	What They Mean
v, —, &	Sentential connectives, defined by truth tables.
○, □, △, ◇, ⌂	Sentential variables, which are place holders, locations into which statements can be plugged.
A, B, C, . . . , Z	Sentential constants, which stand for statements.
(,), [,], {, }	Brackets, intended for punctuation.

9.5 / Calculations in Propositional Logic

The parentheses determine the order of the computations. We make one exception: we will adopt the convention that the dash applies to the nearest letter unless brackets are frequent. Thus $-A \vee B$ will be understood as $(-A) \vee B$ rather than as $-(A \vee B)$.

Let us work a few examples. Suppose that A, B, and C are T, and X, Y, and Z are F.

Example 1: Find the truth value of $-A \vee -B$.

> **Step 1:** Spot the 'main connective,' that is, determine the overall form of the PL sentence. I will call it to your attention with an arrow.

$-A \overset{\swarrow}{\vee} -B$

> **Step 2:** Write the truth values assigned underneath each sentential constant.

$-A \vee -B$
  T   T

> **Step 3:** Working with the smallest compounds, calculate each by the appropriate table. Cross out the old values as you go.

$-A \vee -B$
F T̶   F T̶

> **Step 4:** Repeat step 3 until every connective has a truth value underneath it, and they are all crossed out except one—the main connective. That is your final answer. Circle it.

$-A \vee -B$
F̶ T̶ Ⓕ F̶ T̶

Example 2: $-(-A \vee X) \mathbin{\&} -Z$

> **Step 1:** $-(-A \vee X) \overset{\swarrow}{\mathbin{\&}} -Z$

> **Step 2:** $-(-A \vee X) \mathbin{\&} -Z$
>   T   F   F

> **Step 3:** $-(-A \vee X) \mathbin{\&} -Z$
>   F T̶ F   T F̶

Then: −(−A v X) & −Z

Then: −(−A v X) & −Z

Then: −(−A v X) & −Z

Doing the calculations all on one line by this 'crossing out' technique ensures accuracy and it allows the whole calculation to fit on one line.

EXERCISES 9.5

In each of the following, A, B, and C are T; X, Y, and Z are all F. Calculate the truth values of these compounds. (An asterisk indicates problems that are answered in the back of the book.)

1. −B v B
* 2. −Y v C
3. −−A
4. −−B v Z
5. −Z & A
* 6. A & −−X
7. (A v X) & (Y v B)
8. (B & Z) v −−A
9. (Z & −Z) v (X v Z)
* 10. −(A v Y) & (B v Z)
11. [Z & −(X v A)] & B
12. −(B v Z) & −(X v −Y)
13. −[(B v Z) & −(X v −Y)]
* 14. [Z v (X v X)] v (Y & A)
15. −[(B & C) & −(C & B)]
16. −[(A & B) v −(Z & A)]
17. [Z v (B & X)] & −[(A v B) v Z]

* 18. [B v (Z & X)] v −−[(A & Z) & B]
  19. [X v (Y & Z)] v −[(X v Y) & (A & B)]
  20. {[X v (A v B)] & [−X & (−X v Z)]} & (A v −B)
  21. −{−[A v (B v C)] v −[X v (Y v Z)]}
* 22. −{[A v (B v C)] & [X v (Y v Z)]}
  23. [X & (Y v Z)] & −[(A & B) v (A & C)]
  24. −{[(−A & Z) &(−X & Y)] & −[(A & −B) v −(−Y & −Z)]}
  25. −{−[(B & −C) v (Y & −Z)] & [(−B v X) v (B v −Y)]}

9.6 An Iterative Method of Translation

We have constructed most of PL and have learned how to calculate truth values of compounds in it. We now want to learn how to translate from English to PL. An argument in either language is just a structured set of sentences, so it suffices to learn how to translate sentence by sentence.

Translating simple statements is no problem. Simply pick a capital letter (if one has not been selected already). But compound statements can be tricky. In what follows, we will discuss a few hints about translating types of English sentences, and then a general method (an interative or stepwise method) for translation, which minimizes the possibility of errors. Think of the hints we give as suggestions for translating idioms. An *idiom* is a phrase that cannot be translated literally. For example, the French idiom "mon petit choux" means literally "my little cabbage," but is more accurately translated as "sweetheart" or "darling." The English phrases we will examine are idiomatic from the PL point of view.

We begin with the hints. The first two hints pertain to simple statements, the other hints relate to compound statements.

Hint No. 1: Spell out indexicals. An *indexical* is a term that refers to the context of utterance. For example, the pronoun "I" refers to the person who utters it. Other common indexicals are listed in Table 9.5.

In assigning letters to simple statements, fill in all references. Example:

He likes sushi.

Table 9.5 *Common English Indexicals*

I	There
Me	Then
Myself	Now
Mine	At that time
You	At that point
Yours	Former
Yourself	Latter
He, him	We
She, her	They
It	Those
Hers	That
His	Them
Its	This
Himself, herself	At that place
Itself	Here

should be spelled out completely before symbolization:

R: Renaldo Kupie likes sushi.

Hint No. 2: Clarify vagueness and ambiguity. We saw in Chapter 6 that vagueness and ambiguity can lead to logical fallacies. In symbolization, then, take care to assign letters to atomic sentences (simple statements) that are precise and unambiguous.

Hint No. 3: Watch out for covert negations. Often a statement that appears simple at first glance is really a negation. Some examples:

Desmond lacks the qualities that make nurses great.

(This is the negation "It is not the case that Desmond has the qualities that make nurses great," which is translated as $-D$.)

Rhonda is disinclined to accept Giorgio's word.

(This is the negation "It is not the case that Rhonda is inclined to accept Giorgio's word," translated as $-R$.)

Mookie is unconscious.

(This is the negation "It is not the case that Mookie is conscious," translated as $-M$.)

Although we could symbolize such statements as atomic, they are better symbolized as negations.

Table 9.6 lists the most common English words used to express negation.

Hint No. 4: Watch out for the scope of the dash. Compare the following statements:

1. Maria wrote ten incorrect answers.
2. Maria did not write ten correct answers.

In (1), the negation is applied directly to the predicate: Maria wrote ten answers that were not correct. In (2), the negation applies to the whole statement: it is not the case that Maria wrote ten correct answers. Only the latter is correctly symbolizable as $-F$. A similar caution applies to contraries and contradictories as follows:

Sally is sad.

is not the negation of

Sally is happy.

Hint No. 5: The word 'and' does not always express conjunction. The virtue of PL is that its words are unambiguous. Not so with English. In particular, while the word 'and' usually expresses conjunction, it does not always do so. Consider these two examples:

1. Juan and Gary are Democrats.
2. Juan and Gary are cousins.

The first statement is equivalent to "Juan is a Democrat and Gary is a Democrat"—clearly a conjunction—whereas the second statement

Table 9.6

English Phrase	Symbolization
Not ○ It is not the case that ○ It is false that ○	$-○$
Non- Un- Occurring Dis- inside ○ -less	

is not equivalent to "Juan is a cousin and Gary is a cousin." The point is that they are cousins of each other. In that second statement, the word 'and' is used relationally: to express relation. The ampersand does not capture this use. Thus (1) is translated as *J* & *G*, but (2) is misleadingly translated as *J* & *G*.

Hint No. 6: Many other words in English besides 'and' are used to express conjunction. English has a large vocabulary, capable of expressing many subtle points. PL has a small, precisely defined vocabulary, which is not as expressive. So not surprisingly, something often gets lost in translating English into PL.

Table 9.7 lists some of the most common English phrases or punctuation that express conjunction.

Even though something may get lost in the translation, translate all of them using the ampersand.

For instance, "He whistled while he shaved" means more than that he whistled and that he shaved; the statement asserts that he did both at the same time. But this temporal aspect is lost when we translate into PL. (*W* & *S* is the best we can do.)

The point holds for the word 'and' as well. "Nina hit the vase and it broke" does more than just assert the conjunction of the two facts—it asserts a causal link between them. This does not get reflected in the PL translation (*H* & *B*). This loss of meaning will not trouble us, however, because the arguments we want to assess using PL do not depend on temporal or causal factors.

Hint No. 7: Translate the word 'or' with the wedge. The English word 'or' is ambiguous. It has an **inclusive sense**, as when a counselor says, "You can satisfy the foreign language requirement by taking either Japanese or German." The counselor means that you can satisfy the requirement by taking one or the other, or both. On the

Table 9.7

English Phrase		Symbolization
○ and ▢ ○ yet ▢ ○ still ▢ ○ while ▢ ○ albeit ▢ ○ even though ▢ ○ nonetheless▢ ○ nevertheless▢	○ moreover ▢ ○ although ▢ ○ whereas ▢ ○ however ▢ ○ also ▢ ○ but ▢ ○ ; ▢	○ & ▢

other hand, when the menu says, "You may have soup or salad," it is using the word 'or' in the **exclusive sense** of ruling out (excluding) your having both.

A wedge translates the word 'or,' even though that automatically assumes the word is being used in its inclusive sense. If the English sentence explicitly says, ". . . but not both," we have a way to translate that into PL. For instance, "You may have soup or salad, but not both" is rendered:

$(S \vee L) \mathrel{\&} -(S \mathrel{\&} L)$

(Where S = you may have soup, L = you may have salad.)

This hint is again only heuristic. In the statement:

Either Fred or Tanya could beat Joe.

what is asserted is conjunction, not a disjunction:

Fred could beat Joe, and Tanya could beat Joe.

Hint No. 8: Use a wedge to translate 'unless' (always) and 'without' (usually). The word 'unless' can be translated in various ways, but it is always wise to translate it in the easiest way possible: with a wedge. Thus "John will fail the class unless he does exceptionally well on the final exam" would be best translated as $F \vee W$.

Words that have the same meaning as 'unless' include 'except,' 'save,' 'barring,' 'supposing not,' and 'waiving.' Here are some examples:

The store is always open, except on Sunday.
The store is always open, save on Sunday.
She will succeed, barring interference by others.
She will succeed, supposing no interference by others.
Waiving interference by others, she will succeed.

All these examples should be translated with the wedge.

A similar point holds with respect to the word 'without.' It, too, can be translated in several ways, but the easiest is with the wedge. Thus "Andrea won't go to the party without her sister" is translated as $-A \vee S$.

Table 9.8 summarizes what we have just discussed.

These hints are heuristic tips, as opposed to mechanical rules. As such, they admit exceptions. For example, the sentence:

Mona is going to law school without financial support.

is translated as

$L \ \& -F$

rather than (as the hint suggests)

$L \ v -F$

We must understand what we are translating to make sure that our translation captures the initial statement correctly.

Hint No. 9: The occurrence of the words 'either' and 'both' indicates how to bracket the PL sentence. How a PL sentence is bracketed is crucial. $(A \ v \ B) \ \& \ C$ is quite different from $A \ v \ (B \ \& \ C)$ (at times the first may be T, yet the second may be F). The words 'either' and 'both' can help you decide how to bracket or, in other words, how to spot the main connective.

Consider these two sentences, both composed of the same words:

1. Either Tibor and Maria or Sue will go.
2. Tibor and either Maria or Sue will go.

The only difference between (1) and (2) is where the word 'either' is placed, but it is a big difference. Sentence (1) asserts the disjunction:

$(T \ \& \ M) \ v \ S$

whereas (2) asserts the conjunction:

$T \ \& \ (M \ v \ S)$

Table 9.8

English Phrase		Symbolization
○ or □	○ save □	
○ unless □	○ barring □	○ v □
Unless ○, □	○ except □	
○ without □	○ waiving □	

9.6 / An Iterative Method of Translation

The word 'either' signals that the material which follows it and comes before the 'or' is to be bundled together by brackets.

A similar point holds regarding the word 'both.' Compare:

1. Both Fred or Tibor and Sue will go.
2. Fred or both Tibor and Sue will go.

(1) is translated as

$(F \vee T) \,\&\, S$

(2) as

$F \vee (T \,\&\, S)$

with the placement of the word 'both' making the difference in bracketing.

Hint No. 10: Translate "neither ○ nor □" as $-○ \,\&\, -□$. The phrase "neither ○ nor □" habitually causes problems. Consider the claim:

You will get neither a Porsche nor a Ferrari for Christmas.

What claim is being made? One way to put it is:

You are not getting a Porsche and you are not getting a Ferrari for Christmas.

Picking the letters P and F for the simple statements "You will get a Porsche" and "You will get a Ferrari," respectively, we can render this as $-P \,\&\, -F$. But we can also equivalently say:

It is not the case that you are getting either a Porsche or a Ferrari for Christmas.

which can be rendered $-(P \vee F)$. We may choose to translate "neither ○ nor □" as $-○ \,\&\, -□$ or as $-(○ \vee □)$, but a good idea is to choose one way and always stick with that practice. We will accordingly always symbolize "neither ○ nor □" as $-○ \,\&\, -□$.

But both $-(P \,\&\, F)$ and $-P \vee -F$ are bad translations. The first says that you will not get both the Porsche and the Ferrari—but it leaves it open that you might get one of them. $-P \vee -F$ is bad, since it says you will either not get the Porsche or not get the Ferrari. But that leaves open the chance that you might get one of them.

The reason the phrase "neither ○ nor □" is an idiom in PL is that we don't have a connective which allows us to symbolize the phrase directly (that is, in the form ○ | □). The reason we don't have such a connective (the 'nor' connective) is explained in Section 9.13.

Hint No. 11: When the words 'both' and 'not' occur together in a sentence, the one that occurs first is usually the main connective. Consider these examples:

1. Bush and Clinton will not both be elected.
2. Bush and Clinton will both not be elected.

(1) states the trivial fact that any presidential election can only have one winner. We translate this as:

$-(B \mathbin{\&} C)$

(2) says more—it says that Bush will lose and Clinton will lose. It is translated:

$-B \mathbin{\&} -C$

Hint No. 12: Restore proper clause order before translating. English evolved as a spoken language. A consequence is that we often encounter clauses put at the end of the sentence for rhetorical emphasis. For example, an instructor might express the statement:

You will fail unless you study.

as:

Unless you study, you will fail.

to dramatize what will happen to the student if that student does not study. Before translating such a sentence, restore the logical clause order:

○ unless □

9.6 / An Iterative Method of Translation

In PL, the wedge connects two sentences together; a string of the form:

v ○ □

makes no sense.

Indeed, several other English words are often found in front of the clauses they connect:

While ○, □

Even though ○, □

Unless ○, □

Without ○, □

In every case, we should bring the end clause out front before symbolizing. This will be especially important in Section 9.7.

So much for hints. Let us talk now about a general method for translating from English to PL. Keep in mind that while computation is something that can be done mechanically, translation involves insight and understanding. But the following approach to translation can maximize the chances of success.

The stepwise method of translating involves doing the translation in steps or *iteratively*: translating only one connective at a time. Recall our earlier discussion in Chapter 1 (Section 1.2) of heuristics. We found that, as a general rule of problem-solving, we should try to tackle a big problem in stages. This rule lies behind the method we discuss now.

Suppose we are asked to symbolize the compound:

Unless both SANDRA and DOLORES don't show up, the PARTY will be a success.

Step 1: Select some constants. The choice is entirely ours, except that we must be consistent—a capital letter must stand for one and only one simple statement. To make things easier, I have capitalized some words in these examples to indicate which letters to choose. In our example:

S = Sandra shows up.

D = Dolores shows up.

P = The party will be a success.

Step 2: Spot the main connective, restore clause order if necessary, and then symbolize just that one connective. Put the remaining English in brackets.

[The party will be a success] v [both Sandra doesn't show up and Dolores doesn't show up.]

(Fill in omitted words whenever you can.)

Step 3: Repeat step 2 on each bracketed part. Whenever you hit a simple statement, replace it by the corresponding capital letter.

[P] v [(Sandra does not show up) and (Dolores does not show up)]

P v [$-$(Sandra shows up) and $-$(Dolores shows up)]

P v [$-S$ & $-D$]

Do not try to solve it all at once. Break the problem down into parts.

Example: Either it isn't true both that the WARREN Report is biased and that Garrison's case is PLAUSIBLE, or it isn't at all true that both OSWALD was not guilty and that he was COMMUNIST-inspired.

Step 1: W = The Warren Report is biased.
P = Garrison's case is plausible.
O = Oswald was guilty.
C = Oswald was communist–inspired.

Step 2: [It is not true both that the Warren Report is biased and that Garrison's case is plausible.] v [It is not true both that Oswald was not guilty and that he was communist-inspired.]

Step 3: [The Warren Report is biased and Garrison's case is plausible] v $-$[Oswald was not guilty and Oswald was communist-inspired.]

Step 4: $-(W \text{ \& } P) \text{ v } -(-O \text{ \& } C)$

Example: Either Juan will take Mary to the dance and Fred will take Billie Mae, or else Juan will take Billie Mae and Fred will take Mary, but one thing's sure—Juan won't take Fred.

Step 1: J = Juan takes Mary to the dance.
F = Fred takes Billie Mae to the dance.

9.6 / An Iterative Method of Translation

H = Juan takes Billie Mae to the dance.
L = Fred takes Mary to the dance.
K = Juan takes Fred to the dance.

Our artificial language is so sparse that it cannot show tenses—"Juan takes Mary to the dance," "Juan took Mary to the dance," and "Juan will take Mary to the dance" are all symbolized the same way.

Step 2: {Either (Juan takes Mary and Fred takes Billie Mae) or (John takes Billie Mae and Fred takes Mary)} and (It is not the case that John takes Fred.)

Step 3: {[(Juan takes Mary) and (Fred takes Billie Mae)] or [(Juan takes Billie Mae) and (Fred takes Mary)]} and −K.

Step 4: {(J & F) v (H & L)} & −K

EXERCISES 9.6

Symbolize the following sentences. Pick sentential constants suggested by the words in capital letters. (An asterisk indicates problems that are answered in the back of the book.)

   Example: Unless FRED objects, KELLY will go to the dance.

becomes:

   KELLY will go to the dance unless FRED objects.

which is symbolized:

   K v F

1. Either Paolo won't go to COLLEGE or else he will go to NIGHT school.
* 2. JACQUE or KAREN will buy the tacos.
3. ALI won't marry Sue and neither will BILL.
4. YVETTE won't help Andrea unless Andrea's SISTER doesn't help Andrea.
5. Without YVETTE helping her, ANDREA won't pass the exam.
* 6. Either JACQUE and BILL both went up the hill or else they both didn't.
7. Either JACQUE and BILL went up the hill, or else Jacque went without Bill.
8. Unless KIM fails to fix the car, GARY and BINDHU will drive to Barstow.

9. Either KIM or both TOM and BINDHU will fix the car.

* 10. Neither will KEVIN fail to win the race nor will TED not fail to pass the exam.

11. It's not true that ANDREA will go to the dance without her HUSBAND.

12. Neither JAGUARS nor PUMAS nor TIGERS are found in Las Vegas.

13. SAMIRA and DAVE won't both pass the exam.

* 14. SAMIRA and DAVE will both not pass, unless they both study. (A = Samira studies, B = Dave studies)

15. Unless Cornelia fails to find a JOB, her future will not be both POOR and UNCERTAIN.

16. Without either TED or BUBBA on the team, both the COACH and the OWNER will be disappointed.

17. RICHARD will either go without KELLY or else without LORI.

* 18. RICHARD and GEORGE won't fail to not win.

19. Unless SINGH and KELLY fail the test, YVETTE will not fail.

20. It's not true that unless SINGH and LORI don't both fail, they will both pass.

21. Even though MARIA liked the dance, neither SUE nor TRACY liked it.

* 22. Unless the RAMS win while the COWBOYS win, the LIONS will win.

23. However reluctantly, YOSHI paid the bill.

24. It's not the case that neither YOSHI nor ALFRED will not fail to pass the test.

25. Either neither YOSHI nor SAM went, or else neither KAREN nor MADONNA went.

9.7 The Conditional

We have designed PL to express negation, conjunction, and disjunction. Another type of compound we need to consider is the **conditional**. As in Section 9.4, we will look at a sample statement, use it to figure out a plausible table definition of our new PL connective, and then adopt it without qualification.

Consider this statement, a promise made by Samira to another: "If I go to the store, I'll buy you a sandwich." Is she promising to buy a sandwich? No. Is she promising to go to the store? No. She is promising to buy the sandwich conditional upon her going to the store. View the matter this way. In what situation can we say the promise has been broken, and when is it not broken?

Table 9.9 sums this up.

9.7 | The Conditional

Table 9.9

Did She Go to the Store?	Did She Buy the Sandwich?	Was the Promise Kept?
Yes	Yes	Yes
Yes	No	No
No	Yes	Yes
No	No	Yes

In the case where she does not go to the store, her promise is still kept—since she did not promise to go to the store.

We introduce a new connective, the 'arrow'(→), to translate the phrase "if A then B," and define it in Table 9.10.

We call what comes before the arrow 'the antecedent,' what comes after it is 'the consequent.' As in Section 9.6, we use the 'crossing out' method of calculation. Spot the main connective, write the assigned truth values underneath the respective letters, and calculate one connective at a time in accordance with the bracketing, crossing out as the calculation proceeds.

Example: If A, B, and C are T, and X, Y, and Z are F, what is the truth value of $-[A \rightarrow (B \mathbin{\&} Z)]$?

Step 1: $-[A \rightarrow (B \mathbin{\&} Z)]$ — Main connective
          T   T   F

Step 2: $-[A \rightarrow (B \mathbin{\&} Z)]$
          T   T̶ F T̶

Step 3: $-[A \rightarrow (B \mathbin{\&} Z)]$
          T̶ F T̶ T̶ T̶

Table 9.10

○	□	○ → □
T	T	T
T	F	F
F	T	T
F	F	T

Step 4: $-[A \rightarrow (B \& Z)]$ Final answer
         T F F F F F

EXERCISES 9.7

Calculate the truth values of the following compounds, using this assignment of truth values: A, B, and C are T; X, Y, and Z are F. (An asterisk indicates problems that are answered in the back of the book.)

1. $A \rightarrow B$
* 2. $-A \rightarrow -B$
3. $(A \& X) \rightarrow Z$
4. $(A \& B) \rightarrow (Z \vee A)$
5. $-(A \& B) \rightarrow (Z \vee -B)$
* 6. $(A \rightarrow Z) \vee (C \rightarrow X)$
7. $(A \rightarrow X) \rightarrow Y$
8. $(X \rightarrow A) \rightarrow Y$
9. $(X \rightarrow X) \rightarrow (A \rightarrow Z)$
* 10. $(A \rightarrow A) \rightarrow (X \rightarrow X)$
11. $-[(X \rightarrow Y) \rightarrow (A \rightarrow X)]$
12. $-(-X \rightarrow -Y) \rightarrow -(-A \rightarrow -X)$
13. $[(X \rightarrow Y) \vee (A \rightarrow B)] \rightarrow (X \rightarrow C)$
* 14. $(C \rightarrow X) \rightarrow [(X \rightarrow Y) \vee (A \rightarrow B)]$
15. $-\{X \rightarrow [(Y \rightarrow Z) \vee (Z \rightarrow A)]\}$
16. $-\{A \rightarrow [(B \rightarrow C) \vee (B \rightarrow Z)]\}$
17. $-A \rightarrow [(B \rightarrow C) \vee (B \rightarrow Z)]$
* 18. $-X \rightarrow [(Y \rightarrow Z) \vee (Z \rightarrow A)]$
19. $-\{[A \rightarrow (Z \vee X)] \rightarrow [(Z \vee X) \rightarrow A]\}$
20. $-\{-[(-A \rightarrow A) \rightarrow (-Z \vee -X)] \rightarrow -[(-Z \vee -X) \rightarrow -A]\}$

9.8 Translating Conditionals

We turn now to translating conditionals from English to PL. As in Section 9.6, we will give you a couple of hints about translating conditionals and then discuss several examples done by the stepwise method.

Hint No. 1: Always translate "if ○ then □" as ○ → □. Just as many senses are attached to the words 'or' and 'and' in English, so "if ○ then □" has many meanings, or put another way, many types of conditionals exist.

Consider these examples:

1. If Freda is rich and lazy, then she is rich.
2. If Freda is a bachelor, then she is unmarried.
3. If you cut Freda's head off, then she will die.
4. If Freda goes to the store, then she will buy you a kimono.
5. If Freda is a good swimmer, then I'm King of Scotland.

(1) is a logical implication. It asserts that the consequent follows logically from the antecedent. (2) is a definitional implication. It asserts that the consequent is part of the definition of the key terms in the antecedent. (3) is a causal conditional. It asserts that the event mentioned in the consequent will be caused by the event mentioned in the antecedent. We have already discussed (4), promisorial implication. The tie between the antecedent and the consequent is a matter of a promise. (5) is a material implication—there is no tie between the antecedent and the consequent.

How does PL view the difference between these? It does not differentiate among them at all. All are translated as ○ → □, even though something gets lost in translation. As an example, consider:

If you toss this book against the wall, then it will turn into a vampire bat.

We would translate this into PL as $B \to P$. Now suppose that you do not, as a matter of fact, toss this book against the wall. Then B is F. So $B \to P$ is T, by our truth-table definition of the arrow. Yet the English statement "The book is a vampire bat" is clearly F! The moral of the story: PL cannot accurately express causal connections between events. But since the arguments of interest to us do not depend upon causality, PL will serve us well enough.

Hint No. 2: Become conversant with the idioms used to express the conditional. We saw in Section 9.6 that English has numerous words which express conjunction. English likewise has numerous idioms which express the conditional. Table 9.11 lists the most common of such idioms.

Table 9.11

English Phrase		Symbolization
If ○, then □ If ○, □ Given that ○, □ ○ is sufficient condition for □ In case ○, □ ○ only if □ ○ implies □ Assuming that ○, □ On condition that ○, □ Providing that ○, □	Notice that in the phrases that follow, the clause order is reversed: □ if ○ □ provided that ○ □ in case that ○ □ given that ○ □ is a necessary condition for ○ □ assuming that ○ □ on the condition that ○ □ follows from ○	○ → □

A general rule when dealing with any of these idioms is: try to rephrase the sentence to get something of an explicit "if ○ then □" form, because we know how to translate that. The order of the arrow is all important: ○ → □ may have an entirely different truth value from □ → ○, and translations must reflect the proper logical order of the original.

With this general hint in mind, let us turn to some specific idioms.

The term 'if' should cause no trouble. "If ○, □" is clearly rephrasable as "if ○ then □" and translated as ○ → □. And as was just pointed out, "○ if □" is rephrased as "if □, ○" then translated as □ → ○.

The phrase 'only if' causes trouble. Consider an example:

Students will graduate only if they pass first-year English.

Is the claim being made that if you pass first-year English, then automatically you will graduate? If that were true, then nobody would ever need 4 years to get a degree! Instead, the claim here is that if you want to graduate, then (among other things) you have to take first-year English. So "○ only if □" becomes "if ○, then □," which is translated ○ → □.

The translation of 'only if' will be easier if we remember that the word 'only' generally introduces the consequent of the conditional.

Two other important idioms are 'necessary condition' and 'sufficient condition.'

Suppose people tell you that getting a social security number is a *necessary condition* for your getting a job. (They would probably put it this way: "To get a job, you must have a social security number.") What are they claiming? They are not saying that if you get such a number, you will automatically get a job. They mean that before you can get a job, you will have to get a number. Put another way, if you get a job, then for sure you will have gotten a number (or: if you are going to get a job, then you must first have a number). So: "Your getting a number is a necessary condition for your getting a job" becomes "if you (do) get a job, then you get a number." The last clause is bad grammar—we should say "then you will have gotten a number." But PL cannot express tense. So in general, "○ is a necessary condition for □" is translated as □ → ○.

Next, suppose people tell you that it is a *sufficient condition* for you to get a job that you have a social security number. They are saying that all you need to get a job is a number. That is, "if you get a number, you get a job." So: "○ is a sufficient condition for □" is translated as ○ → □.

Two complications should be considered. First, do not confound 'necessary for' with 'necessary that.' Compare:

1. Education is necessary for success.
2. To get an education, it is necessary that you have money.

(1) can be rephrased as:

Having an education is a necessary condition for success.

and is symbolized accordingly. But example (2) can only be rephrased:

Having money is a necessary condition for getting an education.

The best tip here is to determine what the condition is and what the effect or result of that condition is, then fit them into the templates:

○ is a necessary condition for □.

or

○ is a sufficient condition for □.

as appropriate.

A second complication we discussed in Section 9.6 is clause order inversion. Before translating any idiom, first restore clause order. Thus:

Only if □, ○.
A necessary condition for □ is ○.
A sufficient condition for □ is ○.

should be rephrased as:

○ only if □.
○ is a necessary condition for □.
○ is a sufficient condition for □.

Again, do your translation in steps.

Example: Given that you have taken either French or Latin, you may take philosophy and Greek.

    Step 1: A = You take French.
                     B = You take Latin.
                     C = You take philosophy.
                     D = You take Greek.

    Step 2:   [You have taken French or you have taken Latin.]→ [You may take Philosophy and you may take Greek.]

    Step 3: $(A \vee B) \rightarrow (C \ \& \ D)$

Example: Were you to take the Jason tour of beautiful Barstow, you would be educated while being thrilled.

    Step 1: A = You take the Jason tour.
                     B = You are educated.
                     C = You are thrilled.

    Step 2: [You take the tour.] → [You are educated and you are thrilled.]

    Step 3: $A \rightarrow (B \ \& \ C)$

Before turning to the exercises, let us return to an earlier point in Section 9.6. We said that 'unless' and 'without' are best translated with the wedge, but can be translated in another way. The other way involves the arrow. For example:

You will fail unless you study.

symbolized as:

$F \vee S$

can be rephrased as:

If you don't study, you will fail.

symbolized as:

$-S \rightarrow F.$

Again, 'without' can be symbolized with an arrow:

Lavondra won't go without her sister.

translated as:

$-L \vee S$

can be rephrased as:

If Lavondra goes, then her sister will go also.

and symbolized as:

$L \rightarrow S.$

These hints are, as usual, heuristic. The word 'if' does not always signal implication. For instance:

Michelle wants to know if her car is ready.

expresses the nonconditional statement that Michelle wants to know whether her car is ready.

EXERCISES 9.8A

Translate the following, using sentential constants that the capitalized words suggest. (An asterisk indicates problems that are answered in the back of the book.)

1. If the RAMS win, DALLAS will go to the Super Bowl.
* 2. Only if the RAMS win will DALLAS go to the Super Bowl.

3. If the RAMS and the CHARGERS win, then the STEELERS and DALLAS will go to the Super Bowl.

4. The STEELERS and the VIKINGS will go to the Super Bowl, assuming that the CHARGERS win.

5. Unless the CHARGERS win, the STEELERS and the VIKINGS will go to the Super Bowl.

* 6. If the STEELERS go to the Super Bowl, neither the RAIDERS nor DALLAS will go.

7. Unless the RAMS win, if DALLAS wins the PACKERS will go to the Super Bowl.

8. If the RAIDERS win, then only if the JETS win will the BEARS go to the Super Bowl.

9. If neither the PACKERS nor the JETS win, the REDSKINS will fail to go to the Super Bowl.

* 10. Only if the STEELERS fail to win and the JETS win will the CHIEFS go to the Super Bowl.

11. The RAMS will win if and only if the PACKERS win.

12. If the RAIDERS win while DALLAS doesn't win, then the CHIEFS will fail to go to the Super Bowl.

13. The RAIDERS and the STEELERS will both not go to the Super Bowl, if the Raiders WIN.

* 14. If the CHARGERS fail to win, then only if the STEELERS fail to go to the Super Bowl will the RAIDERS go to the Super Bowl.

15. The CHARGERS must win if the STEELERS are to go to the Super Bowl; unless of course, the RAIDERS go.

16. It is neither true that if the RAMS win, then DALLAS will go to the Super Bowl nor that if the CHARGERS win, then the STEELERS will go.

17. Only if you are RICH will you be HAPPY, if you are MATERIALISTIC.

* 18. KELLY will go, given that JONI will not go if SAM goes.

19. Only if only if only if SAM goes will JONI go will KIM go, will ALI stay.

20. If Flukey is RICH only if she wears ALLIGATOR shirts, then assuming that she wears Army BOOTS, she must be POOR.

EXERCISES 9.8B

Translate the following, using sentential constants that the capitalized words suggest. (An asterisk indicates problems that are answered in the back of the book.)

9.8 / Translating Conditionals

1. Acetone being ADDED is a sufficient condition for the mixture to EXPLODE.

* 2. Acetone being ADDED is a necessary condition for the mixture to EXPLODE.

3. Adding HYDROCHLORIC acid is not a sufficient condition for the METAL to dissolve.

4. Adding HYDROCHLORIC acid is neither a necessary nor a sufficient condition for the METAL to dissolve.

5. To get the METAL to dissolve, it is necessary to add SULFURIC acid.

* 6. It is necessary, for the METAL to dissolve, that you add SULFURIC acid.

7. It is sufficient, for the METAL to not corrode, that you keep it away from ACIDS.

8. Being IMMERSED in acid and SUBJECTED to intense radiation is neither necessary nor sufficient to DISSOLVE the metal.

9. If being IMMERSED in acid is a sufficient condition for the METAL to dissolve, then adding ACETONE is necessary for the mixture to EXPLODE.

* 10. It isn't necessary for CRYSTALS to form that ACID be added.

11. If adding ACID isn't sufficient for the mixture to EXPLODE, then it isn't necessary, either.

12. Only if it is necessary to add ACID to get the METAL to dissolve will we add acid.

13. It is neither necessary nor sufficient for the MIXTURE to explode that we fail to add ACETONE.

* 14. It is false that not failing to add ACETONE is not sufficient for the MIXTURE to fail to explode.

15. If it is not either a necessary or sufficient condition for the mixture to EXPLODE that ACETONE be added, then we ought to add SULFURIC acid.

EXERCISES 9.8C

The following sentences are taken from Durk Pearson and Sandy Shaw, Life Extension. *Translate each using capital letters that the capitalized words suggest. (An asterisk indicates problems that are answered in the back of the book.)*

1. If people LIVED a lot longer, we would have a severe POPULATION explosion.

* 2. There would be an INCREASE in population only if people REPRODUCED at a higher rate than the reduced death rate.

3. This SCENARIO would be a serious problem if LIFE extension prolonged mainly the old-age part of life.

4. Life extension methods described in this book EXTEND the vigorous productive part of life and can improve MENTAL as well as PHYSICAL functions.

5. To INCREASE our life expectancy, it is not necessary to solve completely all aging systems PROBLEMS at once.

* 6. Neither POWER nor RICHES nor GENETIC luck are necessary in an effective LIFE extension plan.

7. If a person LIVES long enough, he or she will USE up all the brain's memory storage capacity and be unable to ABSORB any new experiences.

8. Memory may be improved by NUTRIENTS and OTHER chemicals, and there are also a VARIETY of creativity-increasing substances that have been developed but these are not YET widely available . . .

9. . . . if patients in pain are given a SUGAR pill and TOLD it is morphine, 40 percent of those will experience RELIEF . . .

* 10. A knowledge of STATISTICS is definitely not necessary to the UNDERSTANDING of the material in this book.

11. Some FAMILIARITY with statistics is required to INTERPRET original scientific papers.

12. If you have used BARBITURATES for years, do not suddenly DISCONTINUE their use, or death by CONVULSIONS can occur.

13. PHENYLALANINE can significantly reduce the requirement for sleep, but watch out for the development of INSOMNIA if you use too MUCH phenylalanine.

* 14. Neither DURK nor SANDY have high blood pressure.

15. If you wish to USE the treatment, you will either have to be a SCIENTIST or have a PRESCRIPTION from your doctor.

16. Unfortunately, despite its very low TOXICITY and its reported frequent partial EFFECTIVENESS against balding, the companies offering this chemical cannot SELL it for this purpose.

17. If you eat raw PINEAPPLE or papaya, watch out for SORE corners of the mouth. (A = You eat papaya.)

* 18. If you have essential HYPERTENSION or are on a LOW sodium diet, use POWDERED limestone, DOLOMITE, or some OTHER low sodium antacid instead of BAKING soda . . .

19. If you feel you cannot get to SLEEP without a sleeping PILL (assuming you cannot find TRYPTOPHAN), you might want to try taking 1/2 the normal nightly sleeping pill DOSE in conjunction with swallowing one to three 1-milligram tablets of HYDERGINE . . .

20. It is not necessary for all aging mechanisms to be IDENTIFIED and biochemically UNDERSTOOD in order to GET useful information from animal experiments.

9.9 The Biconditional

To assert a **biconditional** is to say that if ○, then □, and furthermore that if □, then ○. For instance, a committee might tell a graduate student who has completed all the requirements for the Ph.D. except the dissertation: "You will get your degree if and only if you submit an acceptable dissertation." One way is to translate it as: (A → B) & (B → A). But let us use a new connective, the double arrow (↔), to abbreviate that conjunction of conditionals. We define it accordingly in Table 9.12.

Table 9.12 captures the notion of biconditionality, as well as the notion of "having the same truth value," the notion logicians call *material equivalence*. This is because the double-arrow compound is T when the components have the same truth value, and it is F when the components have different truth values.

This suggests a test. If the double arrow expresses material equivalence (having the same truth value), and only two truth values (T and F) are permissible, then the double arrow should have a table the same as that for (A & B) v (−A & −B). This is because that disjunction says: "Either both components are true or both are false." And indeed it has in Table 9.13.

That we can introduce the double arrow as a defined symbol brings up a question. How many symbols do we need to introduce primitively (that is by means of a truth table) to give us enough connectives to define all other possible connectives? The surprising answer is one.

Calculating truth values is done just as before, and it is easy once the table is memorized. Translation is likewise not difficult once you are familiar with the idioms listed in Table 9.14.

Table 9.12

A	B	(A → B) & (B → A)	Definition A ↔ B
T	T	T T T T T T T	T
T	F	T F F F F T T	F
F	T	F T T F T F F	F
F	F	F T F T F T F	T

Table 9.13

A	B	A ↔ B	(A & B)	v	(−A & −B)
T	T	T	T T T	T	F T F F T
T	F	F	T F F	F	F T F T F
F	T	F	F F T	F	T F F F T
F	F	T	F F F	T	T F T T F

Table 9.14

English Phrase	Symbolization
○ if and only if □ ○ when and only when □ ○ exactly when □ ○ just in case □ If ○, and only if ○, □ ○ is a necessary and sufficient condition for □	○ ↔ □

The most common idiom expressing biconditionality is 'if and only if.' But people often say 'only if' when they mean 'if and only if.' For example, a father might say to his child, "You can go out to play only if you clean your room." Strictly speaking, he has laid down only one of possibly many conditions the child must meet to be allowed to go out to play. But the father probably would mean (and the child would expect) that if the child cleaned her room, she would be allowed out.

A similar point holds with respect to 'if.' Some people say 'if' when they mean 'if and only if.' For instance, you might say to a mechanic, "Replace the transmission if it is bad." Logically speaking, you have not ruled out replacing the transmission if it is not bad. But again, you would mean (and the mechanic would assume automatically) that the transmission is to be replaced if and only if it is bad.

EXERCISES 9.9A

Compute the truth values of the following biconditionals, given that A, B, *and* C *are all* T, *while* X, Y, *and* Z *are all* F. *(An asterisk indicates problems that are answered in the back of the book.)*

9.9 / The Biconditional

1. $A \leftrightarrow A$
* 2. $A \leftrightarrow B$
3. $A \leftrightarrow X$
4. $(A \& X) \leftrightarrow (B \& Y)$
5. $(A \vee X) \leftrightarrow (B \vee Y)$
* 6. $(X \vee A) \leftrightarrow (Y \vee B)$
7. $(X \& A) \leftrightarrow (Y \& B)$
8. $-(X \vee A) \leftrightarrow (-X \vee A)$
9. $-(A \& B) \leftrightarrow (-A \& -B)$
* 10. $(--A \vee -B) \leftrightarrow -(A \& B)$
11. $-(X \& Y) \leftrightarrow (-X \& -Z)$
12. $(A \leftrightarrow X) \leftrightarrow (X \& Y)$
13. $(X \leftrightarrow Y) \leftrightarrow (A \& B)$
* 14. $(-X \leftrightarrow Y) \leftrightarrow (-C \vee -Z)$
15. $[(X \to Y) \to Z] \leftrightarrow (-Z \vee -Z)$
16. $[A \to (Y \to Z)] \leftrightarrow (-A \& -A)$
17. $[X \to (Y \to X)] \leftrightarrow [X \to (X \to Y)]$
* 18. $[A \to (Y \to X)] \leftrightarrow [A \to (A \to Y)]$
19. $[(X \& A) \to (Y \to A)] \leftrightarrow [(X \& Z) \to (A \to B)]$
20. $[(A \& B) \to (B \to C)] \leftrightarrow [(X \& A) \leftrightarrow (A \to B)]$
21. $[-(A \& B) \to -(B \to C)] \leftrightarrow [(X \& A) \leftrightarrow (-A \to Z)]$
* 22. $[-(A \& B) \leftrightarrow (-B \leftrightarrow C)] \leftrightarrow [(X \& A) \leftrightarrow (A \to B)]$
23. $-[(A \& B) \leftrightarrow (B \leftrightarrow -C)] \leftrightarrow [(X \to A) \leftrightarrow (A \to B)]$
24. $-[(A \& -B) \leftrightarrow -(B \vee -C)] \leftrightarrow [(X \to A) \leftrightarrow (A \to B)]$
25. $\{-[A \leftrightarrow (B \leftrightarrow Z)] \leftrightarrow [(-Z \& C) \leftrightarrow (X \& Y)]\} \leftrightarrow [--A \leftrightarrow (--B \leftrightarrow --Z)]$

EXERCISES 9.9B

Symbolize the following sentences, using letters suggested by capitalized words. (An asterisk indicates problems that are answered in the back of the book.)

1. You will be RICH if and only if you are TALENTED.
* 2. You will be neither RICH nor HAPPY if and only if you are CROOKED.
3. Being FRIENDLY is a necessary and sufficient condition for being LIKED.

4. INTELLIGENCE is neither a necessary nor a sufficient condition for WEALTH.

5. She gets MAD just when WOMBATS are discussed.

* 6. The report is PRINTED exactly when the operator pushes the red BUTTON.

7. A necessary and sufficient condition for you to not fail the course is that you not fail to STUDY. (P = You pass the course.)

8. It is not true that you will PASS the class if and only if you BRIBE the teacher.

9. A necessary and sufficient condition for SUCCESS is that you neither LIE, CHEAT, MURDER, nor listen to BARRY Manilow.

* 10. It will RAIN when and only when the BAROMETER falls, supposing that the barometer WORKS correctly.

9.10 The Assessment of Arguments

We have constructed PL. Its vocabulary consists of the sentential constants, connectives, and brackets. But a language consists of more than words—we need grammatical rules for putting the words together. The rules of "construction" for PL are easy to state and are probably clear already.

1. Any sentential constant is a grammatical sentence of PL.
2. If p is a grammatical PL sentence, then so is $-p$.
3. If p and q are grammatical PL sentences, then so are $p \rightarrow q$, $p \lor q$, and $p \ \& \ q$.
4. No other expressions are sentences of PL.

We have also learned how to calculate the truth values of compounds and how to translate compounds from English to PL. Having mastered these skills, we want to combine them in a method for determining validity, the method of truth tables.

This method grows directly out of the definition of validity, plus two facts: (1) no argument can be infinitely long, and (2) any statement (simple or compound) is one of two things—T or F. Both those facts are open to dispute, and alternative logics have been devised that do not involve assuming them. Such logics do not concern us here, and they can only be understood after we have mastered the classical logic we are discussing.

9.10 / The Assessment of Arguments

Recall the definition of *validity*: an argument is valid if it is impossible for the premises to be T and the conclusion F.

Proving that something is impossible is generally hard to do. To prove it is possible to run a 3-minute mile, all you have to do is point to one person who has done it. To prove it impossible, however, you cannot simply point to all the people who have tried but failed to run the mile in 3 minutes, because somebody may succeed in doing it a century from now.

Similarly, to prove an argument is invalid—to prove that it is possible for the premises to be T and the conclusion F—you need only point out a case in which that happens. (By *case* we mean an assignment of truth values to constants.) But how can we prove validity?

Because the number of statements in an argument is finite and each component is either T or F, the number of cases, the number of possible assignments of truth values, is finite. Thus we can list them and check things mechanically.

A truth table of an argument has three parts, as in Table 9.15.

After listing the possible assignments of truth values to components, we compute the truth values of the premises and conclusion, and then we see whether a case occurs where the premises turn out to be T and the conclusion F.

Let us work through some examples.

Example 1: Assess this argument.

> If Amos is in favor of war as a solution, then he is not in favor of a diplomatic solution. Either he is in favor of war as a solution or he is not in favor of a diplomatic solution. So he doesn't favor a diplomatic solution.

One of the nice things about PL is that it reveals the logical structure of compounds quite clearly. The PL translation of this argument is:

1. $W \rightarrow -D$
2. $W \vee -D$

∴ $-D$

Table 9.15

List of possibilities	Computations of the truth values of the premises	Computations of the truth values of the conclusion

Then we set up the table. What letters are used as the building blocks of this argument? Just *W* and *D*. So we have to list all possible truth-value assignments to those two letters.

W	D
T	T
T	F
F	T
F	F

Then we make one column for each of the premises, and one for the conclusion. We separate basic letters, the premises and the conclusion, each by double lines to avoid confusion.

W	D	W → −D	W v −D	−D
T	T			
T	F			
F	T			
F	F			

Now we compute the truth values of the premises and conclusion under each assignment. If we ever see the premises all being T while the conclusion is F, we can immediately conclude that the argument is invalid. If we check every possibility and discover that in none of them are the premises T and the conclusion F, then we know that the argument is valid.

W	D	W → −D	W v −D	−D
T	T	T F F T	T T F T	F T
T	F	T T T F	T T T F	T F
F	T	F T F T	F F F T	F T
F	F	F T T F	F T T F	T F

We conclude that the argument is valid, since we have checked all possibilities and found it impossible for the premises to be T and the conclusion F.

Example 2:   Assess the (already symbolized) argument:

1. *A* v *B*
2. −*A* → *B*

∴ *B*

Since only two building blocks (*A* and *B*) are here, only four possibilities occur: both T, both F, or one T and the other F.

A	B	¬A → B	A v B	B
T	T	F T T	T T T	T
T	F	F T F	T T F	F
F	T			
F	F			

We do not need to finish the table, since in the second row (case) the premises are all T and the conclusion F. This proves that the argument is invalid.

The method of truth tables is both a method of discovery of validity or invalidity (a 'decision procedure') and a method of proving what has been discovered (a 'proof procedure').

Consider another example, this time involving three blocks. In such arguments there are three independent letters, each able to be either T or F, so there are $2 \times 2 \times 2 = 8$ possibilities. With n letters, there are 2^n possibilities.

Example 3:

1. $X \to Y$
2. $A \to X$

$\therefore A \to Y$

A	X	Y	X → Y	A → X	A → Y
T	T	T	T T T	T T T	T T T
T	T	F	T F F	T T T	T F F
T	F	T	F T T	T F F	T T T
T	F	F	F T F	T F F	T F F
F	T	T	T T T	F T T	F T T
F	T	F	T F F	F T T	F T F
F	F	T	F T T	F T F	F T T
F	F	F	F T F	F T F	F T F

An easy way to generate all the possible cases is to first figure out how many rows will occur (2 to the nth power for n letters). Then under the first letter, write half that number of Ts, followed by half Fs. Next, under the second letter, alternate blocks of Ts and Fs in groups of half the previous number. And so on. Thus the list of possibilities for an argument whose premises and conclusion are built up out of five letters (say, *A, B, C, D,* and *E*) will have $2^5 = 32$ rows, and the possibility can be listed as in Table 9.16.

Table 9.16

A	B	C	D	E
16 Ts	8 Ts	4 Ts	2 Ts	T
				F
			2 Fs	T
				F
			2 Ts	T
		4 Fs		F
			2 Fs	T
				F
	8 Fs	4 Ts	2 Ts	T
				F
			2 Fs	T
				F
		4 Fs	2 Ts	T
				F
			2 Fs	T
				F
16 Fs	8 Ts	4 Ts	2 Ts	T
				F
			2 Fs	T
				F
		4 Fs	2 Ts	T
				F
			2 Fs	T
				F
	8 Fs	4 Ts	2 Ts	T
				F
			2 Fs	T
				F
		4 Fs	2 Ts	T
				F
			2 Fs	T
				F

EXERCISES 9.10A

Use the truth tables to determine the validity or invalidity of the following arguments (already translated into PL). (An asterisk indicates problems that are answered in the back of the book.)

1.
    1. $A \to B$
    2. A
    $\therefore B$

* 2.
    1. $A \to B$
    2. $-A$
    $\therefore -B$

3.
    1. $A \to B$
    2. $-B$
    $\therefore -A$

4.
    1. $X \vee Y$
    2. $-X$
    $\therefore Y$

5.
    1. $X \vee Y$
    2. X
    $\therefore -Y$

* 6.
    1. $M \to N$
    $\therefore M \to (M \& N)$

7.
    1. $M \to N$
    $\therefore M \to (M \vee N)$

8.
    1. $M \to N$
    $\therefore M \to (N \to M)$

9.
    1. $X \to Y$
    2. $Y \to Z$
    $\therefore X \to Z$

* 10.
    1. $X \to Y$
    2. $Y \to Z$
    $\therefore Z \to X$

11.
    1. $X \to (Y \vee Z)$
    2. $-Y \& X$
    $\therefore Z$

12.
    1. $X \to (Y \& Z)$
    2. $-Y \vee X$
    $\therefore Z$

13.
    1. $E \vee F$
    2. $F \to E$
    $\therefore --E$

* 14.
    1. $-X \vee M$
    2. $-M \vee N$
    $\therefore -N \to -X$

15.
    1. $X \to Y$
    2. $-Y$
    3. $-X \to Z$
    $\therefore Z$

16.
    1. $A \to X$
    2. $A \to -X$
    3. A
    $\therefore Q$

17.
    1. $(A \to B) \& (C \to D)$
    2. $A \vee C$
    $\therefore B \vee D$

* 18. 1. $(A \to B)\ \&\ (C \to D)$
    2. $-B \lor -D$
    ∴ $-A \lor -C$

20. 1. X
    ∴ $X \lor Z$

19. 1. $(A \to B)\ \&\ (C \to D)$
    2. $B \lor D$
    ∴ $A \lor C$

EXERCISES 9.10B

Translate each argument into PL, then do a truth table to determine whether it is valid. (An asterisk indicates problems that are answered in the back of the book.)

    1. Life is tough. But if life is tough, Arthur will shoot himself. So Arthur will shoot himself.

* 2. If the experience didn't destroy Friedrich, then it made him strong. But Friedrich is weak. So the experience didn't destroy him.

    3. Either he's on drugs or he's incredibly stupid. If he's on drugs he's incredibly stupid. So he's incredibly stupid.

    4. If I graduate, Daddy will buy me either a Porsche or a Jaguar. But I'm not going to graduate. So I'm getting neither.

    5. If we spend that much money to see that movie, then we're stupid. But we're not stupid. So we're going to spend that much to see that movie.

* 6. If the United States doesn't stop OPEC from raising prices, then the recession won't end. But if the recession continues, then Mexico will be in grave danger. Therefore, if Mexico is not in danger, then America stopped OPEC from raising prices.

    7. Either Clinton or Bush will win the election. But the people haven't forgotten the recession, and only if people have forgotten the recession will Bush win. So Clinton will win.

    8. If rats don't read comics, then they're smarter than undergrad students. If rats are smarter than undergrad students, then the rats should be in grad school. But if rats should be in grad school, then professors should carry more rat-food in their briefcases. So if rats don't read comics, professors should carry more rat-food in their briefcases.

    9. If the Dolphins lose, then the Cowboys and the Raiders will go to the Super Bowl. But if either the Raiders or the Cowboys go to the Super Bowl, then I'll be sure to watch it on TV. So if the Dolphins lose, then I'll be sure to watch the Super Bowl on TV.

* 10. If the Dolphins lose, then neither the Cowboys nor the Raiders will go to the Super Bowl. But if both the Raiders and the Cowboys go to the Super Bowl, then I will be sure to watch it on TV. So if the Dolphins lose, then I'll be sure to watch the Super Bowl on TV.

9.11 Tautologies and the Nature of Validity

We now have an apparatus that does two of the three jobs required of any theory of deduction; it provides a decision procedure and a proof procedure. We take any argument, translate it into PL, set up the table, grind out the answer—and prove it at the same time! But what about the third job any theory of deduction must do, namely, explain what validity is? To this task we now turn.

To explain validity, we need to introduce some concepts. Compare these two statements:

1. Clinton won the 1992 election.
2. Either Clinton won the 1992 election, or he didn't.

(1) is T, but it might have turned out differently—for instance, had his opponent run a more effective campaign. Statements whose truth values depend upon the way the world happens to be, we call **contingent**.

On the other hand, while (2) is T, it could not possibly have turned out otherwise; even if Clinton had lost, (2) would still be T. A statement that cannot possibly be F we call a **tautology**. A tautology is utterly uninformative. Imagine hearing a news bulletin on the night of an election, "News bulletin just in—either Clinton will win or he won't!"

We can use our truth table technique to determine whether a statement is a tautology. We simply check to see if the statement in question is T under all possible assignments of truth values to constants. Thus, whereas a truth table for an argument consists of several columns, one for each premise and the conclusion, a truth table for a statement has just one column for computations.

Look at our example again. "Either Clinton will win or he will not" is translated C v $-C$. A truth table for this example is given in Table 9.17.

Since nothing but Ts are under the main connective, the statement is a tautology. A more complicated example: Is $(M \rightarrow N)$ v

Table 9.17

C	C v −C
T	T T F T
F	F T T F

($M \to -N$) a tautology? The answer is found by constructing a table (Table 9.18) and seeing whether under the main connective there are all Ts.

The statement is indeed a tautology. Had at least one F come under the main connective, it would have been contingent.

What about a statement like "Clinton will win and he will also lose," where the statement is not intended as a metaphor. That statement cannot possibly be T. We call such statements 'self-contradictions.' We can detect a contradiction by looking at the table—if it is a contradiction, there are nothing but Fs under the main connective. Our example translates into $C \mathbin{\&} -C$ and the table is:

C	C & −C
T	T F F T
F	F F T F

Sure enough, only Fs will be found under the main connective.

In summary, if nothing but Ts occur under the main connective, the statement is a tautology. If nothing but Fs occur, it is a contradiction. If at least one T and at least one F occur, it is contingent.

The negation of a contradiction will have only Ts under the main connective, so it will be a tautology. Similarly, the negation of a tautology is a contradiction. The negation of a contingent statement will still be contingent. Can you see why that is so?

Table 9.18

M	N	($M \to N$) v ($M \to -N$)
T	T	T T T T T F F T
T	F	T F F T T T T F
F	T	F T T T F T F T
F	F	F T F T F T F F

9.11 / Tautologies and the Nature of Validity

EXERCISES 9.11A

For each of the following determine whether it is a tautology, a contradiction, or contingent. (An asterisk indicates problems that are answered in the back of the book.)

1. $A \to A$
* 2. $A \to B$
3. $(A \to A) \vee B$
4. $A \to (A \vee B)$
5. $A \to (A \& B)$
* 6. $(A \& B) \to A$
7. $(A \& B) \to (A \vee B)$
8. $(A \to B) \& (B \to A)$
9. $(A \to B) \vee (B \to A)$
* 10. $(A \to B) \to (B \to A)$
11. $(A \& B) \vee (A \& -B)$
12. $[A \to (B \& C)] \to (A \to B)$
13. $A \to [(A \vee B) \vee C]$
* 14. $[A \& (B \vee C)] \to [(A \& B) \vee (A \& C)]$
15. $[(A \& B) \vee (A \& C)] \to [A \& (B \vee C)]$

We need to introduce another concept. Recall that the double arrow expresses the notion of having the same truth value materially, that is, in fact. Now compare these two pairs of sentences:

1. "Clinton was president in 1993." vs. "Gore was vice-president in 1993."
2. "Clinton was president in 1993." vs. "It's not true that Clinton was not president in 1993."

In pair (1), the two statements happen to have the same truth value, but things could have been different: Clinton might have chosen a different running mate. Again, we call that **material equivalence**.

In pair (2), the two statements must have the same truth value; they couldn't possibly have different values. This is more than material equivalence; it is **logical equivalence**. If ○ and □ are logically equivalent, it is impossible for them to have different truth values.

Thus we can prove two statements are equivalent by showing they have the same truth values in all cases.

An even better way to prove logical equivalence between two statements is to connect them with a double arrow, and then do one single truth table on the resultant biconditional to see if it is a tautology. After all, if ○ and □ have the same truth values in all possible cases, then the biconditional formed between them will always be T.

Consider our earlier example in this section. Are the statements C and $--C$ logically equivalent? Just connect them by a double arrow and see if you get nothing but Ts under the main connective:

C	$C \leftrightarrow --C$
T	T T T F T
F	F T F T F

How about the pair $-A \,\&\, -B$ and $-(A \vee B)$? (They better be equivalent—they are the two ways we indicated earlier translated "neither A nor B.")

A	B	$(-A \,\&\, -B) \leftrightarrow -(A \vee B)$
T	T	F T F F T T F T T T
T	F	F T F T F T F T T F
F	T	T F F F T T F F T T
F	F	T F T T F T T F F F

We will employ the symbol ⇔ to indicate logical as opposed to material equivalence. That is, the logical equivalence:

○ ⇔ □

holds just in case

○ ↔ □

is a tautology.

As logical equivalence should not be confounded with material equivalence, so one should not confound logical equivalence with synonymy (sameness of meaning). Two expressions can be logically equivalent without being synonymous.

For example:

1. Alfred is friendly.
2. Alfred is friendly, or else Billie Mae is aggressive and not aggressive.

(1) and (2) certainly do not mean the same thing, for (2) talks about Billy Mae and (1) does not. But they are logically equivalent (as one can verify by constructing the appropriate table).

On the other hand, if two expressions are synonymous, then they must be logically equivalent. Can you see why that claim is correct?

Similarly, two expressions may be materially equivalent (that is, may happen to have the same truth value) without being logically equivalent, but if two expressions are logically equivalent, they must be materially equivalent. Table 9.19 summarizes these points.

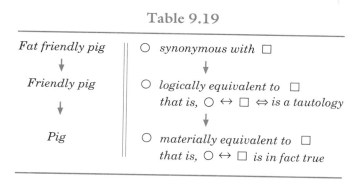

Table 9.19

EXERCISES 9.11B

Determine which of the following biconditionals are tautologies (thus determining whether the expressions flanking the double arrows are logically equivalent). (An asterisk indicates problems that are answered in the back of the book.)

1. $A \leftrightarrow (A \& A)$
* 2. $A \leftrightarrow (A \vee A)$
3. $(A \& B) \leftrightarrow (B \& A)$
4. $(-A \vee B) \leftrightarrow (A \rightarrow B)$
5. $(A \vee B) \leftrightarrow (B \vee A)$
* 6. $(A \rightarrow B) \leftrightarrow (B \rightarrow A)$

7. $-(A \mathbin{\&} B) \leftrightarrow (-A \lor -B)$
8. $[A \lor (B \lor C)] \leftrightarrow [(A \lor B) \lor C]$
9. $[A \mathbin{\&} (B \mathbin{\&} C)] \leftrightarrow [(A \mathbin{\&} B) \mathbin{\&} C]$
* 10. $[A \leftrightarrow B] \leftrightarrow [(A \mathbin{\&} B) \lor (-A \mathbin{\&} -B)]$
11. $(A \to B) \leftrightarrow (-B \to -A)$
12. $(A \to B) \leftrightarrow (-A \to -B)$
13. $[A \to (B \mathbin{\&} C)] \leftrightarrow [(A \to B) \to C]$
* 14. $[A \to (B \mathbin{\&} C)] \leftrightarrow [A \to (B \to C)]$
15. $[(A \mathbin{\&} B) \to C] \leftrightarrow [A \to (B \to C)]$
16. $[(A \mathbin{\&} B) \to C] \leftrightarrow [(A \to B) \to C]$
17. $[A \leftrightarrow B] \leftrightarrow [-(A \mathbin{\&} B) \lor -(-A \mathbin{\&} -B)]$
* 18. $[(A \leftrightarrow B) \leftrightarrow C] \leftrightarrow [A \leftrightarrow (B \leftrightarrow C)]$
19. $[A \mathbin{\&} (B \leftrightarrow C)] \leftrightarrow [(A \mathbin{\&} B) \leftrightarrow (A \mathbin{\&} C)]$
20. $[A \leftrightarrow B] \leftrightarrow [(A \to B) \mathbin{\&} (B \to A)]$

9.12 Validity Explained

We are now in position to fit in the last piece of the puzzle. Look at the one-premise argument:

1. P
∴ C

and consider the conditional formed by taking the premise as antecedent and the conclusion as consequent: $P \to C$. Clearly, if the argument is valid, then it is impossible for P to be T and C to be F. But since that is the only case where $P \to C$ is F, it follows that if the argument is valid, then the associated conditional is a tautology. By *associated conditional* we mean the conditional formed by taking the conjunction of the premises as antecedent and the conclusion as consequent.

The point can be generalized to arguments of any number of premises.

Consider:

1. P_1
2. P_2
∴ C

and the associated conditional $(P_1 \,\&\, P_2) \to C$. If the argument is valid, it is impossible for P_1 and P_2 to be T while C is F. But that would be the only case where the associated conditional would be F; hence the associated conditional is a tautology.

So the notion of validity is reducible to the notion of a tautology. But tautologies are empty. This leads to the explanation of what validity really is: in a valid argument, the conclusion asserts nothing that is not already contained in the premises. A valid argument is simply the unpacking of information contained in the premises.

In the same way that we distinguished between material and logical equivalence, so we must distinguish between material and logical implication:

$$A \Rightarrow B$$

will be taken to mean that

$$A \to B$$

is a tautology. We can again use truth tables to decide the matter.
Example: Show $(X \,\&\, Y) \Rightarrow X$

X	Y	$(X \,\&\, Y) \to X$
T	T	T T T T T
T	F	T F F T T
F	T	F F T T F
F	F	F F F T F

EXERCISES 9.12

Show the following logical implications. (An asterisk indicates problems that are answered in the back of the book.)

1. $X \Rightarrow (X \vee Y)$
* 2. $(X \,\&\, Y) \Rightarrow (Y \,\&\, X)$
3. $(X \vee Y) \Rightarrow (Y \vee X)$
4. $[(X \to (Y \,\&\, X)] \Rightarrow Y$
5. $[X \to (Y \to Z)] \Rightarrow [(X \,\&\, Y) \to Z]$
* 6. $[(X \,\&\, Y) \,\&\, Z] \Rightarrow [Z \,\&\, (Y \,\&\, Z)]$

7. $[A \mathbin{\&} (B \mathbin{\&} C)] \Rightarrow C$
8. $[-A \mathbin{\&} (B \mathbin{v} A)] \Rightarrow B$
9. $[-B \mathbin{\&} (A \rightarrow B)] \Rightarrow -A$

9.13 Variants of Propositional Logic

In this section, we examine other ways of setting up propositional logic and along the way gain additional insight into why translation from English to PL can be tricky.

To begin, other versions of PL allow connectives different from those we have. For example, we could have introduced a connective $\otimes$ to express *exclusive-or*, defined in the following table:

p	q	$p \otimes q$
T	T	F
T	F	T
F	T	T
F	F	F

Such a symbol would allow for easier translation of exclusive-or statements. For example:

You may have either soup or salad, but not both.

which, with our version of PL, is symbolized:

$(S \mathbin{v} L) \mathbin{\&} -(S \mathbin{\&} L)$

could be symbolized more directly as:

$S \otimes L$

Again, some versions of PL allow a connective $\downarrow$ to symbolize *neither-nor*, defined as:

p	q	$p \downarrow q$
T	T	F
T	F	F
F	T	F
F	F	T

9.13 / Variants of Propositional Logic

Such a connective allows for a more direct translation of neither-nor statements. For instance, the statement:

You will receive neither a Porsche nor a Jaguar for Christmas.

symbolized in our version of PL as:

$-P \& -J$

or as:

$-(P \lor J)$

could be symbolized more briefly as:

$P \downarrow J$

Like the dagger function defined previously is the Sheffer stroke |, which expresses the notion of disjunctive negation, as in:

He will either not go to the dance or not go to the store.

The table for the Sheffer stroke is:

p	q	$p \mid q$
T	T	F
T	F	T
F	T	T
F	F	T

Whereas in our version of PL we would symbolize the example as:

$-D \lor -S$

we can symbolize more economically with the stroke as:

$D \mid S$

Finally, some versions of PL allow a symbol to express clauses of the form:

p, if q

(called Horn clauses). We can use a backward arrow defined as follows.

p	q	$p \leftarrow q$
T	T	T
T	F	T
F	T	F

Again, having such a connective would simplify some symbolizations. For example:

Bindhu will fail if he doesn't study.

would be symbolized:

$F \leftarrow -S$

rather than as in our version of PL:

$-S \rightarrow F$

(in which the order of the clauses in the PL sentence is different from that in the English). The same holds for:

Taking aspirin is a necessary condition for getting rid of a headache.

which can be symbolized using the Horn symbol as:

$A \leftarrow H$

rather than:

$H \rightarrow A$

as in our version of PL.

You might ask why we didn't just introduce the previous connectives and still others for all the tricky idioms we meet in English. The answer is that to introduce more connectives into PL makes it easier to confound them in practice and much harder to remember their table definitions. A trade-off occurs here: the fewer the connectives, the easier it is to keep them in mind, but the more difficult symbolization becomes.

EXERCISES 9.13A

Assume that A, B, and C are T, while X, Y, and Z are F. Determine the truth values of the following components. (An asterisk indicates problems that are answered in the back of the book.)

1. $A \downarrow X$
* 2. $-(X \otimes X)$
3. $-X \mid -X$
4. $-(-A \downarrow -X)$
5. $A \downarrow (X \mid X)$
* 6. $(A \rightarrow X) \& (A \leftarrow X)$
7. $-(X \leftarrow Y) \otimes Z$
8. $X \vee (A \otimes A)$
9. $-[A \otimes (B \leftarrow X)]$
* 10. $(A \downarrow A) \downarrow (X \downarrow X)$
11. $A \leftarrow -(A \leftrightarrow B)$
12. $-A \mid -(A \downarrow B)$
13. $-A \leftarrow (B \mid -B)$
* 14. $--(A \leftrightarrow B) \mid B$
15. $B \downarrow (B \downarrow (X \downarrow X))$
16. $A \leftarrow (B \downarrow C)$
17. $X \leftrightarrow (B \downarrow -A)$
* 18. $-[(A \mid B) \downarrow (C \mid X)]$
19. $(-A \downarrow -B) \mid (X \leftarrow Y)$
20. $-[X \downarrow (-X \mid -A)] \vee A$

EXERCISES 9.13B

Translate the following using $\leftarrow$, $\downarrow$, $\mid$, and $\otimes$ where appropriate. Select letters suggested by capitalized words. (An asterisk indicates problems that are answered in the back of the book.)

1. Neither FRED nor YOSHI will fail to attend.
* 2. She will DANCE or SING, but not both; unless, of course, she is PAID double.
3. He hates CHILDREN if they don't BEHAVE.
4. Ted will neither RUN nor HIDE if he can AVOID it.
5. Either SUE or ANGELICA won't go.
* 6. Neither WEALTH nor POWER are necessary for HAPPINESS.
7. She is either not BRIGHT or not AWARE if she is not dating FRED.
8. Neither MARIE nor ANGELICA will go if GARY does.
9. Getting a DEGREE is neither necessary nor sufficient for SUCCESS.
* 10. Neither FRED nor TED nor ED will go.

We have looked at the variety of connectives available in versions of PL. Let us turn now to methods of punctuation. Our version of PL uses parentheses for punctuation, but other versions employ dots for that purpose. For instance, the sentence:

$A \& (B \vee C)$

would be rendered:

A & . *B* v *C*

where the dot operates away from the main connectives to indicate grouping. Similarly, the sentence:

(*A* & *B*) v (*C* & *D*)

would be dot punctuated as:

A & *B* .v. *C* & *D*

Surprisingly, we can formulate PL in such a way to dispense with punctuation symbols altogether and to rely on position alone to disambiguate sentences. A sentence is in *prefix form* when the connective for one or more components appears in front of that or those components. That is, instead of writing:

A & *B*

we would write:

& *A B*

We can distinguish between:

A & (*B* v *C*)

and

(*A* & *B*) v *C*

without parentheses by where we place the connectives:

& *A* v *B C*

vs.

v & *A B C*

In prefix sentences, the main connective always occurs at the beginning.

Now, why did we develop our version of PL to be *infix* (that is, where binary connectives occur between rather than in front of the components) instead of prefix in structure? We did so because English, our metalanguage, has sentences that are usually infix in form, and also because the mathematics we learned in school has infix expressions (such as 2 + 2) rather than prefix ones (+ 2 2).

However, that our version of PL is purely infix in orientation explains another aspect of the difficulty we experience in symbolizing English. English, while primarily infix in orientation, is not purely infix. That is, while most compound sentences in English are like:

> José likes cheese, although he does not like pizza.

many have prefix form such as:

> Unless José likes cheese, he won't like pizza.

This inconsistency in English grammatical structure leads, as we saw, to confusion in symbolization, especially when combined with idioms for which our version of PL has no immediate symbol. For example, to symbolize:

> A necessary condition for happiness is wealth.

in our version of PL, we must first restore infix clause order:

> Wealth is a necessary condition for happiness.

and then, because we do not have the Horn symbol ← to express the idiom "○ is a necessary condition for □" directly as ○ ← □, we must change the clause order again:

> $H \rightarrow W$

Why not allow prefix as well as infix expression in PL? That is, why not allow such expressions as:

> (A & B) & v BC

Again, the answer is that we want to keep our system as simple as possible to prevent confusion. To either adopt a purely infix or a purely prefix approach to PL is more elegant, and given the familiarity of infix structure, we chose that approach. So constructing a logic can often be a matter of choice.

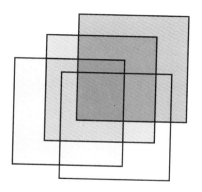

10

Natural Deduction

10.1 Drawbacks of the Method of Truth Tables

In Chapter 9, we learned a method for discovering and proving validity in arguments. But the truth-table method has two major drawbacks. First, it is very cumbersome. The clearly valid argument:

$A \to B$
$B \to C$
$C \to D$
$D \to E$
$E \to F$
$F \to G$
$G \to H$
$H \to I$
$I \to J$
$J \to K$
───────
$A \to K$

would require a table of 2,048 rows! We can do such tables by computer but not by hand.

Worse, the truth-table method cannot be generalized much beyond propositional logic (PL). So we are better off devising a method that mirrors the thought processes we naturally use to determine validity. We call this new method accordingly, *the method of natural deduction.*

The basic idea is this. Suppose we know ahead of time that an argument is indeed valid. Then we will have proven it valid if we can go, step by step, from 'the premises to the conclusion' (from P to C), provided that each small step is clearly valid. But that means we must agree ahead of time on what steps are 'clearly valid.' For that reason, we have to single out step-making rules, called *inference rules*, which we can prove to be valid, and which we will all agree to use exclusively in constructing any proofs.

We can look at proof construction as a game. In a game such as poker, bridge, or chess, rules are laid down, which define the game, and which no player is free to break. For example, in bridge, each player is dealt 13 cards at the beginning of the hand. Similarly, our inference rules are what define the proof game. If they are misapplied or if other rules not on our list are applied, then the proof is not acceptable. Different textbooks present different sets of inference rules—nothing is magical about ours. The point is that, first, we agree on some set of rules (subject to restrictions) and only then can proof construction take place.

Given some set of inference rules, $I_1, I_2, \ldots, I_n$, a proof of an argument will have the form:

1. P $\lfloor C$
2. Step 1
3. Step 2
   .
   .
   .

$r + 1$. Step r, which has to be C

We put the conclusion in a half-open box opposite the last premise, to make it clear at what we must eventually arrive. We must justify each step by citing one and only one inference rule. The justification will always take the form of citing the inference rule used to get that line, plus the numbers of the preceding lines used:

1. P $\lfloor C$
2. Step 1 $1, I_i$
3. Step 2   Some lines, some I_i
4. Step 3   Some lines, some I_i

.
.
.

r + 1. Step r, which has to be C   Some lines, some I_i

We will examine several examples in Section 10.3.

It takes only a few minutes to explain the rules of chess, but that does not mean we can play the game well right off. We have to practice and learn the fine art of strategy: what rules to use and when to use them. We will discuss tips on strategy as we proceed, but keep in mind that the method of natural deduction is not a mechanical, grind-it-out method as was the method of truth tables.

10.2 The Notion of an Inference Rule

The crucial concept in the method of natural deduction is the notion of an *inference rule*. To understand what that term means, we need to first learn what an 'argument form' is. Consider the following arguments:

$$\begin{array}{lll} 1.\ A \to B & 1.\ -X \to -Z & 1.\ (R\ \&\ S) \to (L \text{ v } Z) \\ 2.\ B & 2.\ -X & 2.\ R\ \&\ S \\ \hline \therefore\ B & \therefore\ -Z & \therefore\ L \text{ v } Z \end{array}$$

While the specifics of each argument differ, they all share a common form or structure:

1. ○ → □
2. ○
___
∴ □

This pattern or form we call **argument form**, which is an array of variables, connectives, and brackets (but no constants) that has the property that when we replace the variables by statements, we get an argument as the result. The following are argument forms:

$$\begin{array}{lll} 1.\ (\bigcirc\ \&\ \square) \text{ v } \triangle & 1.\ \bigcirc \text{ v } (\square \to \triangle) & 1.\ \bigcirc \\ \hline \therefore\ \square & 2.\ -\bigcirc & \therefore\ \bigcirc \\ & \hline \therefore\ \square \to \triangle & \end{array}$$

The following are not argument forms:

1. ○ & A   ○   1. →○ □
―――――   2. ○
∴ A   ―――――
                                ∴ □

The first is not, because it has constants mixed in. The second is not, because when you replace the variable by statements, you get only a statement as a result. The third is not, because when you replace variables by statements, the first premise is ill-formed.

We can define, in a simple fashion, the notion of an argument being an instance of a given argument form or of its 'having' that form. An argument α is an instance of argument form β if and only if it is possible to get from β to α by the uniform substitution of statements for variables. 'Uniform' means that the same variable is replaced by the same statement throughout. We can prove that a given argument is an instance of a given form by simply showing in a table the substitution.

Example: Show that the argument:

1. $(A \& B) \leftrightarrow C$
2. C
―――――――
∴ $(A \& B)$

is an instance of the form:

1. ○ ↔ □
2. □
―――――
∴ ○

Answer: Exhibit the table of substitution:

○	A & B
□	C

Sure enough, if we replace every ○ by the statement $A \& B$ and every □ by C, we get the argument from the form.

10.2 / *The Notion of an Inference Rule*

Example: Show that:

1. $A \to -B$
2. B
   ∴ A

is not an instance of the form:

1. $\bigcirc \to \square$
2. $\square$
   ∴ $\bigcirc$

Answer: To get the first premise to fit the argument form, we must either plug in A for $\bigcirc$ and $-B$ for $\square$, or else plug in $A \to -B$ for $\square$. But if we try the first option, then the second premise isn't obtained (it's B, not $-B$); and if $\square$ is replaced by $A \to -B$, we don't get the first premise (it's $A \to -B$, not $\bigcirc \to (A \to -B)$).

Example: Show that:

1. $-A \to (B \text{ v } C)$
2. $(B \text{ v } C) \to (D \text{ \& } E)$
   ∴ $-A \to (D \text{ \& } E)$

is an instance of:

1. $\bigcirc \to \square$
2. $\square \to \triangle$
   ∴ $\bigcirc \to \triangle$

Answer: Simply exhibit the table of substitution:

$\bigcirc$	$-A$
$\square$	$B \text{ v } C$
$\triangle$	$D \text{ \& } E$

A given argument may be an instance of many different forms. For example, the argument:

10 / *Natural Deduction*

    1. $-A \vee B$
    2. $--A$
    ―――――
    $\therefore B$

is an instance of the following forms:

    1. ○   1. ○ v □   1. ○ v □
    2. □   2. △   2. −○
    ――――   ――――   ――――
    $\therefore$ △ $\therefore$ ◇ $\therefore$ △

    1. ○ v □   1. ○ v □   1. ○ v □
    2. −−△   2. −−△   2. −○
    ――――   ――――   ――――
    $\therefore$ ◇ $\therefore$ □ $\therefore$ □

We leave it to the reader to exhibit the tables of substitution and to find the other forms of which the argument is an instance.

    Validity applies to argument forms as well as arguments. We can apply truth tables to discover validity. For example, the argument form:

    1. ○ → □
    2. −□
    ――――
    $\therefore$ −○

can be shown valid by a table [true (T); false (F)]:

○	□	○ → □	−□	−○
T	T	T	F T	F T
T	F	F	T F	F T
F	T	T	F T	T F
F	F	T	T F	T F

If an argument form is valid, any instance of it will also be valid. However, the converse is not T: if an argument form is invalid, it doesn't necessarily follow that a given instance of it is invalid. For example, the argument:

    1. $A \to B$
    2. A
    ――――
    $\therefore B$

is an instance of the valid form:

10.2 / The Notion of an Inference Rule

1. $\bigcirc \to \square$
2. $\bigcirc$
   ∴ $\square$

but also of the invalid forms:

1. $\bigcirc$
2. $\square$
   ∴ $\triangle$

1. $\bigcirc \to \square$
2. $\triangle$
   ∴ $\square$

(and others). However, an invalid argument has to be an instance of at least one invalid form, no matter how many valid forms it has.

We can now define precisely the notion of an inference rule. An **inference rule** is a simple, yet valid argument form. Simplicity is admittedly a relative matter, but clearly an argument form such as:

1. $\bigcirc \to \square$
2. $\bigcirc$
   ∴ $\square$

is simple, while one such as:

1. $\bigcirc \to (\square \to (\triangle \to \Diamond))$
2. $\bigcirc \,\&\, \square \,\&\, \triangle$
   ∴ $\Diamond$

is not. But validity is not a relative matter, of course, and it is required that our inference rules must all be valid.

EXERCISES 10.2A

For each of the following arguments, (1) identify four forms of which it is an instance, and give the tables of substitution; (2) indicate which of the forms (if any) are valid; and (3) give the truth tables. (An asterisk indicates problems that are answered in the back of the book.)

1.
1. 1. $A \rightarrow -B$
2. B
   ———
∴ $-A$

* 2. 1. $M \vee L$
2. $-L$
   ———
∴ $-M$

3. 1. $A \& -B$
   ———
∴ $A \vee -B$

4. 1. $A \rightarrow (B \rightarrow C)$
2. $B \& A$
   ———
∴ C

5. 1. $M \& (N \vee R)$
2. $-N$
   ———
∴ $-R$

* 6. 1. —————— A
   ———
∴ A

EXERCISES 10.2B

For each of the following argument forms, give at least five different instances using the same constants A, B, *and* C. *(An asterisk indicates problems that are answered in the back of the book.)*

1. 1. $\bigcirc \leftrightarrow \square$
2. $-\bigcirc$
   ———
∴ $\square$

* 2. 1. $\bigcirc \& (\square \rightarrow \triangle)$
2. $\square$
   ———
∴ $\bigcirc$

3. 1. $\bigcirc \leftrightarrow (\square \leftrightarrow \triangle)$
2. $\bigcirc \leftrightarrow \triangle$
3. $\bigcirc$
   ———
∴ $\square$

4. 1. $\bigcirc \vee (\square \vee \triangle)$
2. $-\bigcirc \& -\square$
   ———
∴ $\triangle$

5. 1. $\bigcirc \rightarrow \square$
2. $\square \rightarrow \triangle$
   ———
∴ $\bigcirc \rightarrow \triangle$

* 6. 1. $-(-\bigcirc \vee -\square)$
   ———
∴ $\bigcirc \vee \square$

10.3 The First Four Inference Rules

We will introduce our rules of inference slowly, starting with just four, taking time to adjust to them. These rules are all 'one-way,' in the sense that from the premise(s) one may move on to the conclusion, but not the other way around.

The first rule is *simplification* (Simp). If at any point during the proof game we have a conjunction, we are entitled (permitted, that is, not forced) to write on the next line the first conjunct. Symbolically:

10.3 / *The First Four Inference Rules*

Simp

1. ○ & □
   ∴ ○

Be clear on what Simp allows. The rule allows the following steps:

1. A & B
   ∴ A

1. (A v B) & L
   ∴ A v B

1. [(A → L) v Z] & (P v Q)
   ∴ (A → L) v Z

Each of the previous arguments are instances of the argument form called Simp. On the other hand, the following are not instances of Simp:

1. 1. A & B
      ∴ B

2. 1. (A → B) v (C v D)
      ∴ A → B

3. 1. (A & B) & C
      ∴ A

(1) is not, because our rule of Simp only allows us to simplify to the first conjunct, not the second. (2) is not, because Simp only allows us to step to a conjunct, not a disjunct. And (3) is not, because we skipped a step—we really applied Simp twice.

The rule of Simp is a valid argument form, as we can prove by a truth table (Table 10.1).

The second rule is *conjunction* (Conj). At any point during the proof game, we may put together any two preceding lines, using an ampersand (&).

Table 10.1

○	□	○ & □	○
T	T	T	T
T	F	F	T
F	T	F	F
F	F	F	F

Conj

1. ○
2. □
∴ ○ & □

The order in which the premises of an inference rule occur is irrelevant, so in the case of Conj (but not Simp!) we could just as well write:

1. □
2. ○
∴ ○ & □

Here are some genuine instances of Conj.

1. A
2. B
∴ $A \& B$

1. $L \& Z$
2. $T \vee L$
∴ $(T \vee L) \& (L \& Z)$

1. $L \rightarrow X$
2. $L \rightarrow Z$
∴ $(L \rightarrow X) \& (L \rightarrow Z)$

Here are some arguments that are not instances of Conj, although they are valid.

1. A
2. B
∴ $A \rightarrow B$

10.3 / *The First Four Inference Rules*

1. $L \,\&\, Z$
2. $T \vee L$

∴ $(T \vee L) \,\&\, (L \vee Z)$

1. $L \to X$
2. $L \to Z$

∴ $L \to (X \,\&\, Z)$

The first is not, because we can only conjoin with an ampersand, not an arrow. The second is not, because the second conjunct in the conclusion is different from the first premise. And the third is not, because Conj only applies to whole lines. The third inference rule is *modus ponens* (MP). If at some point during the game we discover that ○ → □ and later discover that ○ is indeed true, then we may infer—write down as the next step—□. The order in which the preceding lines occur is irrelevant—we may discover ○ first, then go on to discover ○ → □. The rule of MP is obvious: if the teacher tells you that absolutely if you pass the final, you will pass the class, and a month later you take the final and indeed pass it, you can logically conclude that you will pass the class (supposing the teacher's mind has not changed.)

Symbolically:

1. ○
2. ○ → □

∴ □

Some instances of MP are:

1. $A \to Z$
2. A

∴ Z

1. $L \vee M$
2. $(L \vee M) \to (K \vee R)$

∴ $K \vee R$

1. $(A \to B) \to Z$
2. $[(A \to B) \to Z] \to (T \to Z)$

∴ $T \to Z$

These are not instances of MP:

1. 1. $A \rightarrow B$
   2. B
   $\therefore A$

2. 1. $(A \rightarrow B) \rightarrow C$
   2. A
   $\therefore C$

3. 1. $(Z \mathbin{\&} X) \rightarrow L$
   2. $X \mathbin{\&} Z$
   $\therefore L$

(2) and (3) are not MP, because the substitution into the variable must be uniform. (1) is so often confused with MP that logicians have a label for it: 'the fallacy of affirming the consequent.' MP is valid; affirming the consequent is not, as Table 10.2 shows.

Our fourth rule is *addition* (Add). This rule says that we may disjoin, that is, add with a wedge (v), any statement we wish, to any statement already proved. Symbolically:

Table 10.2

(Table for MP)

○	□	○ → □	○	□
T	T	T	T	T
T	F	F	T	T
F	T	T	F	F
F	F	T	F	F

(Table for the fallacy of affirming the consequent)

○	□	○ → □	□	○
T	T	T	T	T
T	F	F	F	T
F	T	T	T	F	← Falsifying case
F	F	T	F	F

10.3 / The First Four Inference Rules

Add

1. ○
   ───────
   ∴ ○ v □

Such an inference rule is valid, because if ○ is true, then (by the definition of the wedge) ○ v (anything) is true.

Instances of Add include:

1. *A*
   ───────
   ∴ *A* v *B*

1. (*A* v *B*) → *C*
   ───────
   ∴ [(*A* v *B*) → *C*] v *E*

1. *A* v *B*
   ───────
   ∴ (*A* v *B*) v (*A* v *B*)

But the following are not instances of Add:

1. 1. *A*
      ───────
      ∴ *B* v *A*

2. 1. *A* → *B*
      ───────
      ∴ *A* → (*B* v *C*)

3. 1. *A* v *B*
      ───────
      ∴ (*B* v *A*) v *D*

(1) is not, because the order of the variables has been changed. (2) is not, because we can only add to whole statements other whole statements. (3) is not, because we can only substitute for variables uniformly; what gets plugged into the first variable in the premise must get plugged into that variable throughout.

EXERCISES 10.3

Identify each of these short arguments as either an instance of Simp, Conj, MP, Add, or else none of those. (An asterisk indicates problems that are answered in the back of the book.)

1. 1. $(A \& B) \to C$
   $\therefore A \to C$

* 2. 1. $(A \lor E) \lor L$
   $\therefore [(A \lor E) \lor L] \lor S$

3. 1. $(M \& S) \lor Q$
   $\therefore (M \& S) \lor (Q \lor T)$

4. 1. $(M \to S) \to (R \to Z)$
   2. $M \to S$
   $\therefore R \to Z$

5. 1. $A \lor B$
   $\therefore (A \lor B) \lor C$

* 6. 1. $L \to Z$
   2. $Z \to L$
   $\therefore (Z \to L) \& (L \to Z)$

7. 1. $M \to A$
   2. M
   $\therefore A \lor Z$

8. 1. $L \to (T \lor X)$
   2. L
   $\therefore L \& [L \to (T \lor X)]$

9. 1. $(A \leftrightarrow Z) \leftrightarrow (R \lor X)$
   $\therefore [(A \leftrightarrow Z) \leftrightarrow (R \lor X)] \lor [(A \leftrightarrow A) \leftrightarrow (Z \lor X)]$

* 10. 1. $[(A \leftrightarrow Z) \& R] \leftrightarrow [R \leftrightarrow (A \leftrightarrow Z)]$
   $\therefore (A \leftrightarrow Z) \& R$

11. 1. $(A \leftrightarrow B) \leftrightarrow C$
    2. $[(A \leftrightarrow B) \leftrightarrow C] \to [A \leftrightarrow B]$
    $\therefore B \leftrightarrow A$

12. 1. $(M \to R) \to C$
    2. $[(M \to R) \to C] \to (L \lor Z)$
    $\therefore L \lor Z$

10.4 Direct Proofs Using the Four Inference Rules

Rather than introduce the other seven inference rules now, we will try a few simple proofs to get our feet wet. The four rules we have are:

Simp

1. $\bigcirc \& \square$
$\therefore \bigcirc$

Conj

1. $\bigcirc$
2. $\square$
$\therefore \bigcirc \& \square$

10.4 / Direct Proofs Using the Four Inference Rules

MP

1. ○ → □
2. ○
   ―――――
   ∴ □

Add

1. ○
   ―――――
   ∴ ○ v □

Think of these rules as devices for tearing apart compound statements and rebuilding new components repeatedly until the desired conclusion is reached. Simp and MP both allow the detachment of chunks from larger compounds. Conj allows pieces to be put back together. Add allows the introduction of material not given initially. These rules go only one way, and they only apply to whole lines. Let us look at some simple proofs and then make a few general comments about strategy.

Consider this argument:

1. $A \& (X \text{ v } Y)$
2. B $\underline{/ A \& B}$

We want to get a conjunction $A \& B$. Clearly, we can get $A \& B$ by using Conj—but only if we have managed to get A and B first. Here, B is already "dealt to us at the beginning of the hand," that is, given as a premise. To get A, we must tear apart the only line in which A occurs, namely, line 1. The rule for detaching a conjunct from a conjunction being Simp, the proof of this argument has the form:

1. $A \& (X \text{ v } Y)$
2. B $\underline{/ A \& B}$
3. A 1, Simp
4. $A \& B$ 3, 2, Conj

This is indeed a proof, because it is a sequence of lines, each one of which is either a premise or else follows from preceding lines by the application of an inference rule, and of which the last line is the conclusion.

Consider next:

1. $A \to B$
2. $A \& R$ $\underline{|B \& A}$

Here again, we are asked to derive a conjunction $B \& A$. We can use Conj, but that requires us to first get B and then get A. B occurs only in premise 1. To get B out of $A \to B$, we must use MP. But to do that, MP requires us to have the antecedent A. That occurs in line 2, and to get it, we use Simp. The proof is:

1. $A \to B$
2. $A \& R$ $\underline{|B \& A}$
3. A 2, Simp
4. B 3, 1, MP
5. $B \& A$ 3, 4, Conj

Two comments regarding strategy can be made. Let us put them as 'rules':

Rule of Strategy No. 1: To get a conclusion of the form $\alpha \& \beta$ try getting α, then β, then conjoining them.

Rule of Strategy No. 2: If you do not see how to do the proof right away, try thinking backwards. Look at the conclusion as a pattern, and figure out what inference rule (or rules) could give rise to such a pattern. Then let the rule do the work for you; let the rule tell you what sort of things to look for next.

Consider:

1. $(A \lor B) \to C$
2. A $\underline{|C}$

Try thinking backwards. We want C. C only occurs in line 1, which is $(A \lor B) \to C$. Do we have a rule that allows us to get C from $(A \lor B) \to C$? Yes, MP—but MP tells us that we need $A \lor B$. But line 2 only gives us A. So our first step will be to get from A to $A \lor B$.

3. $A \lor B$ 2, Add
4. C 1, 3, MP

Consider:

1. $A \& B$
2. $C \& D$ $\underline{|(A \& C) \lor R}$

10.4 | *Direct Proofs Using the Four Inference Rules*

Working backwards, we see that the conclusion is (A & C) v R, yet R occurs nowhere in the premises. This suggests Add: get A & C, then add R. Here the inference rule tells us what to do: we must first get A & C. To do that, we must get A, then C, then conjoin them:

1. A & B
2. C & D |(A & C) v R
3. A   1, Simp
4. C   2, Simp
5. A & C   3, 4, Conj
6. (A & C) v R   5, Add

Rule of Strategy No. 3: To get a conclusion of the form α v β, try getting α, then adding β.

Proofs often can be tricky for those who do not see clearly how general the rule of Add is. If instead of the previous problem we were given:

1. A & B
2. C & B |(A & C) v (A & C)

the proof would have been the same—yet some would have been confused by the fact that we have to add (A & C) to itself.

EXERCISES 10.4

Construct proofs for the following valid arguments. (An asterisk indicates problems that are answered in the back of the book.)

1.   1. A → B
     2. A |B v C

* 2.   1. A & B
     2. A → C |C

3.   1. X → L
     2. X
     3. S & M |S & L

4.   1. A
     2. A → B
     3. B → C |C

5.   1. X & Z
     2. (X v Z) → R |R

* 6.   1. M
     2. (M v X) → (Z & A)
     3. (Z v A) → D |D

7.   1. M & Z
     2. (M v Z) → Q
     3. (Q v T) → R |R

8.   1. A & L
     2. A → Z
     3. M & P |M & (A & Z)

9. 1. $A \,\&\, Q$
   2. $(A \text{ v } Z) \to Z$
   3. $(Z \text{ v } Q) \to B$ $\lfloor B$

* 10. 1. $A \,\&\, Z$
   2. $B \,\&\, T$
   3. $(A \,\&\, B) \to Q$ $\lfloor Q$

11. 1. $(A \,\&\, B) \,\&\, (L \,\&\, E)$
    2. $(A \text{ v } M) \to Z$
    3. $(A \text{ v } B) \to T$ $\lfloor (T \,\&\, Z) \text{ v } M$

12. 1. $A \,\&\, [B \,\&\, (C \,\&\, E)]$
    2. $A \to (R \to Z)$
    3. $R \,\&\, M$
    4. $(A \,\&\, Z) \to (L \text{ v } P)$
    5. $[(L \text{ v } P) \text{ v } A] \to Q$ $\lfloor R \,\&\, Q$

10.5 The Next Five Inference Rules

The next five inference rules are **logical equivalences**: they are rules that allow the substitution for an expression of another to which it is logically equivalent. Consider the rule of *double negation* (DN): for a double-negated statement $--\bigcirc$ we may substitute the positive statement $\bigcirc$ and vice versa: for any $\bigcirc$ we may substitute $--\bigcirc$.

Diagrammatically:

$\bigcirc \Leftrightarrow --\bigcirc$

Why have such a rule? Because an argument such as:

1. $A \to B$
2. $--A$

$\therefore B$

is valid but could never be proven unless we had a rule that allowed us to erase the double negation.

While the first four inference rules are 'one-way' only, these new rules are 'two-way,' that is, logical equivalences. From $\bigcirc$ we may infer $--\bigcirc$, and from $--\bigcirc$, we may infer $\bigcirc$. But while from $\bigcirc \to \square$ and $\bigcirc$ we may infer $\square$, we cannot infer $\bigcirc \to \square$ and $\bigcirc$ from $\square$.

Another difference between one-way and two-way inference rules is that the one-way rules apply only to whole lines. Thus:

1. $A \to (B \,\&\, C)$

$\therefore A \to B$

10.5 / *The Next Five Inference Rules*

is not an acceptable application of Simp. But you may apply any two-way rule to part of a line. Thus:

1. $A \rightarrow B$

∴ $A \rightarrow --B$

would be an acceptable application of DN.

The next two-way rule is *commutation* (Comm).

○ & □ ⇔ □ & ○
○ v □ ⇔ □ v ○

Both are Comm.

The following are instances of Comm:

1. $A \vee (B \& C)$

∴ $(B \& C) \vee A$

1. $A \rightarrow (Z \vee L)$

∴ $A \rightarrow (L \vee Z)$

1. $(A \leftrightarrow Z) \& (L \vee R)$

∴ $(L \vee R) \& (A \leftrightarrow Z)$

The following arguments are not instances of Comm:

1.   1. $A \vee (B \& C)$

  ∴ $(C \& B) \vee A$

2.   1. $A \rightarrow B$

  ∴ $B \rightarrow A$

3.   1. $A \leftrightarrow (Z \vee L)$

  ∴ $(Z \vee L) \leftrightarrow A$

(1) is not, because two commutations were done at once. (2) is not, because the connective we are commuting around is an arrow. (3) is

not, because the connective we are commuting around is a double arrow. (We can only commute around a wedge or ampersand.)

The third two-way rule is again a familiar one, *association* (Assoc), which has two forms:

$\bigcirc \mathbin{\&} (\square \mathbin{\&} \triangle) \Leftrightarrow (\bigcirc \mathbin{\&} \square) \mathbin{\&} \triangle$
$\bigcirc \vee (\square \vee \triangle) \Leftrightarrow (\bigcirc \vee \square) \vee \triangle$

It should be obvious to you what counts and what does not count as instances of Assoc.

Equivalence (Equiv) is a rule that allows us to unpack a biconditional into its conjoined conditionals.

$\bigcirc \leftrightarrow \square \Leftrightarrow (\bigcirc \to \square) \mathbin{\&} (\square \to \bigcirc)$

If we want to move from $\bigcirc \leftrightarrow \square$ to $(\square \to \bigcirc) \mathbin{\&} (\bigcirc \to \square)$, we must use Comm along the way.

The final two-way rule is *De Morgan's Laws* (DM), which allow us to convert wedges to ampersands and ampersands to wedges.

$-(\bigcirc \vee \square) \Leftrightarrow -\bigcirc \mathbin{\&} -\square$
$-(\bigcirc \mathbin{\&} \square) \Leftrightarrow -\bigcirc \vee -\square$

The first form of DM is just the *neither-nor* equivalence. DM makes sense: to say that "it is not the case that either $\bigcirc$ or $\square$" is just to say that $\bigcirc$ is F and $\square$ is F. And to say that "$\bigcirc$ and $\square$" is F is just to say that either $\bigcirc$ is F or $\square$ is F.

The following are instances of DM:

1. $-(-A \mathbin{\&} B)$
   ―――――――
   ∴ $--A \vee -B$

1. $-(A \to B) \vee -(B \vee A)$
   ――――――――――――――
   ∴ $-[(A \to B) \mathbin{\&} (B \vee A)]$

1. $-(--L \mathbin{\&} -Z)$
   ―――――――――
   ∴ $---L \vee --Z$

1. $-[(A \to -B) \mathbin{\&} (--Z \vee L)]$
   ―――――――――――――――
   ∴ $-(A \to -B) \vee -(--Z \vee L)$

10.5 / The Next Five Inference Rules

    1. $--(M \lor -Z)$
    ―――――――――
    ∴ $-(-M \,\&\, --Z)$

The following are not instances of DM:

1. 1. $-(A \lor -B)$
    ―――――――
    ∴ $-A \lor --B$

2. 1. $-(-M \lor -N)$
    ――――――――
    ∴ $M \,\&\, N$

3. 1. $-(M \lor -N)$
    ―――――――
    ∴ $-M \,\&\, -N$

(1) is not an instance of DM, because DM converts the denial of a wedge into an ampersand; (2) skips two DN steps; (3) omits the second dash that should be in front of N in the conclusion.

EXERCISES 10.5

Identify which two-way rules (if any) the following arguments show. (An asterisk indicates problems that are answered in the back of the book.)

1. 1. $-A \to B$
    ∴ $B \to -A$

* 2. 1. $-A \,\&\, -B$
    ∴ $-B \,\&\, -A$

3. 1. $-A \to -(B \,\&\, Z)$
    ∴ $A \to (-B \,\&\, -Z)$

4. 1. $--A \to -(Z \,\&\, R)$
    ∴ $A \to (-Z \lor -R)$

5. 1. $-A \to (-B \to -C)$
    ∴ $(-A \leftrightarrow -B) \to -C$

* 6. 1. $(--A \lor -B) \,\&\, C$
    ∴ $--A \lor (-B \,\&\, C)$

7. 1. $---------Z$
    ∴ $----------Z$

8. 1. $--(-A \lor --B)$
    ∴ $----(A \,\&\, -B)$

9. 1. $--L \leftrightarrow (Z \to X)$
    ∴ $[--L \leftrightarrow (Z \to X)] \,\&\,$
       $[(Z \to X) \to --L]$

* 10. 1. $(---A \leftrightarrow L) \,\&\, M$
    ∴ $[(---A \to L) \,\&\,$
      $(L \to ---A)] \,\&\, M$

10.6 Proofs Using the Nine Inference Rules

We now have a total of nine inference rules. From the point of view of the rules of the game, these nine rules are equal: they are laws the player is not free to break. But from the point of view of strategy, they are not equal. Instead, these are more like one boss and eight helpers—the boss rule, the Santa Claus among the elves, is MP. Strategically, most proofs involve using the 'transforming rules' (Simp, Conj, Add, DN, DM, Comm, Assoc, and Equiv) to set up one or more MP moves. We call this 'setting up the chain,' since we are using the transforming rules to set up chain reasoning. Diagrammatically:

One-Chain Proof

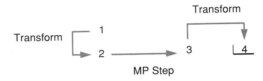

Consider this example:

    1. $-(C \text{ v } B)$
    2. $-B \to D$ $\underline{D}$

This problem is a one-chain proof: we need D; we are given $-B \to D$. To set up the chain, we need to get $-B$. We obviously cannot get $-B$ out of $-B \to D$; after all, it would be ridiculous to conclude from the fact that if I were rich, I would buy a Porsche, that I am rich. The only other premise is $-(C \text{ v } B)$; so there must be some way to use the transforming rules to get $-B$ (what we need to set up the chain) from $-(C \text{ v } B)$. Only one rule seems plausible: DM. So we will try it. Proofs are not like truth tables—we have no guarantee of success.

    3. $-C \ \& -B$ 1, DM

Now we see how to get $-B$: just commute line 3 and simplify.

    4. $-B \ \& -C$ 3, Comm
    5. $-B$ 4, Simp

We can chain reason to get the conclusion:

    6. D 5, 2, MP

10.6 / Proofs Using the Nine Inference Rules

Let us try another one-chain proof.

    1. $Z \& A$
    2. $-(-Z \vee -A) \to \underline{|T}$

We can get T from $-(-Z \vee -A) \to T$, but only if we first get $-(-Z \vee -A)$ by itself—and we cannot use line 2. The only alternative is to transform line 1 into what we need to set up the chain.

    3. $--(Z \& A)$ 1, DN (we need dashes)
    4. $-(-Z \vee -A)$ 3, DM (we needed to turn the
                                         ampersand into a wedge)

Having set the chain up, we can deal the coup de grace:

    5. T 2, 4, MP

Proofs can involve multiple chaining (two chains, three chains, four chains, etc.). But these are no trouble if you are setting up a chain. Diagrammatically:

Two-Chain Proof

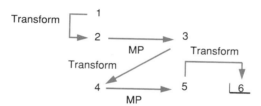

Example:

    1. $-A$
    2. $-(A \& X) \to B$
    3. $(B \vee Z) \to Q$ $\underline{|Q}$

General idea: Transform line 1 to chain to get B, add Z, then chain to get Q.

    4. $-A \vee -X$ 1, Add (we need X)
    5. $-(A \& X)$ 4, DM
    6. B 5, 2, MP
    7. $B \vee Z$ 6, Add
    8. Q 7, 3, MP

If it is not immediately clear how to transform, try applying DM to the antecedent of any conditional where DM might apply, to see if some pattern emerges. For example, in the previous problem, we might have started by applying DM to the antecedent of line 2:

4. $(-A \text{ v} -X) \to B$

Then it would have been clear what we must add to line 1 to get the antecedent:

5. $-A \text{ v} -X$ 1, Add
6. B 5, 4, MP
7. $B \text{ v } Z$ As before
8. Q As before

Here is a triple-chain proof:

1. $-A$
2. $-A \leftrightarrow Z$
3. $Q \leftrightarrow Z$
4. $(Q \& -A) \to M$ $\underline{M}$

Before we begin the chain, we need to get single arrows from the double arrows, pointing in the right direction:

5. $(-A \to Z) \& (Z \to -A)$ 2, Equiv
6. $-A \to Z$ 5, Simp
7. $(Q \to Z) \& (Z \to Q)$ 3, Equiv
8. $(Z \to Q) \& (Q \to Z)$ 7, Comm
9. $Z \to Q$ 8, Simp
10. Z 6, 1, MP
11. Q 9, 10, MP
12. $Q \& -A$ 1, 11, Conj
13. M 12, 4, MP

Here again are the rules of strategy:

1. To get $\alpha \& \beta$, get α, then get β, then Conj.
2. To get $\alpha \text{ v } \beta$, either get α and add β, or (since we have the rule of Comm) get β and add α.
3. To get β having first gotten $\alpha \to \beta$, set up the chain, that is, try to get α by transforming other lines.

EXERCISES 10.6

Prove the following arguments. (An asterisk indicates problems that are answered in the back of the book.)

10.6 / Proofs Using the Nine Inference Rules

1. 1. $A \to B$
   2. $-(-A \text{ v} -C)$ $\underline{|B}$
* 2. 1. $A \leftrightarrow B$
   2. $-(-A \text{ v} -D)$ $\underline{|B}$
3. 1. $-(C \text{ v } A)$
   2. $-A \to -(B \text{ v } C)$ $\underline{|-B}$
4. 1. $-A$
   2. $-(B \text{ \& } A) \to Q$ $\underline{|Q}$
5. 1. $-(B \text{ v } C)$
   2. $-B \to -(-R \text{ \& } -S)$
   3. $(R \text{ v } S) \to Z$ $\underline{|Z}$
* 6. 1. $-(D \text{ \& } E) \to E$
   2. $-D$ $\underline{|E \text{ v } Q}$
7. 1. $--A \leftrightarrow -X$
   2. A $\underline{|-X}$
8. 1. $-(Z \text{ v } R)$
   2. $-R \to M$ $\underline{|M}$
9. 1. $-(M \text{ v } -R)$
   2. $----R \leftrightarrow T$ $\underline{|T}$
* 10. 1. $-(M \text{ v } -Z)$
   2. $[(-L \text{ v } (X \text{ v } Z)] \to Q$ $\underline{|Q}$
11. 1. $-[X \text{ v } (Y \text{ v } Z)]$
    2. $-(Z \text{ \& } L) \to S$ $\underline{|S}$
12. 1. $A \text{ v } (B \text{ v } C)$
    2. $[(A \text{ v } B) \text{ v } C] \to -(M \text{ v } T)$
    3. $(-T \text{ v } T) \to P$ $\underline{|P}$
13. 1. A
    2. $-(-Z \text{ \& } -A) \to F$
    3. $(F \text{ v } -F) \to W$ $\underline{|X \text{ v } W}$
* 14. 1. $A \text{ v } L$
    2. $[R \text{ v } (L \text{ v } A)] \to Z$
    3. $Z \to P$
    4. $--P \to Q$ $\underline{|Q}$
15. 1. W
    2. $(W \text{ v } W) \to (W \text{ \& } Z)$ $\underline{|(W \text{ \& } Z) \text{ \& } W}$
16. 1. $-(-M \text{ v } -Q)$
    2. $--(M \text{ \& } Q) \leftrightarrow B$ $\underline{|B \text{ v } B}$

17. 1. (A & B) & (C & D)
    2. [A & [B & (C & D)]] → (E & F)
    3. [[A & (C & B)] & D] → (F & G) /[G & (F & F)] & E

10.7 The Rule of Conditional Proof

To make our set of rules complete (capable of proving every valid PL argument), we need two more rules. These rules are the *conditional proof* (CP) and *reductio ad absurdum* (Reductio). We begin with CP.

Often we will 'entertain a proposition,' that is, consider a proposal, to see what would happen as a consequence if it were adopted. Someone suggests that we all see a movie. You think to yourself, "OK, suppose I go to the movie. But then I won't have time tonight to study for the big test tomorrow. And if I don't study for that test, I won't pass it." You temporarily assumed (for the sake of argument) that a statement was T (that you were going to the movie) and then deduced from it conclusions. At the end of all this, you confidently assert: "If I go to the movie, then I will fail the test tomorrow."

The rule of CP codifies this mental process. It says that you may temporarily assume any statement during a proof, so long as that assumption is discharged before the end of the proof. In the previous example, you did not conclude, "I will fail the test." Instead, you concluded, "If I go to the movie, I will fail the test." An assumption is discharged by drawing a bent arrow beneath some subsequent line after the temporary assumption, with the arrow pointed to the assumption, and underneath it a conditional with the temporary assumption as the antecedent and the subsequent line as the consequent.

Symbolically:

```
 ┌→ ○
 │  .
 │  .
 │  .
 │  □
 └─
   ○ → □
```

10.7 / The Rule of Conditional Proof

As an example, consider:

 1. $A \rightarrow B$
 2. $-(-B \mathbin{\&} -C) \rightarrow E$ $\underline{A \rightarrow E}$
 3. A Assume
 4. B 1, 3, MP
 5. $--B$ 4, DN
 6. $--B \lor --C$ 5, Add
 7. $-(-B \mathbin{\&} -C)$ 6, DM
 8. E 2, 7, MP

We have no right to quit at this point, because A was not given to us as a premise; we assumed it only temporarily. So we must discharge the assumption:

 1. $A \rightarrow B$
 2. $-(-B \mathbin{\&} -C) \rightarrow E$ $\underline{A \rightarrow E}$
 3. A Assume
 4. B 1, 3, MP
 5. $--B$ 4, DN
 6. $--B \lor --C$ 5, Add
 7. $-(-B \mathbin{\&} -C)$ 6, DM
 8. E 2, 7, MP

 9. $A \rightarrow E$ 3–8, CP

Line 9 does not assert that A is T or that E is T; only that *if* A were T then E would also be T. Thus while every line from 3–8 depends upon the assumption that A is T, line 9 is beyond the range of that assumption. Line 9 does not depend on the temporary assumption. To assert a conditional is not to assert that the antecedent is T. We indicate this by discharging the assumption with a bent arrow and justifying line 9 by saying that we used the sequence of lines 3–8 and the rule of CP.

The rule of CP is similar to Add in this regard: it allows anything to be picked from "out of the blue," but in practice what should be chosen is dictated by the context of the proof. We can use CP several times in a proof, either one after another,

 Step
 Step
 Step

 Step

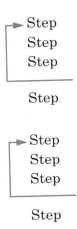

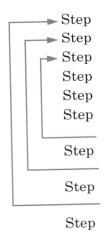

or we may nest assumptions as follows:

As an example of the first, consider:

1. $A \to B$
2. $--B \to D$
3. $(X \to Y) \& (--Y \to Z)$ $\lfloor (X \to Z) \& (A \to D)$

We already know that to prove ○ & □, we generally get ○, then □, then conjoin. But in this case, ○ and □ are both conditionals. A rule of strategy is evident: to get something of the form ○ → □, try assuming ○, getting □, and then boxing it out by CP.

10.7 | *The Rule of Conditional Proof*

```
  ┌─► 4. X                          Assume
  │   5. X → Y                      3, Simp
  │   6. (− −Y → Z) & (X → Y)       3, Comm
  │   7. − −Y → Z                   6, Simp
  │   8. Y → Z                      7, DN
  │   9. Y                          4, 5, MP
  │  10. Z                          8, 9, MP
  └─────────────────────────────────────────
     11. X → Z                      4–10, CP
```

After an assumption has been discharged, you cannot use it or any of the lines that depend upon it, that is, that lie within the box.

```
  ┌─► 12. A                         Assume
  │   13. B                         12, 1, MP
  │   14. − −B                      13, DN
  │   15. D                         14, 2, MP
  └─────────────────────────────────────────
     16. A → D                      12–15, CP
     17. (X → Z) & (A → D)          11, 16, Conj
```

As an example of nested boxes, consider:

1. $(A \mathbin{\&} B) \to C$
2. A Assume
3. B Assume
4. $A \mathbin{\&} B$ 2, 3, Conj
5. C 1, 4, MP

6. $B \to C$ 3–5, CP (discharges the second assumption)

7. $A \to (B \to C)$ 2–7, CP (discharges the first assumption)

The arrows should never cross each other. A proof that looked like:

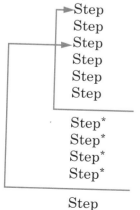

Step
Step
Step
Step
Step
Step

Step*
Step*
Step*
Step*

Step

would automatically be invalid, since some of the lines in the second box (the ones with stars) might well depend on the first assumption, even though it was supposed to have been discharged.

EXERCISES 10.7

Prove the following arguments. (An asterisk indicates problems that are answered in the back of the book.)

1. 1. $A \rightarrow (B \rightarrow C)$ $B \rightarrow (A \rightarrow C)$ *   6. 1. $A \rightarrow X$
* 2. 1. $A \rightarrow Q$ $(A \& B) \rightarrow Q$ 2. $X \rightarrow Y$
   3. 1. $X \rightarrow A$ 3. $Y \rightarrow Z$ $A \rightarrow Z$
       2. $(A \vee R) \rightarrow L$ $X \rightarrow L$ 7. 1. $A \rightarrow (B \& C)$ $(A \rightarrow B) \& (A \rightarrow C)$
   4. 1. $A \rightarrow (B \& C)$ $A \rightarrow B$ 8. 1. $A \rightarrow (A \rightarrow C)$ $A \rightarrow C$
   5. 1. $A \rightarrow (B \& C)$ 9. 1. $A \rightarrow L$ $(B \& A) \rightarrow (B \& L)$
       2. $(B \vee Z) \rightarrow R$ $A \rightarrow R$ *   10. 1. $M \rightarrow T$ $M \rightarrow (\neg M \rightarrow T)$

10.8 Reductio

The last rule, *reductio ad absurdum* ("reduction to the absurd"), also involves making assumptions. Suppose we want to prove that some statement ○ is F. We may do this by assuming temporarily that it is T, and then showing that such an assumption leads to a contradiction. This is a common method of proof in mathematics.

Symbolically:

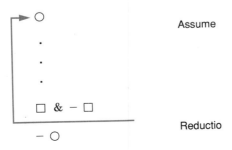

Assume

Reductio

10.8 / Reductio

With Reductio, we try to tease out of previous lines any contradiction we can—it does not matter which.

An easy way to see the point of Reductio is to recall that we can represent an argument by an associated conditional:

1. P

$\therefore C$

associates with $P \rightarrow C$.

If the argument is valid, we know that $P \rightarrow C$ is a tautology, hence it is T. But if C is a contradiction, C is F. The only way $P \rightarrow C$ could be T when C is known to be F is for P to be F. So he rule of Reductio makes sense. If we validly infer a contradiction from the starting point $\bigcirc$, $\bigcirc$ must have been F all along.

Consider an example:

1. $A \vee B$
2. $-A$ $\underline{\quad B}$

In Reductio, we assume the opposite of what we want to show, and go for a contradiction. For this reason, Reductio is often called 'indirect proof.'

   → 3. $-B$ Assume
      4. $-A \,\&\, -B$ 2, 3, Conj
      5. $-(A \vee B)$ 4, DM
      6. $(A \vee B) \,\&\, -(A \vee B)$ 1, 5, Conj

We have gotten our contradiction, so we apply the rule.

7. $--B$ 3–6, Reductio
8. B 7, DN

In addition to the rules of strategy listed in Section 10.6 is another: if utterly stuck in a proof, try doing it by Reductio—assume the negation of the conclusion and try to deduce a contradiction.

EXERCISES 10.8

Prove the following arguments. Use Reductio where appropriate. (An asterisk indicates problems that are answered in the back of the book.)

1. 1. $A \to B$
    2. $-B$ ⊢ $-A$

* 2. 1. $X \to T$
    2. $X \to -T$ ⊢ $-X$

3. 1. $A \leftrightarrow B$
    2. $-A$ ⊢ $-B$

4. 1. X
    2. $-X$ ⊢ Z

5. 1. $A \to B$
    2. $C \to -B$ ⊢ $C \to -A$

* 6. 1. $A \to (B \& C)$ ⊢ $-(A \& -(B \& C))$

7. 1. $A \to B$
    2. $-(A \& B)$ ⊢ $-A$

8. 1. $-A \& -B$
    2. $C \to A$
    3. $D \to B$ ⊢ $-(C \vee D)$

9. 1. $A \to X$
    2. $X \to Y$
    3. $Y \to -X$ ⊢ $-A$

* 10. 1. $A \to (Z \to X)$
    2. $A \to (Z \to -X)$ ⊢ $A \to -Z$

10.9 Proofs Using the Complete Set of Rules

We are now able to construct a proof for any valid PL argument—theoretically able, that is. But to do proofs, we must keep in mind strategy as well as the 11 rules.

Here is a summary of the rules of strategy.

1. Before taking any steps, think.
2. Focus upon the conclusion first and try to work backwards to see how the chain reasoning will flow.
3. To get α v β, use the direct approach. Try getting one side and adding the other.
4. To get α & β, use the direct approach. Try getting each side separately and then conjoin them.
5. To get α → β, use the conditional proof approach. Assume α and try to show that β follows.
6. To get α ↔ β, show α → β and then β → α and conjoin.
7. To get −α, use Reductio. Go for any contradiction you can find.
8. When stuck, try Reductio.

EXERCISES 10.9

Prove the following arguments any way you can. (An asterisk indicates problems that are answered in the back of the book.)

1. 1. $-A$
2. 1. $A \to B$
3. 1. C
4. 1. $A \to B$
   2. $C \to -B$
5. 1. $D \to (E \lor F)$
   2. $-E \& -F$
6. 1. $(Q \lor R) \to S$
7. 1. $(G \to H) \to I$
   2. $-G \lor H$
8. 1. $H \to (I \& J)$
9. 1. $T \to U$
   2. $T \to V$
10. 1. $A \to B$
11. 1. $-A \lor B$
12. 1. $-A \to (B \to C)$
13. 1. $[(M \& N) \& O] \to P$
    2. $Q \to [(O \& M) \& N]$
14. 1. $(J \lor K) \to L$
    2. $L \to J$
15. 1. $M \to N$
    2. $M \to (N \to O)$
16. 1. $(X \lor Y) \to (X \& Y)$
    2. $(X \lor Y) \to -(X \& Y)$
17. 1. $[H \lor (I \lor J)] \to (K \to J)$
    2. $L \to [I \lor (J \lor H)]$

    $\underline{A \to C}$
    $\underline{A \to (B \lor C)}$
    $\underline{D \to C}$

    $\underline{C \to -A}$

    $\underline{\vdash -D}$
    $\underline{Q \to S}$

    $\underline{I}$
    $\underline{H \to I}$

    $\underline{T \to (U \& V)}$
    $\underline{\vdash -A \lor B}$
    $\underline{\vdash -B \to -A}$
    $\underline{A \lor (-B \lor C)}$

    $\underline{\vdash -Q \lor P}$

    $\underline{L \to L}$

    $\underline{M \to O}$

    $\underline{\vdash -X}$

    $\underline{(L \& K) \to J}$

18. 1. $-A \lor B$
19. 1. $A \to B$
20. 1. $H \to P$
    2. $(H \& P) \to M$
    3. $-(H \& M)$
21. 1. $P \to (C \to N)$
    2. $(N \lor R) \to T$
22. 1. $(S \lor T) \to M$
    2. $(M \lor T) \to [S \to (P \to C)]$
    3. $S \& P$
23. 1. $(A \& B) \to -C$
    2. $C \lor O$
    3. $A \& B$
24. 1. $D \to E$
    2. $-(D \& E)$
25. 1. $(H \to X) \& (R \to Y)$
26. 1. $G \to F$
    2. $F \to -P$
    3. P
27. 1. $(H \lor -H) \to G$
28. 1. $G \to (A \& -A)$
29. 1. $P \to O$
    2. $-O$
    3. $(T \lor C) \to (V \& P)$

    $\underline{A \to B}$
    $\underline{(A \& C) \to (B \& C)}$

    $\underline{\vdash -H}$

    $\underline{\vdash P \to (C \to T)}$

    $\underline{\vdash P \to C}$

    $\underline{\vdash Q}$

    $\underline{\vdash -D}$
    $\underline{(H \& R) \to (X \& Y)}$

    $\underline{\vdash -G}$

    $\underline{\vdash G}$

    $\underline{\vdash -T}$

10.10 Other Inference Rules

Our set of 11 rules is complete (capable of proving any valid PL argument), but if we add more rules to our set, proofs can often be shortened considerably. In this section, we mention a few of the most common rules not already in our set.

The rule of *modus tollens* (MT) is:

1. $\bigcirc \to \square$
2. $-\square$

∴ $-\bigcirc$

We can prove this by our existing rules as follows:

1. $A \to B$
2. $-B$
    $|-A$ Assume for contradiction
3. A
4. B 1, 3, MD
5. $B \& -B$ 4, 2, Conj

6. $-A$

The rule of *disjunctive syllogism* (DS) is:

1. $\bigcirc \vee \square$
2. $-\bigcirc$

$\therefore \square$

We can prove this by our rules as follows:

1. $A \vee B$
2. $-A$
3. $-B$ Assume for a contradiction
4. $-A \& -B$ 2, 3, Conj
5. $-(A \vee B)$ 4, DM
6. $(A \vee B) \& -(A \vee B)$ 1, 5, Conj

7. B

The rule of *hypothetical syllogism* (HS) is:

1. $\bigcirc \to \square$
2. $\square \to \triangle$

$\therefore \bigcirc \to \triangle$

Proof:

1. $A \to B$
2. $B \to C$
    $\underline{A \to C}$ Assume to show C
3. A
4. B 1, 3, MP
5. C 2, 4, MP
6. $A \to C$ 3–5, CP

10.10 / Other Inference Rules

Two useful equivalences are implication and contraposition. *Implication* (Imp) is:

$$\bigcirc \to \square \Leftrightarrow -\bigcirc \vee \square$$

Proof (of each logical implication):

1. $A \to B$	$\underline{\vert -A \vee B}$
2. $-(-A \vee B)$	Assume for a contradiction
3. $--A \,\&\, -B$	2, DM
4. $A \,\&\, -B$	3, DN
5. A	4, Simp
6. B	5, 1, MP
7. $-B \,\&\, A$	4, Comm
8. $-B$	7, Simp
9. $B \,\&\, -B$	6, 8, Conj
10. $-A \vee B$	2–9, Reductio

1. $-A \vee B$	$\underline{\vert A \to B}$
2. A	Assume to show B
3. $--A$	2, DN
4. B	1, 2, DS
5. $A \to B$	2–4, CP

Contraposition (Contra) is:

$$\bigcirc \to \square \Leftrightarrow -\square \to -\bigcirc$$

Proof:

1. $A \to B$	$\underline{\vert -B \to -A}$
2. $-B$	To show $-A$
3. $-A$	1, 2, MT
4. $-B \to -A$	2, 3, CP

1. $-B \to -A$	$\underline{\vert A \to B}$
2. A	To show B
3. $--A$	2, DN
4. $--B$	3, 1, MT
5. B	4, DN
6. $A \to B$	2–5, CP

The rules of MT, DS, Imp, Contra, and HS often serve to convert conditional and indirect proofs to direct proofs. For example, consider the argument:

 1. $A \to -(M \lor Z)$ $\underline{|Z \to -A}$

By our original set of rules, we would prove it thus:

 1. $A \to -(M \lor Z)$ $\underline{|Z \to -A}$
 2. Z To show $-A$
 3. A For a contradiction
 4. $-(M \lor Z)$ 3, 1, MP
 5. $-M \& -Z$ 4, DM
 6. $-Z \& -M$ 5, Comm
 7. $-Z$ 6, Simp
 8. $Z \& -Z$ 2, 7, Conj
 9. $-A$ 3–8, Reductio
 10. $Z \to -A$ 2–9, CP

But we can prove this more directly with the new rules:

 1. $A \to -(M \lor Z)$ $\underline{|Z \to -A}$
 2. Z To show $-A$
 3. $--(M \lor Z) \to -A$ 1, Contra
 4. $(M \lor Z) \to -A$ 3, DN
 5. $Z \lor M$ 2, Add
 6. $M \lor Z$ 5, Comm
 7. $-A$ 4, 6, MP
 8. $Z \to -A$ 2–7, CP

Again, prove:

 1. $-(M \lor S) \to H$
 2. $H \to -T$ $\underline{|T \to (M \lor S)}$

With only our initial set of rules:

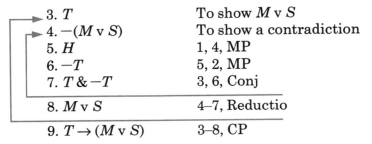

10.10 / *Other Inference Rules*

We can prove this more directly using the new rules:

    3. $-(M \lor S) \to -T$ 1, 2, HS
    4. $T \to (M \lor S)$ 3, Contra

Finally, consider:

    1. $-A \leftrightarrow -(L \& X)$ $\lfloor A \to X$

We can prove this by our initial set of rules:

    2. A To show X
    3. $-X$ For a contradiction
    4. $-X \lor -L$ 3, Add
    5. $-(X \& L)$ 4, DM
    6. $-(L \& X)$ 5, Comm
    7. $[-A \to -(L \& X)] \& [-(L \& X) \to -A]$ 1, Equiv
    8. $[-(L \& X) \to -A] \& [-A \to -(L \& X)]$ 7, Comm
    9. $-(L \& X) \to -A$ 8, Simp
  10. $-A$ 6, 9, MP
  11. $A \& -A$ 2, 10, Conj
  12. X 3–11, Reductio
  13. $A \to X$ 2–12, CP

We can prove this by the new set of rules more directly:

    2. $[-A \to (L \& X)] \& [-(L \& X) \to -A]$ 1, Equiv
    3. $[-(L \& X) \to -A] \& [-A \to -(L \& X)]$ 2, Comm
    4. $-(L \& X) \to -A$ 3, Simp
    5. $A \to (L \& X)$ 4, Contra
    6. A To show X
    7. $L \& X$ 5, 6, MP
    8. $X \& L$ 7, Comm
    9. X 8, Simp
  10. $A \to X$ 6–9, CP

EXERCISES 10.10A

The following are rules of inference found in other versions of PL. Prove them by using our old set of rules. (An asterisk indicates problems that are answered in the back of the book.)

1. Absorption (Abs):
    1. $A \to B$ $\quad\quad\vdash A \to (A \& B)$
* 2. Constructive dilemma (CD):
    1. $A \lor C$
    2. $A \to B$
    3. $C \to D$ $\quad\quad\vdash B \lor D$
3. Destructive dilemma (DD):
    1. $-B \lor -D$
    2. $A \to B$
    3. $C \to D$ $\quad\quad\vdash -A \lor -C$
4. A second form of equivalence:
    $A \leftrightarrow B \Leftrightarrow (A \& B) \lor (-A \& -B)$
5. Tautology (Taut):
    $A \Leftrightarrow (A \& A)$
    $A \Leftrightarrow (A \lor A)$
* 6. Exportation (Exp):
    $A \to (B \to C) \Leftrightarrow (A \& B) \to C$
7. Distribution (Dist):
    $A \& (B \lor C) \Leftrightarrow (A \& B) \lor (A \& C)$
    $A \lor (B \& C) \Leftrightarrow (A \lor B) \& (A \lor C)$

The trick to remembering distribution is to keep in mind that the connective closest to A on one side remains closest to it on the other side.

EXERCISES 10.10B

Redo the proofs from Exercises 10.9 using the new rules wherever possible.

10.11 Proving Invalidity

So far we have a technique for quickly proving validity, supposing that we are convinced ahead of time that the argument is valid. But as a practical matter, how do we first discover whether the argument is valid or not? We could do a long truth table to begin with—but in that case, why bother with natural deduction?

A quicker method of proving invalidity would be useful. This method is best called 'the short truth-table method,' since it grows out of the method we discussed in Chapter 9. The short truth-table method amounts to trying to discover immediately the falsifying case (the assignment of truth values to constants that results in the premises all being T and the conclusion F) in the long table.

Consider a sample argument:

1. $A \to B$
2. $A \to C$
   ∴ $B \lor C$

Is it possible to come up with an assignment of truth values to constants that causes the premises to be T and the conclusion F? Set up a substitution table:

A	B	C

For the conclusion to be F, both B and C must be assigned the value F (that is the only way a disjunction can be F).

A	B	C
	F	F

But we are also trying to get an assignment that makes the premises T, and that means we must assign to A the value F, since if we assign it T, $A \to B$ becomes F. Thus if we try the assignment

A	B	C
F	F	F

we can show that the premises are T and the conclusion F, indicating invalidity.

Now for a more involved example. Again, we seek an assignment of values to constants that makes the premises all T and the conclusion F:

1. $A \leftrightarrow B$
2. $-(Z \vee M)$
3. $-M \rightarrow Q$
4. $Q \rightarrow A$
   ∴ $-B \mathbin{\&} Z$

We set up the substitution table:

A	B	M	Z	Q

To make the conclusion F, however, we must consider three possibilities, since the conjunction is F when $-B$ is T and Z is F,

A	B	M	Z	Q
	F		F	

when $-B$ is F and Z is F,

A	B	M	Z	Q
	T		F	

and when $-B$ is F and Z is T.

A	B	M	Z	Q
	T		T	

We must accordingly go through our stepwise reasoning process on three parallel tracks. We consider how to make the first premise T (assign A and B the same truth value):

A	B	M	Z	Q
F	F		F	

or

A	B	M	Z	Q
T	T		F	

or

A	B	M	Z	Q
T	T		T	

Next, we consider how to make the second premise T: $Z \lor M$ must be F, so Z and M must be F. This eliminates the third track:

A	B	M	Z	Q
F	F	F	F	

or

A	B	M	Z	Q
T	T	F	F	

Finally, for the third premise to be true, M can't be T while Q is F. This is accomplished on the first track by assigning T to Q, on the second track by assigning T to Q as well:

A	B	M	Z	Q
F	F	F	F	T

or

A	B	M	Z	Q
T	T	F	F	T

Under the first assignment, the last premise $Q \to A$ is F, so we don't have a falsifying case. Under the second case, however, $Q \to A$ is indeed

T, so we do have a falsifying case. One such case is all we need to prove invalidity.

EXERCISES 10.11A

Use the short truth-table method to prove the invalidity of these arguments. (An asterisk indicates problems that are answered in the back of the book.)

1.
   1. $A \rightarrow B$
   2. $B \rightarrow C$
   $\therefore -A \rightarrow C$

*2.
   1. $-A \rightarrow E$
   2. $A \vee E$
   $\therefore E$

3.
   1. $(A \& B) \rightarrow C$
   2. $A \vee B$
   $\therefore C$

4.
   1. $A \rightarrow B$
   2. $B \rightarrow C$
   $\therefore -A \rightarrow -C$

5.
   1. $A \rightarrow (B \rightarrow C)$
   2. $A \rightarrow B$
   $\therefore C$

*6.
   1. $A \leftrightarrow B$
   2. $-B \leftrightarrow -C$
   $\therefore A \leftrightarrow -C$

7.
   1. $(A \rightarrow B) \& (C \rightarrow D)$
   2. $B \vee D$
   $\therefore A \vee C$

8.
   1. $L \rightarrow (R \rightarrow S)$
   2. $A \vee L$
   3. $B \rightarrow -A$
   4. $(B \& M) \& R$
   $\therefore -S$

9.
   1. $X \rightarrow (Y \& S)$
   2. $C \rightarrow -Y$
   3. $D \rightarrow -S$
   4. $C \vee D$
   $\therefore X$

*10.
   1. $L \rightarrow -(X \vee Y)$
   2. $(X \rightarrow M) \rightarrow H$
   3. $(Y \rightarrow N) \rightarrow J$
   $\therefore -H \vee -J$

Matters stand thus. If we have decided that a given argument is valid, we can then use natural deduction to prove it valid. On the other hand, if we have decided that it is invalid, we can quickly prove it invalid by the short truth-table method. But what if we have no idea to begin with whether the given argument is valid or invalid? The answer is that we can use the short truth-table method as a method of discovery of validity, though not as a method of proof.

Consider this argument:

1. $-(A \vee B)$

$\therefore A \leftrightarrow B$

Suppose we do not have an idea of whether it is valid or invalid. We can seek a falsifying case as before. To make the conclusion F, we must assign differing truth values to A and B:

A	B
T	F

A	B
F	T

Yet if we assign T to A and F to B, then $-(A \lor B)$ is F. This is not a falsifying case. Similarly, if we assign F to A and T to B, the premise $-(A \lor B)$ is F. Again, we don't get a falsifying case. In no way can we assign truth values to the constants to make the premise T and the conclusion F, so the argument must be valid.

Be careful: when we use the short truth-table method and figure out that the given argument is valid, we have not thereby proved it. We prove the argument valid by going on to construct either a long table or else a proof by natural deduction.

EXERCISES 10.11B

For each of the following, use the short-table method to discover whether the argument is valid or invalid. If valid, construct a proof of it. (An asterisk indicates problems that are answered in the back of the book.)

1. 1. $A \to (B \lor C)$
   2. $-C$
   ∴ $-B \to -A$

* 2. 1. $A \to (B \lor C)$
   2. $C \to B$
   ∴ $A \to B$

3. 1. $A \lor (B \to C)$
   2. $A \lor B$
   ∴ C

4. 1. $A \to (D \to G)$
   2. $D \to (G \to L)$
   ∴ $A \to L$

5. 1. $(Q \lor M) \to P$
   2. $(M \lor P) \to X$
   ∴ $-X \to -Q$

* 6. 1. $A \to (N \& R)$
   ∴ $A \to (R \lor S)$

7. 1. $A \to (X \to -C)$
   2. $(M \to X) \& (E \to A)$
   3. $G \lor C$
   4. $(I \to G) \& (H \to J)$
   5. $I \leftrightarrow -D$
   ∴ $E \leftrightarrow F$

8. 1. $X \leftrightarrow (Z \leftrightarrow R)$
   2. $R \leftrightarrow (M \leftrightarrow X)$
   3. $M \leftrightarrow -Z$
   ∴ $X \leftrightarrow Z$

9. 1. $X \leftrightarrow (L \lor S)$
   2. $L \leftrightarrow M$
   3. $S \leftrightarrow M$
   ∴ $-X \leftrightarrow -M$

* 10. 1. $R \lor [T \lor (S \lor X)]$
   2. $-R \& (-T \& -S)$
   ∴ $Q \to -X$

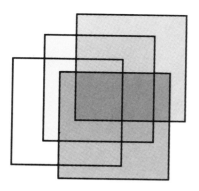

11

Logic Graphs and Set Theory

11.1 Overview

In Chapters 9 and 10, we developed a logic adequate to analyze statements and arguments involving the logical constants 'and,' 'or,' 'not,' and so on, which are the truth-functional connectives.

In this chapter, we will develop logical techniques to deal with the logical constants 'all' and 'some.' These we called *quantifiers*, and accordingly, the logical language we are going to construct is called *quantificational logic* (QL).

Quantifiers quantify: they specify how much of a group or collection of things is being talked about. In this chapter, we will develop our intuitive grasp of the concepts about groups of things, or what we call **sets**. This area of logic is called **set theory**, and our approach is to explore set theory using **logic graphs**, graphical devices for representing sets.

Logicians in the nineteenth century developed logic graphs. John Venn and Leonhard Euler developed the earliest forms of logic graphs—called accordingly, 'Venn diagrams' and 'Euler circles.' Writer, mathematician, and logician Charles Dodgson, who, under the name Lewis Carroll, wrote *Alice in Wonderland*, developed later in the nineteenth century the logic graphs we shall use. More recently, computer scientists have devised other logic graphs, which we will discuss in Section 11.13.

Figure 11.1

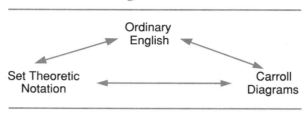

Thus in this chapter we deal with three types of discourse: ordinary English, set theoretic notation, and Carroll diagrams. We will learn to move back and forth among these three techniques for representing sets. Figure 11.1 summarizes this point.

Keep in mind the distinction between a description or specification of a group, and a statement about groups. "Purple people" is a phrase that picks out a group, whereas "Purple people are seldom encountered in Puerto Vallarta" is a statement about a group—a statement to the effect that few members of that group are to be found in Puerto Vallarta. "Solid gold mountains" is a phrase that picks out a group; "Solid gold mountains don't exist" is a statement about that group, namely, that it is empty. In the first part of this chapter we will focus on set descriptions; in Section 11.7 we will shift our discussion to statements about sets.

11.2 The Concept of a Set

We will use the word 'set' to mean a collection of objects or items. The objects can be of any sort: pigs, people, pals, Porsches, prime numbers, prime ribs, puzzles, principles, and plutocrats. These entities or ideas may not have anything in common. We will use the = sign and braces to name sets, as follows:

$M = \{\text{Fred, Ted, Ed}\}$

Read this as "the set M is (identical with) the set consisting of Fred, Ted, and Ed."

We have two ways to define a set: by enumeration and by giving a defining property. Defining a set by *enumeration* means explicitly listing the names of the members of that set. Examples:

$N = \{1, 3, 5, 7\}$

$L = \{\text{cheese, wine, loaf of bread}\}$

$P = \{\text{Paul Pig, Suzie Pig, Alphonse Pig}\}$

The order in which the members are listed is unimportant. Of course, defining a set by enumeration of its members is most practical when the set is small.

Much more flexible is definition of a set by using a *defining property*. That is, we pick out the set by mentioning the kind of things the members are. We use the notation:

$\{x \mid x \text{ has property } P\}$

which is read, "the set of x such that x has property P." Examples:

$B = \{x \mid x \text{ is a prime number}\}$
$Q = \{x \mid x \text{ is a food popular among Hungarians}\}$
$R = \{x \mid x \text{ is a white rabbit chased by Alice}\}$
$S = \{x \mid x \text{ is a unicorn}\}$

Two sets will be of special interest to us: the universal set (denoted Σ) and the empty set (denoted Λ). The **universal set** is the set of everything being talked about at a given time: the set of all pigs, the set of all movies, or the set of all restaurants. If the discussion indicates no other universal set, then it is taken to be the whole universe of actual objects. The **empty set** is the set containing no elements (often expressed by empty braces { }).

11.3 Basic Set Operations

Three basic operations or ways of manipulation can be performed on sets to create other sets. These operations are union, intersection, and complement.

The **union** of two sets is the set composed of all the elements of the two component sets. We use the symbol $\cup$ to denote the union. So if

$A = \{1, 2, 3, 4\}$

and

$B = \{4, 5, 6\}$

then the union A and B is

$$A \cup B = \{1, 2, 3, 4, 5, 6\}$$

Caution: We do not list the same element twice.

We can define the union of two sets that are themselves characterized by defining properties as follows. If

$$A = \{x \,|\, x \text{ has } P_1\}$$

and

$$B = \{x \,|\, x \text{ has } P_2\}$$

then

$$A \cup B = \{x \,|\, x \text{ has } P_1 \text{ or } x \text{ has } P_2\}$$

The **intersection** of two sets is the set composed of those objects common to both sets. We represent intersection by the symbol $\cap$. Thus if

$$A = \{1, 2, 3, 4\}$$

and

$$B = \{2, 3, 5,\}$$

then

$$A \cap B = \{2, 3\}$$

We can define the intersection of sets A and B characterized by defining properties P_1 and P_2, respectively, as follows:

$$A \cap B = \{x \,|\, x \text{ has } P_1 \text{ and } x \text{ has } P_2\}$$

Finally, the **complement** of a set A is the set of those members of the universal set that are not members of A. We denote the complement of a set A by the bar.

Complement of $A = \overline{A}$

For example, if

$$\Sigma = \{1, 2, 3, 4, 5\}$$

11.3 / Basic Set Operations

and

$$A = \{1, 2, 3\}$$

then

$$\overline{A} = \{4, 5\}$$

If a set A is characterized by defining property P_1, then

$$\overline{A} = \{x \mid x \text{ does not have } P_1\}$$

We used A as a variable in the previous definition. We used geometric symbols for variables in Chapters 9 and 10, but that notation would be confusing here. You should be able to tell when a letter is being used to talk about a particular set and when to talk about any and all sets.

These three basic operations can be combined. For example, suppose:

$$\Sigma = \{1, 2, 3, 4, 5, 6, 7\}$$
$$A = \{2, 4, 6\}$$
$$B = \{1, 3, 4\}$$
$$C = \{5, 6, 7\}$$

Then

$$\overline{A} = \{1, 3, 5, 7\}$$
$$\overline{B} = \{2, 5, 6, 7\}$$
$$(A \cup B) = \{1, 2, 3, 4, 6\} = \{5, 7\}$$
$$(A \cup B) \cup C = \{1, 2, 3, 4, 5, 6, 7\} = \Lambda$$

As we saw with calculating the truth values of long compounds from the truth values of components, the trick is to do the calculation in steps, dictated by brackets. So, for example, using the previous sets, calculate:

$$\overline{(\overline{A} \cup \overline{B})} \cap \overline{\overline{C}}$$

Calculating in stages:

$$\overline{A} = \{1, 3, 5, 7\}$$
$$\overline{B} = \{2, 5, 6, 7\}$$
$$\overline{A} \cup \overline{B} = \{1, 2, 3, 5, 6, 7\} \text{ (eliminate repetition of names of members)}$$

$$\overline{\overline{C}} = \{1, 2, 3, 4\}$$
$$\overline{C} = \{1, 2, 3, 4\} = \{5, 6, 7\}$$

Like double negation, the double complement of a set is equivalent to the original set.

$$(\overline{A \cup B}) \cap \overline{\overline{C}} = \{5, 6, 7\}$$

Again, we obey the heuristic rule that recommends breaking down any complex task into less complex parts. We calculate $\overline{(A \cap B)} \cup \overline{(C \cap \overline{B})}$ by calculating in this order:

$$\overline{(A \cap B)} \cup \overline{(C \cap \overline{B})}$$
$$\begin{array}{cc} 1 & 3 \\ 2 & 4 \\ & 5 \\ & 6 \end{array}$$

EXERCISES 11.3

Given the following basis sets, determine the members of compound sets. (An asterisk indicates problems that are answered in the back of the book.)

$\Sigma = \{$Paula, Freda, Hans, Isaac, Kathy, Luisa, Lauren, Mischa$\}$
$A = \{$Paula, Hans$\}$
$B = \{$Isaac$\}$
$C = \{$Freda, Hans, Luisa, Lauren$\}$

1. $\overline{A}$
* 2. $\overline{B}$
3. $\overline{C}$
4. $(A \cup B)$
5. $(A \cap B)$
* 6. $(A \cup B) \cup C$
7. $A \cup (B \cup C)$
8. $(\overline{A \cup B}) \cup \overline{C}$
9. $\overline{(A \cup B)} \cup C$
* 10. $\overline{(A \cup B)} \cup \overline{(A \cap B)}$

11. $(A \cup \overline{A}) \cup A$
12. $\overline{(A \cup A)}$
13. $(\overline{A} \cup \overline{B}) \cup \overline{\overline{C}}$
* 14. $\overline{A} \cap \overline{(B \cup C)}$
15. $\overline{(A \cap B) \cap \overline{\overline{C}}}$
16. $A \cup \Sigma$
17. $\overline{\overline{A \cap \Sigma}}$
* 18. Λ
19. $\overline{(\Sigma \cup \Sigma)} \cap \Lambda$
20. $(\overline{A \cup \Lambda}) \cap (\overline{\Sigma \cup \Sigma})$
21. $(A \cup \Sigma) \cup [(B \cap \Lambda) \cup (C \cap \overline{C})]$
* 22. $(A \cap \Lambda) \cup [(B \cup \Sigma) \cap (C \cap \overline{\overline{C}})]$
23. $\overline{[(\overline{B} \cup \Sigma) \cap (A \cup \Sigma)] \cup (\overline{C \cup C})}$
24. $[(B \cup \Lambda) \cap (A \cup \Sigma)] \cap \overline{(\overline{C} \cup C)}$
25. $[\overline{(\overline{C} \cup C)} \cap (B \cap \Sigma)] \cup \overline{(\overline{A} \cap \Lambda)}$

11.4 Translating from English to Set Theory Notation

We now have symbols to represent particular sets, to represent the universal and the empty sets, and to represent the three basic set operations. In this section, we want to learn how to translate English set descriptions into set theoretic notation, and vice versa.

We will give several specific hints; as always, these hints are heuristic aids, not hard-and-fast rules.

Hint no. 1: Think of set theoretic notation as a kind of code. Then, at least with the more complicated English expressions, do not translate into the code all at once. Instead, translate first into an intermediate stage, a half-code, half-English stage we will call *pseudocode*. Let us examine some examples, using the dictionary:

$\Sigma = \{x \mid x \text{ is a man}\}$
$B = \{x \mid x \text{ is a bachelor}\}$
$R = \{x \mid x \text{ is rich}\}$
$H = \{x \mid x \text{ is happy}\}$

Example: Symbolize the following:

Rich, happy bachelors.

Step 1: Put the expression into pseudocode.

The set of things x such that x is rich and x is happy and x is a bachelor.

Step 2: Turn the pseudocode into code, that is, set theoretic notation.

$(R \cap H) \cap B$

Hint No. 2: Negation indicates the complement. Words or prefixes like 'not,' 'un-,' and 'non-' generally indicate that the set is the complement of a more basic set. Example: translate:

Unhappy men.

Pseudocode:

The set of x such that x is not happy.

Code:

$\overline{H}$

Hint No. 3: At the level of ordinary language, the word 'and' may indicate union, but it also may indicate intersection. This is another reason to translate first into pseudocode, because in pseudocode 'and' always indicates intersection, while 'or' always represents union. To determine whether the 'and' indicates union or intersection, ask three questions. First, does the English expression refer to a 'larger' or a 'smaller' set than the components? Usually, the intersection of two sets is smaller than the components, whereas the union is larger. Example:

The rich and the happy women.

Pseudocode (we are talking here about a larger set than just rich women or happy women alone):

The set of x such that x is rich or x is happy or both.

'Or' indicates union, so the results should be:

$R \cup H$

11.4 / *Translating from English to Set Theory Notation*

A second question to ask is whether the word 'and' can be replaced by the phrase 'together with.' We often see the word 'and' used this way:

The rich people, and those who are happy.

This means:

The rich, together with the happy.

A third question is whether the word 'and' is preceded by a comma—if it is, probably union is indicated. Example:

Frogs, and dogs.

Pseudocode:

The set of x such that x is either a frog or a dog.

Code:

$F \cup D$

Translating from set theoretic notation back to English is generally easier than going from English to code. But you should again go through the intermediate pseudocode stage. Example:

Code:

$\overline{(R \cup H)}$

Back to pseudocode:

The set of x such that x is not either rich or happy.

English:

People who are neither rich nor happy.

EXERCISES 11.4A

Given the following sets,

$\Sigma = \{x \mid x \text{ is a dog}\}$
$S = \{x \mid x \text{ is smart}\}$

$M = \{x \mid x \text{ is mean}\}$
$H = \{x \mid x \text{ is housebroken}\}$

translate the following English set descriptions into set theory symbols. (An asterisk indicates problems that are answered in the back of the book.)

1. Dogs that aren't smart.
* 2. Dogs that aren't mean.
3. Dogs that are smart or housebroken.
4. Dogs that are neither smart nor housebroken.
5. Smart dogs, and those that are not mean.
* 6. Dogs that aren't smart, and those that aren't housebroken.
7. Smart but mean dogs.
8. Dogs that are smart, mean, but housebroken.
9. Dogs that are not neither smart nor mean.
* 10. Dogs that are neither mean nor smart, together with dogs that are neither housebroken nor smart.

EXERCISES 11.4B

For the same dictionary, translate the following from set notation to English. (An asterisk indicates problems that are answered in the back of the book.)

1. $\overline{S}$
* 2. $\overline{S} \cup \overline{H}$
3. $S \cup \overline{H}$
4. $\overline{S} \cap \overline{H}$
5. $S \cup (H \cap M)$
* 6. $\overline{S} \cap (\overline{H} \cup M)$
7. $\overline{S \cup H}$
8. $\overline{S} \cup \overline{M}$
9. $(S \cup H) \cap M$
* 10. $(S \cap H) \cap M$
11. $S \cap \overline{S}$
12. $S \cup \overline{S}$
13. $(S \cup H) \cup \overline{M}$

14. $(\overline{S} \cap \overline{H}) \cap \overline{\overline{M}}$
15. $\overline{(S \cup H)} \cup \overline{M}$

11.5 Two-Set Carroll Diagrams

Lewis Carroll invented a diagram technique called the **Carroll diagram**, a graphical device for representing sets and statements about sets. The simplest Carroll diagram consists of a rectangle divided in half. The top half will represent the set A, the bottom the complement of A. See Figure 11.2.

Shading the north half represents the set A (Figure 11.3). Shading the south half represents $\overline{A}$ (Figure 11.4).

This technique can be generalized to handle expressions involving combinations of two sets. Supposing the two sets to be A and B, the corresponding Carroll diagram is given in Figure 11.5. The northwest square represents the members of A that are also members of B, that is, $A \cap B$. The northeast region of the diagram represents the members of A that are not members of B, that is, $A \cap \overline{B}$. The southwest square represents the members of the complement of A that are also members

Figure 11.2

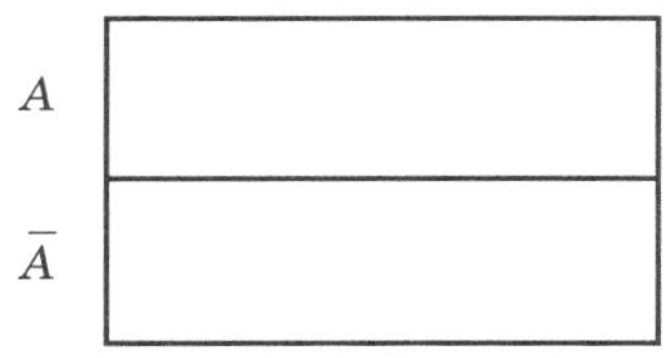

Figure 11.3

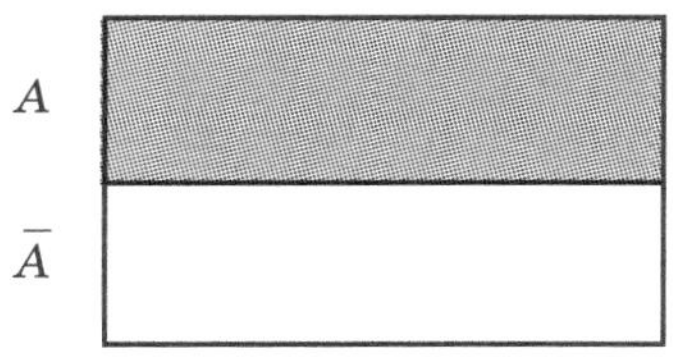

Figure 11.4

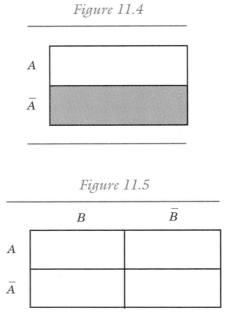

Figure 11.5

of B, that is, $\bar{A} \cap B$. The southeast square represents the members of the complement of A that are also outside of B, that is, $\bar{A} \cap \bar{B}$.

Shading can represent sets. $A \cup B$ is given in Figure 11.6. The entire north half of the diagram represents set A (Figure 11.7). The entire west region represents set B (Figure 11.8). Figure 11.9 represents the set $A \cap B$.

We have before us two levels of discourse. In Section 11.4, we discussed moving back and forth between ordinary English and set

Figure 11.6

$A \cup B = \{x \mid x \text{ is in } A \text{ or } x \text{ is in } B\}$

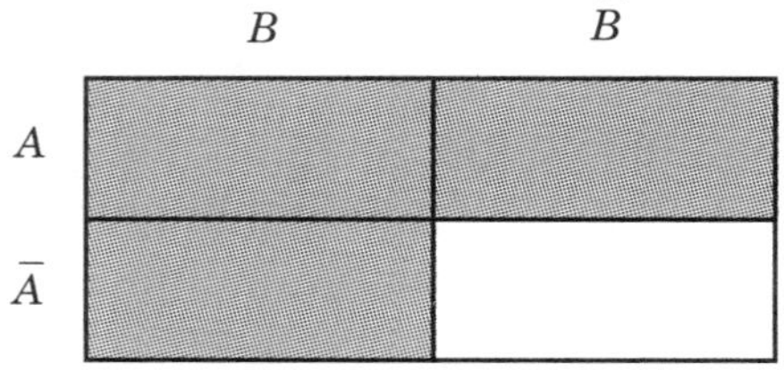

11.5 / Two-Set Carroll Diagrams

Figure 11.7

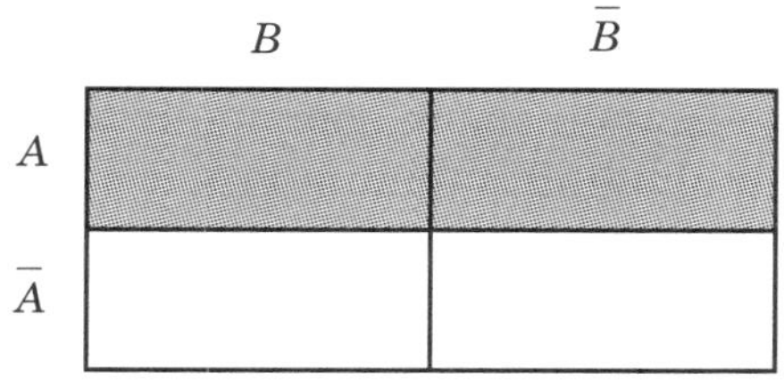

	B	B̄
A	▓	▓
Ā		

Figure 11.8

	B	B̄
A	▓	
Ā	▓	

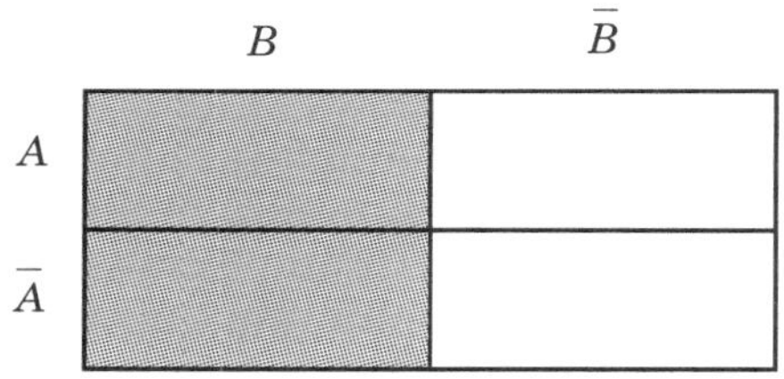

Figure 11.9

$A \cap B = \{x \mid x \text{ is in } A \text{ and } x \text{ is in } B\}$

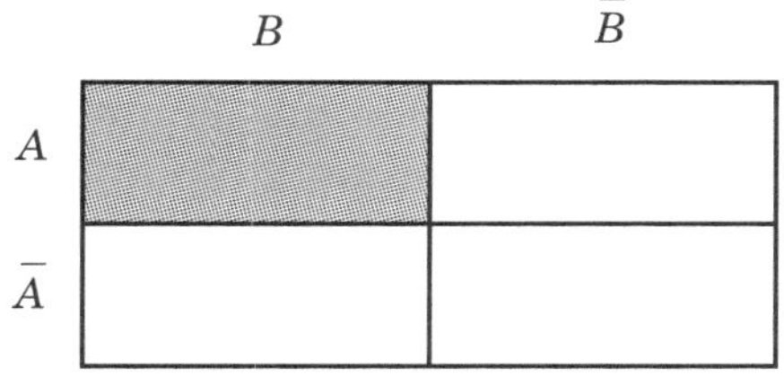

	B	B̄
A	▓	
Ā		

theoretic notation; we will now discuss moving back and forth between set theoretic notation and Carroll diagrams.

Start with moving from a given set theoretic description to a Carroll diagram. The following step-by-step procedure will help.

Step 1: Start with the smallest compound. It is either a union, an intersection, or a complement.

Step 2: If dealing with an intersection, shade completely one of the components, then shade completely the other component.

The area in which the shading overlaps is the intersection. Copy it over with more shading.

Step 3: If dealing with a union, shade completely one of the components, then shade completely the other component. The area of any shading is the union. Copy it over with more shading.

Step 4: If dealing with a complement of a component, shade that component, and then copy the diagram, shading the opposite area.

Step 5: Go to the next largest component.

Example: Draw a Carroll diagram for $\overline{A} \cup \overline{B}$.

Answer: The order of calculation will be:

$\overline{A} \cup \overline{B}$
 1 3 2

Step 1: Shade $\overline{A}$.

	B	$\overline{B}$
A		
$\overline{A}$	shaded	shaded

Step 2: Shade $\overline{B}$.

	B	$\overline{B}$
A		shaded
$\overline{A}$	shaded	shaded (darker)

Step 3: $\overline{A} \cup \overline{B}$ is the entire shaded region. Copy it over.

	B	$\overline{B}$
A		shaded
$\overline{A}$	shaded	shaded

11.5 / Two-Set Carroll Diagrams

Example: Draw a Carroll diagram to represent $\overline{(A \cap B)}$.

Answer: The order of calculation will be:

$A \cap B$ then $\overline{A \cap B}$
   1   2

Step 1: Shade $A \cap B$.

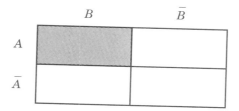

Step 2: To get the complement, shade the opposite region.

The Carroll diagram for $\overline{A \cap B}$ and $\overline{A} \cup \overline{B}$ are the same. This is analogous to De Morgan's law:

$-(A \ \& \ B) \Leftrightarrow (-A \ v -B)$.

De Morgan's law holds in set theory as well as sentential logic.

Example: Draw a Carroll diagram to represent $[(A \cap B) \cup B] \cup A$.

Answer: The order of calculation will be:

$A \cap B$, then $(A \cap B)] \cup B$, then $[(A \cap B) \cup B] \cup A$

Step 1: Shade in $A \cap B$.

Step 2: Shade in $(A \cap B) \cup B$.

Step 3: Shade in $[(A \cap B) \cup B] \cup A$.

We can move now in a step-by-step manner from set theoretic notation to Carroll diagrams. Next, we need to develop the ability to move from a Carroll diagram to a set theoretic description. The same set can have any number of logically equivalent descriptions—as we saw previously with De Morgan's law—so no unique set theoretic description exists for a given Carroll diagram.

The method we will use involves decomposing any Carroll diagram into basic pieces—like the pieces of a jigsaw puzzle. The pieces are the four squares. To develop a set theoretic description of a given Carroll diagram, decompose it into some combination of those basic patterns. Then a correct description results from taking the union of the set theoretic descriptions of the pieces.

Example: Give a correct set theoretic description of:

Answer: This Carroll diagram has the northwest and southeast squares shaded, thus this region is:

$$(A \cap B) \cup (\overline{A} \cap \overline{B})$$

11.5 / Two-Set Carroll Diagrams

Example: To reinforce the idea that there can be any number of equivalent descriptions of the same set, consider this Carroll diagram:

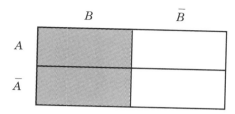

Clearly, the simplest description is just B. But it can also be described as the union of the northwest and the southwest squares:

$$(A \cap B) \cup (\bar{A} \cap B)$$

These two expressions are logically equivalent.

EXERCISES 11.5A

Draw Carroll diagrams to represent the following sets. (An asterisk indicates problems that are answered in the back of the book.)

1. $A \cap B$
* 2. $\overline{A \cap B}$
3. $\bar{A} \cap \bar{B}$
4. $\bar{A} \cup \bar{B}$
5. $A \cap \bar{B}$
* 6. $\overline{\overline{A \cup B}}$
7. $(\overline{A \cap B}) \cup (\bar{A} \cap \bar{B})$
8. $(A \cup B) \cup (A \cap B)$
9. $(A \cup B) \cap (B \cup \bar{A})$
* 10. $(\overline{A \cup B}) \cap (\overline{\bar{B} \cap A})$
11. $[(\bar{A} \cap \bar{B}) \cup B] \cup A$
12. $[(B \cup A) \cup A] \cup A$

EXERCISES 11.5B

Give a set theoretic description to the following Carroll diagrams. (An asterisk indicates problems that are answered in the back of the book.)

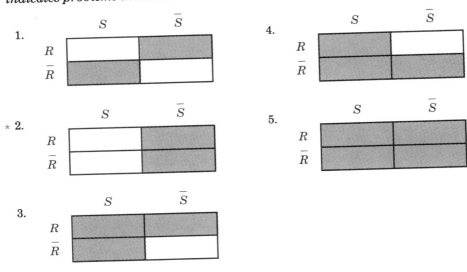

11.6 Three-Set Carroll Diagrams

We can generalize the Carroll diagram technique to accommodate three sets. Let A, B, and C be any three sets. A Carroll diagram for them is given in Figure 11.10.

Figure 11.10

Region 1 represents $(A \cap B) \cap \bar{C}$
Region 3 represents $(A \cap \bar{B}) \cap \bar{C}$
Region 5 represents $(A \cap B) \cap C$
Region 7 represents $(\bar{A} \cap B) \cap C$

Region 2 represents $(A \cap \bar{B}) \cap \bar{C}$
Region 4 represents $(\bar{A} \cap \bar{B}) \cap \bar{C}$
Region 6 represents $(A \cap \bar{B}) \cap C$
Region 8 represents $(\bar{A} \cap \bar{B}) \cap C$

11.6 / Three-Set Carroll Diagrams

As before, we want to go from set theory to Carroll diagrams and back again. Remember, while only one Carroll diagram can be correct for a given set theoretic description, many equivalent set theoretic descriptions may apply to the same Carroll diagram. Things proceed as before.

Example: Construct a Carroll diagram for $A \cup (\overline{B} \cap \overline{C})$.

Answer: Our order of calculation will be: $A \cup (\underset{2}{\overline{B}} \cap \underset{1}{\overline{C}})$

Step 1: Shade $\overline{B}$.

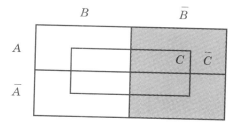

Step 2: Shade $\overline{C}$.

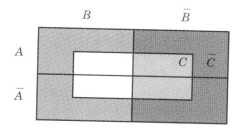

Step 3: Copy over $\overline{B} \cap \overline{C}$, that is, the area of overlap.

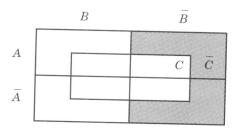

Step 4: Shade in A.

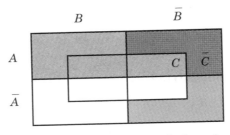

Step 5: Copy over the entire shaded region.

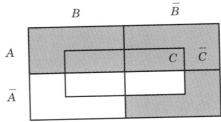

Finding a set theory description is easy: just decompose the figure into some combination of the standard regions.

Example: Find one set theoretic description of:

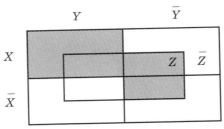

Answer: View this as:

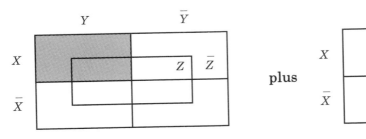

The first region is just:

$X \cap Y$

since it includes both the portion of $X \cap Y$ within Z and the portion outside of Z.

The second region is just:

$$Z \cap \overline{Y}$$

since it includes both the portion of $Z \cap \overline{Y}$ within X and outside of X as well.

So the answer is:

$$(X \cap Y) \cup (Z \cap \overline{Y})$$

Example: Find a set theoretic description of:

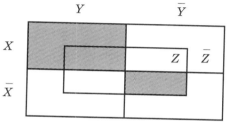

Answer: View this as:

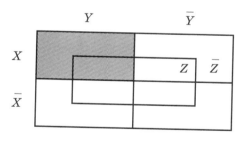 plus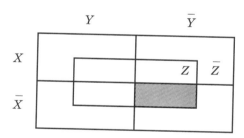

The first is:

$$X \cap Y$$

The second is:

$$Z \cap (\overline{Y} \cap \overline{X})$$

Thus the answer is:

$$(X \cap Y) \cup [Z \cap (\overline{Y} \cap \overline{X})]$$

In review: To go from English to set theory notation or vice versa, use pseudocode as an intermediate step. To go from set theoretic notation to a Carroll diagram, do the shading in stages. To go from a

Carroll diagram to set theoretic description, mentally decompose the Carroll diagram into a combination of the standard pieces, and union together the pieces.

EXERCISES 11.6A

Translate the following from English to set theoretic notation. (An asterisk indicates problems that are answered in the back of the book.)

$\Sigma = \{x \mid x \text{ is a person}\}$
$M = \{x \mid x \text{ is a mugger}\}$
$C = \{x \mid x \text{ is a car thief}\}$
$J = \{x \mid x \text{ likes Michael Jackson}\}$

1. Muggers who like Michael Jackson.
* 2. Muggers, together with car thieves.
3. Car thieves who aren't muggers.
4. Muggers who dislike Michael Jackson.
5. Muggers who either are car thieves or else like Michael Jackson.
* 6. Muggers who are car thieves, along with those who dislike Michael Jackson.
7. People who are neither muggers nor car thieves.
8. Nonmuggers who are not car thieves.
9. Muggers who are car thieves but dislike Michael Jackson.
* 10. Nonmuggers who dislike Michael Jackson, who are car thieves.

EXERCISES 11.6B

Turn the following set theoretic descriptions into colloquial English. (An asterisk indicates problems that are answered in the back of the book.)

$\Sigma = \{x \mid x \text{ is a cat}\}$
$S = \{x \mid x \text{ is Siamese}\}$
$T = \{x \mid x \text{ is a tabby}\}$
$F = \{x \mid x \text{ is friendly}\}$

11.6 / Three-Set Carroll Diagrams

1. F
* 2. $\overline{F}$
3. $F \cap S$
4. $\overline{F} \cap S$
5. $F \cap \overline{S}$
* 6. $F \cap (S \cup T)$
7. $S \cap T$
8. $\overline{S \cup T}$
9. $\overline{F} \cup (S \cap T)$
* 10. $F \cap \overline{(S \cup T)}$
11. $(F \cap S) \cup (F \cap T)$
12. $\overline{(F \cup T) \cup S}$
13. $(\overline{F} \cap \overline{S}) \cap \overline{T}$
* 14. $\overline{(F \cap T)} \cup \overline{(F \cap S)}$
15. $(F \cap T) \cap S$

EXERCISES 11.6C

Draw a Carroll diagram for each of the following. (An asterisk indicates problems that are answered in the back of the book.)

1. $E \cup F$
* 2. $(E \cup F) \cup G$
3. $(\overline{E} \cup \overline{F}) \cap G$
4. $\overline{(E \cup F)} \cap G$
5. $(E \cap F) \cap G$
* 6. $(\overline{E} \cap \overline{F}) \cap G$
7. $\overline{(E \cup F) \cup G}$
8. $(E \cup G) \cap \overline{F}$
9. $(\overline{E} \cup \overline{G}) \cup \overline{F}$
* 10. $\overline{(E \cup F) \cup G}$
11. $\overline{(E \cap F) \cap G}$
12. $\overline{(E \cap F)} \cup G$
13. $\overline{(E \cup F)} \cap \overline{(E \cup G)}$
* 14. $(\overline{F} \cap \overline{G}) \cup (F \cup \overline{E})$
15. $(E \cap F) \cup (G \cap \overline{F})$

EXERCISES 11.6D

Find one correct set theoretic description for each of the following Carroll diagrams. (An asterisk indicates problems that are answered in the back of the book.)

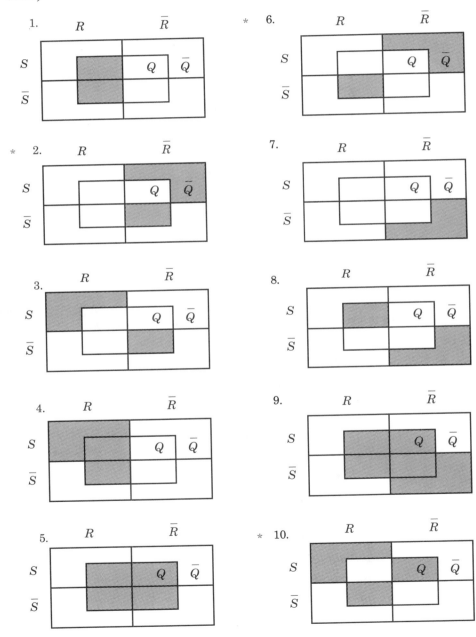

11.7 Statements in Set Theory

So far we have dealt with classes—classes described by English, set theory notation, and Carroll diagrams. Next, we need to consider statements about classes.

There are three basic statements we can make about classes. The first is **set membership**, which is a statement to the effect that a given individual is a member of a given class. We will use the symbol $\in$ to denote membership. Thus to say that an individual x belongs to set A, we would write:

$x \in A$

Examples:

Fred $\in$ {Fred, Sam, Kelly}
Ronald Reagan $\in \{x \mid x$ is a U.S. president$\}$
Fish $\in \{x \mid x$ is a protein-rich food$\}$

The second basic statement about sets is set inclusion. **Set inclusion** is any statement to the effect that one set is included in another set. We will use the symbol $\subseteq$ to denote inclusion. We express the statement that set A is included in set B thus:

$A \subseteq B$

Examples:

$\{1, 2, 3\} \subseteq \{1, 2, 3, 4, 5\}$
$\{a, b, r\} \subseteq \{a, b, c, e, f, g, r\}$
$\{10, 20, 30\} \subseteq \{x \mid x$ is a multiple of 10$\}$

We can define set inclusion by set membership. One set is contained in another if and only if every member of the first is an element of the second. That is, $A \subseteq B \Leftrightarrow$ for any x, if $x \in A$, then $x \in B$.

The third basic statement about sets is set equality. **Set equality** is any statement to the effect that two sets are identical. We will use the = sign thus:

$A = B$

Examples:

$\{1, 2, 3\} = \{1, 2, 3\}$
$\{x \mid x$ is an even number$\} = \{x \mid x$ is a number divisible by 2 without remainder$\}$

$\{a, b, c\} = \{c, a, b\}$

We can define set equality by set inclusion:

$A = B \Leftrightarrow A \subseteq B$ and $B \subseteq C$

The previous expression is equivalent to:

$A = B \Leftrightarrow x \in A$, if and only if $x \in B$

We will allow ourselves a number of derived symbols as well. These symbols and their meanings are listed in Table 11.1.

As we did in Chapter 9, we should distinguish between contingently true set statements and logically true set statements. The following statements are examples of contingently true set statements:

Ronald Reagan $\in \{x \mid x$ is a U.S. president$\}$
Gary Jason $\notin \{x \mid x$ is a U.S. president$\}$
$\{a, b, c\} \subseteq \{x \mid x$ is a letter of the alphabet used in the USA$\}$

The first two are contingent, because being president is contingent upon the vagaries of electoral politics. The third is contingent, because we might have had the Cyrillic alphabet (if by some twist of history Russia had controlled North America).

The following are examples of logically true set statements:

$a \in \{a\}$
$A \subseteq \Sigma$
$A = A$

Table 11.1

Set Theory Statement	Meaning
$x \in B$	Individual x is an element of set A.
$A \subseteq B$	Set A is contained in set B.
$A = B$	Set A is identical to set B.
$x \notin A$	x is not an element of A.
$A \not\subseteq B$	A is not contained in B.
$A \neq B$	A and B are not identical.
$A \subset B$	A is strictly contained in B, i.e., $A \subseteq B$ but $A \neq B$.

11.7 / Statements in Set Theory

The first is a tautology, in that any individual is a member of the set consisting of that individual. The second is a tautology, in that any set is contained in the universal set. And the third example merely states that A is identical to itself.

The following are a few common tautological set equalities. You should think them through to verify their tautologousness.

$$A = A$$
$$A \cup \overline{A} = \Sigma$$
$$A \cap \Lambda = \Lambda$$
$$A \cap \Sigma = A$$
$$\overline{A \cup B} = \overline{A} \cap \overline{B}$$

$$\Lambda \neq \Sigma$$
$$A \cap \overline{A} = \Lambda$$
$$A \cup \Lambda = A$$
$$A \cup \Sigma = \Sigma$$
$$\overline{A \cap B} = \overline{A} \cup \overline{B}$$

EXERCISES 11.7

State whether each statement is true (T) or false (F). (An asterisk indicates problems that are answered in the back of the book.)

$A = \{x \mid x \text{ is an even integer}\}$
$B = \{x \mid x \text{ is a movie star}\}$
$C = \{x \mid x \text{ is bigger than } 10\}$
$D = \{x \mid x \text{ is a man}\}$

1. $1 \in A$
* 2. $2 \in A$
3. $\{2, 4, 6\} \subseteq A$
4. $A = B$
5. Marlon Brando $\in B$
* 6. $B = D$
7. $A \neq C$
8. $17 \in C$
9. $\{x \mid x \text{ is divisible by 4 without remainder}\} \not\subseteq A$
* 10. $2 \in D$
11. $12 \in (A \cap C)$
12. $\{12, 13, 14\} \subseteq (A \cap C)$
13. $(A \cap C) \neq A$
* 14. $\{12, 13, 14\} \subseteq (B \cup C)$
15. $\{12, 13, 14\} \subseteq (A \cup C)$
16. Dustin Hoffman $\in (B \cap D)$

17. $13 \notin (A \cap A)$
* 18. $\{12, 13, 18\} \in (A \cup C)$
19. $\{\text{Marlon Brando, Bo Derek}\} \in (B \cup D)$
20. Bo Derek $\notin (B \cap D)$

11.8 Categorical Statements

In this chapter we set out to develop a mechanism to assess **arguments** whose validity turns on the logical constants 'all' and 'some.' We are now in a position to do just that. Since arguments are built up out of statements, we will first turn our attention to them.

It is traditional to divide quantifier statements into four types: **A, I, E,** and **O**. These are called **categorical statements**.

The first statement form is termed *universal affirmatives*, or **A** statements:

$\mathbf{A}SP$ = All S are P.

where S is the *subject* class and P is the *predicate* class.

Examples:

> All pigs are fat.
> All people are lonely.
> All movie stars are vain.

(Note that by using a predicate such as 'fat' that we are using it as a set-defining property.)

The second type of categorical statement is called *particular affirmative*, or an **I** statement. An **I** statement is any statement of the form:

$\mathbf{I}SP$ = Some S is P.

Examples:

> Some men are happy.
> Some dogs are friendly.
> Some mushrooms are poisonous.

The third statement form is termed 'universal negative,' or **E** statement:

$\mathbf{E}SP$ = No S are P.

Examples:

No cats are loyal.
No frogs are shy.
No house is safe.

Fourth, we distinguish 'particular negative,' or **O** statements:

OSP = Some S is not P.

Examples:

Some fish is not fresh.
Some people are not friendly.
Some cats are not evil.

It is traditional to classify these four forms on the basis of 'quality' or 'quantity.' The *quantity* of a categorical statement is either particular or universal, depending on whether the predicate is attributed to some or all of the subject class. **I** and **O** statements are particular, while **A** and **E** statements are universal.

The *quality* of a statement is affirmative or negative, depending on whether the predicate is affirmed of the subject, or denied of it. **A** and **I** statements are affirmative, while **E** and **O** are negative.

Earlier it was pointed out that categorical statements can be viewed as statements about classes. There are two equivalent ways of doing this: (1) We can view the categorical statements as statements of class inclusion, or (2) we can view them as statements of class equality.

We can express the categorical statements as statements about set inclusion (Table 11.2). We can also express the categorical statements as statements of class equality (Table 11.3). We can use Carroll diagrams to represent categorical statements. We draw a two-set diagram and use an O to represent the assertion that a given area is empty. We will use an X to indicate that a given area has elements in it. If we do not want to assert that there are elements in a given region, and yet we do not want to assert that it is empty, we will leave that area blank.

Thus we can represent the **A** form "All S is P" as in Figure 11.11. The Carroll diagram represents the statement of class equality:

A$SP \Leftrightarrow (S \cap \overline{P}) = \Lambda$

Table 11.2

Statements of Set Inclusion

All S is P.	$S \subseteq P$ (Every member of S is a member of P.)
No S is P.	$S \subseteq \overline{P}$ (Every member of S is a member of the complement of P, i.e., is outside of P.)
Some S is P.	$S \not\subseteq \overline{P}$ (Not every member of S is a member of the complement of P.)
Some S is not P.	$S \not\subseteq P$ (Not every member of S is also a member of P.)

Table 11.3

Statements of Class Equality

All S is P.	$(S \cap \overline{P}) = \Lambda$ (The intersection of S and the complement of P is empty.)
No S is P.	$S \cap P = \Lambda$ (The intersection of S and P is empty.)
Some S is P.	$S \cap P = \Lambda$ (The intersection of S and P is not empty.)
Some S is not P.	$S \cap \overline{P} = \Lambda$ (The intersection of S and the complement of P is not empty.)

The Carroll diagram in Figure 11.12 represents the **E** statement "No S is P." This grows out of the statement of class equality:

E$SP \Leftrightarrow (S \cap P) = \Lambda$

The Carroll diagram in Figure 11.13 represents the **I** statement "Some S is P," which is equivalent to the statement of class equality:

I$SP \Leftrightarrow (S \cap P) \neq \Lambda$

Similarly, the Carroll diagram in Figure 11.14 represents the **O** statement "Some S is not P," which is equivalent to:

O$SP \Leftrightarrow (S \cap \overline{P}) \neq \Lambda$

11.8 / Categorical Statements

Figure 11.11

	P	$\bar{P}$
S		O
$\bar{S}$		

Figure 11.12

	P	$\bar{P}$
S	O	
$\bar{S}$		

Figure 11.13

	P	$\bar{P}$
S	X	
$\bar{S}$		

Figure 11.14

	P	$\bar{P}$
S		X
$\bar{S}$		

EXERCISES 11.8

*For each of the following, symbolize it as one of the four categorical statement forms (**A, I, E,** and **O**), then state its quality and quantity, and use a Carroll diagram to represent it. (An asterisk indicates problems that are answered in the back of the book.)*

M = All mammals class
B = All brown things class
C = All cats class

Example:

All cats are mammals

is **A**CM
Quality = affirmative
Quantity = universal

Carroll diagram:

	M	$\bar{M}$
C		O
$\bar{C}$		

1. All mammals are cats.
* 2. All cats are brown.
3. No cats are brown.
4. Some cats are brown.
5. Some cats are not mammals.
* 6. Some mammals are not brown.
7. No brown things are cats.
8. No mammals are cats.
9. Some cats aren't brown.
* 10. All brown things are cats.

11.9 The Square of Opposition

To better explore the relations between the four categorical statement types, put them at each corner of a square ('oppose them') as in **Figure 11.15**.[1]

[1] This section may be omitted without loss of continuity.

11.9 / The Square of Opposition

Figure 11.15

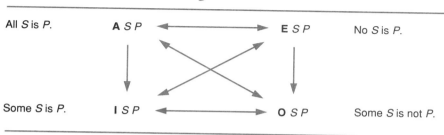

We begin with the diagonals. If All *S* is *P*, what can we say about the statement that Some *S* is not *P*? Clearly, that it is false. If all *S* is *P*, the things in *S* are all inside *P*, so none can be outside it. Similarly, if some *S* is not *P*, then that means that the statement "All *S* is *P*" must be false. Let us introduce the technical term 'contradictories' thus: two statements are **contradictories** if they cannot both be true (T) and cannot both be false (F).

Where subject and predicate are the same in both statements, the **A***SP* and **O***SP* statements are contradictories.

Look next at any **E** statement and its associated **I** statement. Again, we see contradiction. If no *S* is *P*, then no *S* thing is in the class of *P* things—so the statement that some *S* things are in the class of *P* must be F. Similarly, if some *S* is *P*, then the claim that no *S* is *P* must be F.

Next, consider the relation between an **A***SP* claim and its associated **E***SP* claim—for example, "All men are Republicans" and "No men are Republicans." Clearly, it is possible for both statements to be F (if, in this example, some men were Republicans but some were Democrats). But the two statements cannot both be true at the same time. When statements can both be F but cannot both be T, we call them **contraries**.

At the bottom of the square are **I***SP* and **O***SP*. Consider for example: "Some women are magicians" and "Some women are not magicians." Can both these statements be T? Certainly—so they aren't contradictories or contraries. But they can't both be F: if "Some women are magicians" is F, then "No women are magicians" follows (**I***SP* and **E***SP* are contradictories). But if we assume that there are women (that the class of women is not empty), then "No women are magicians" will entail that "Some women are not magicians"—which is just **O***SP* in this case. Upshot: **I***SP* and **O***SP* cannot both be F. When two statements are such that they can both be T but cannot both be F, we say they are **subcontraries**.

In the previous discussion, we made an important presupposition: that the classes the subject terms refer to are not empty. This *subject existence assumption* somewhat limits the applicability of this logical system, but makes its presentation easier.

Let us return to the square of opposition, working under the subject existence assumption. Consider the relation between **A**SP and **I**SP. If "All women are magicians," then (since we assume that there are, in fact, women) it follows that some women are magicians. The relation between **A**SP and **I**SP is one of **subalternation**: the universal implies the particular statement ("the superaltern"). This holds as well between **E**SP and **O**SP.

Putting these points into the square:

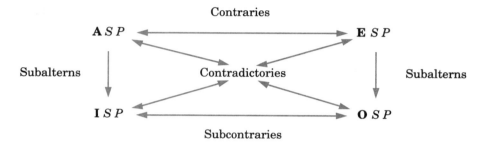

We can also develop a table of 'immediate inferences' (a list of valid one-premise argument forms):

ASP T $\Rightarrow$ **E**SP F, **I**SP T, **O**SP F
ESP T $\Rightarrow$ **A**SP F, **I**SP F, **O**SP T
ISP T $\Rightarrow$ **E**SP F, **A**SP undetermined, **O**SP undetermined
OSP T $\Rightarrow$ **A**SP F, **E**SP undetermined, **I**SP undetermined
ASP F $\Rightarrow$ **O**SP T, **E**SP undetermined, **I**SP undetermined
ESP F $\Rightarrow$ **I**SP T, **A**SP undetermined, **O**SP undetermined
ISP F $\Rightarrow$ **A**SP F, **E**SP T, **O**SP T
OSP F $\Rightarrow$ **A**SP T, **E**SP F, **I**SP T

or, put as argument forms:

1. **A**SP
∴ −(**E**SP)

1. **A**SP
∴ (**I**SP)

1. **A**SP
∴ −(**O**SP)

1. **E**SP
∴ −(**A**SP)

1. **E**SP
∴ −(**I**SP)

1. **E**SP
∴ (**O**SP)

11.10 / Translating Natural Language into Categorical Statements

$$\frac{1.\ \mathbf{I}SP}{\therefore -(\mathbf{E}SP)} \qquad \frac{1.\ \mathbf{O}SP}{\therefore -(\mathbf{A}SP)} \qquad \frac{1.\ -\mathbf{A}SP}{\therefore (\mathbf{O}SP)}$$

$$\frac{1.\ \mathbf{E}SP}{\therefore (\mathbf{I}SP)} \qquad \frac{1.\ -\mathbf{I}SP}{\therefore -(\mathbf{A}SP)} \qquad \frac{1.\ -\mathbf{I}SP}{\therefore (\mathbf{E}SP)}$$

$$\frac{1.\ -\mathbf{I}SP}{\therefore (\mathbf{O}SP)} \qquad \frac{1.\ -\mathbf{O}SP}{\therefore (\mathbf{A}SP)} \qquad \frac{1.\ -(\mathbf{O}SP)}{\therefore -(\mathbf{E}SP)}$$

$$\frac{1.\ -\mathbf{O}SP}{\therefore (\mathbf{I}SP)}$$

EXERCISES 11.9

For each of the immediate inference argument forms previously discussed, plug in a subject and a predicate to get a particular argument. Be sure that the subject class is nonempty. (Afterwards, try substituting for S the empty class unicorns, to see which forms remain valid.)

11.10 Translating Natural Language into Categorical Statements

English sentences do not always have an overt **A**, **I**, **E**, and **O** structure. A few tips may help you symbolize accurately.

Hint No. 1: Watch out for inverted word order. Sentences such as "Men are all sex maniacs" are rhetorically effective but logically deceiving. That sentence is best paraphrased as "All men are sex maniacs" and translated accordingly **A***MS*.

Hint No. 2: Watch out for omitted quantifiers. Some sentences omit the quantifiers altogether. "Cats are sly" usually means "All cats are sly," which is **A***CS*. On the other hand, the sentence "Children are present" means "There are children present," which is **I***CP* ("Some children are present").

Hint No. 3: Watch out for omitted copulas. Sentences in ordinary language often omit the copula (some tense of the verb *to be*), as for example:

All people want money.

This can be paraphrased:

All people are money-wanters.

which is **A**P*M*.

Hint No. 4: Remember that English has many quantifiers besides 'all' and 'some.' Table 11.4 lists the more common ones.

Please note that something may get lost in the translation, as, for instance, in viewing:

Almost all dogs are friendly.

as:

Some dogs are friendly.

We can formulate a step-by-step method of translation, in this case, of putting statements expressed in ordinary language into normal form.

[Quantifier] [Subject class term] [copula *is* or *are*] [Predicate class term]

Step 1: Figure out what the subject is.

Step 2: If the verb is not 'are' or 'is,' substitute for it a synonymous phrase involving 'are' or 'is.'

Step 3: Determine what the predicate is, i.e., what class is claimed to contain all, some, or none of the subject class.

Step 4: Determine the quantifier.

Table 11.4

Anybody		Many	
Anything		Most	
Every(body)		Almost all	
Everything	→ All	A	→ Some
Each		An	
Nothing		A few	
Only		A lot	
		There is (are)	

EXERCISES 11.10

State each of the following in categorical form, state its quality and quantity, and then construct a Carroll diagram for it. (An asterisk indicates problems that are answered in the back of the book.)

1. People are strange.
* 2. Cats can't dance.
3. All that glitters isn't gold.
4. Women are all kind.
5. Police are nearby.
* 6. Almost all burglars are clever.
7. A few burglars are clever.
8. Only politicians love elections.
9. A lot of politicians love elections.
* 10. People hate losers.
11. Each cat is unique.
12. There are happy bachelors.
13. Dogs aren't selective.
* 14. Dogs aren't around.
15. Many dogs are happy.

11.11 Using Carroll Diagrams to Assess Arguments

We can use Carroll diagrams to assess arguments composed of categorical statements. A **syllogism** is a two-premise argument, and a **categorical syllogism** is a syllogism made up of categorical statements.

Of particular interest to us will be categorical syllogisms in standard form. A categorical syllogism is in *standard form* when: (1) its premises and conclusion contain three and only three terms. The predicate term of the conclusion is the *major term*; the subject term of the conclusion is called the *minor term*; the term that appears in both premises is called the *middle term*. (2) The premise containing the major term—called the *major premise*—is listed first. (The second premise contains the minor term, and so is called the *minor premise*.)

Here is an example of a categorical syllogism in standard form:

1. No Marxists are trusting.
2. Some socialists are trusting.

∴ Some socialists are not Marxists.

Here, 'Marxists' is the major, 'socialists' the minor, and 'trusting (people)' the middle term.

From now on, by 'syllogism,' we will understand a categorical syllogism in standard form.

In Section 11.8 we employed two-set Carroll diagrams to represent the four categorical statements. A syllogism has three class terms (the major, the minor, and the middle), so we must employ a three-set diagram to assess a syllogism.

With the following steps, we can employ the marking technique discussed earlier to test for validity.

Step 1: Set up a three-set Carroll diagram with sets labeled accordingly—major on the side, middle on top, and minor inside.

Step 2: Diagram universal premises before particular ones, using O to indicate empty regions.

Step 3: In symbolizing a particular statement, use the X to represent the existence of individuals of a certain type, and put the X on the borderline of the term that is not mentioned in that statement. In symbolizing a universal statement, use two Os.

Step 4: Examine the resulting diagram to see if it captures the conclusion. If it does, the argument is valid. Otherwise, the argument is invalid.

Example: Test the following argument for validity.

1. All people love money.
2. All Communists are people.

∴ All Communists love money.

Step 1: Draw a Carroll diagram.

1. All P are M.
2. All C are P.

∴ All C are M.

Step 2: Both the premises are universal, so we may as well do them in order. Premise 1 asserts that all P are M, so we put O inside the region of P that is outside M.

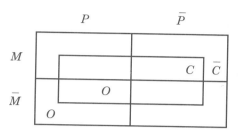

Note that we must put Os in both the inner and outer region of the southwest square. This is because:

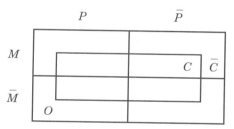

only asserts that $(P \cap \overline{M}) \cap \overline{C}$ is empty, and

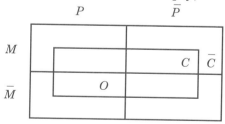

only asserts that $(P \cap \overline{M}) \cap C$ is empty.

We diagram next the second premise (that all C are P).

Step 3: Does the diagram represent the conclusion that all C are $\overline{M}$? Yes, because that portion of the inner square inside the $\overline{M}$ region has O in it. So the argument is valid.

Example: Assess the argument:

1. No men are honest.
2. No women are men.

∴ So no women are honest.

This argument is:

1. No M are H.
2. No W are M.

∴ No W are H.

where M is the middle, W the minor, and H the major term.

Step 1: Set up the Carroll diagram.

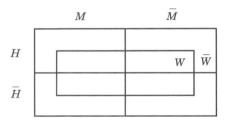

Step 2: Both premises are universal, so we can fill them in any order.

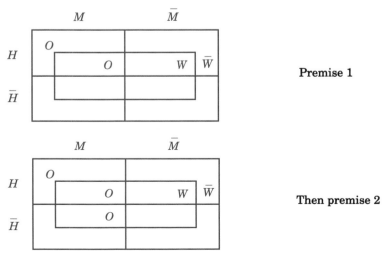

Premise 1

Then premise 2

Step 3: Does this capture the conclusion? "No W are H" is represented by Os in the region $W \cap H$, but only part of the $W \cap H$ region has O in it. So the argument is invalid.

Example: Assess the argument:

1. Some cats are sly.
2. Some cats are not friendly.

∴ So some sly things are not friendly.

The argument here is:

1. Some C are S.
2. Some C are not F.

∴ Some S are not F.

Step 1:

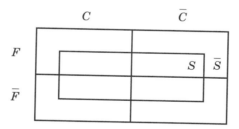

Step 2:

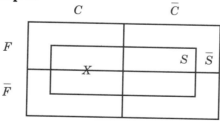

Premise 1

We put the X inside the S box and inside the C region, but since premise 1 does not give us any information regarding F, we must put the X on the borderline. (The X can be put anywhere on the borderline, as long as it occurs in the $C \cap S$ region.)

We must put the X in the $C \cap \overline{F}$ region, but on the S borderline, since premise 2 says nothing one way or the other

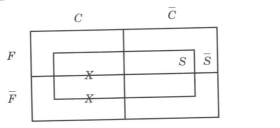

Premise 2

about S. (Again, the X can occur anywhere on the borderline, as long as it is in the $C \cap F$ region.)

Step 3: The conclusion requires that an X occur clearly inside the $S \cap F$ region. But the diagram does not reflect that, so the argument is invalid.

Example: Assess the argument:

1. Some men are rich.
2. Some clowns are men.

∴ So some clowns are rich.

The argument here is:

1. Some M is R.
2. Some C is M.

∴ Some C is R.

Step 1:

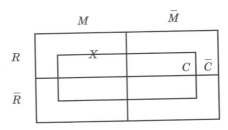

We put an X inside the $M \cap R$ region on the C borderline.

Step 2:

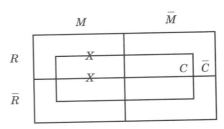

11.11 / Using Carroll Diagrams to Assess Arguments

We put an X inside the $M \cap C$ region on the R borderline.

Step 3: The conclusion requires that an X occur clearly inside the $C \cap R$ region. Since the diagram doesn't show that, the argument is invalid.

EXERCISES 11.11

For each of the following syllogisms: (a) Put the argument in standard form; (b) Identify the major, minor, and middle terms; and (c) Use the Carroll Diagram technique to determine whether or not the argument is valid. (An asterisk indicates problems that are answered in the back of the book.)

1. No rock stars are stable, so no bachelors are stable, since all rock stars are bachelors.

* 2. Some boys dance and some girls dance, so some boys are girls.

3. Some reggae is mellow, since all disco is reggae and some mellow music is disco.

4. Every hustler is self-centered, so a few self-centered people are not honest, since some hustlers aren't honest.

5. All British cars are delicate, but no delicate cars are suitable as police cars, so no British cars are suitable as police cars.

* 6. All Communists favor rent-control and so do all Democrats, so all Democrats are Communists.

7. Some high-protein dishes are low in sodium, and since low-sodium foods are good for you, some high-protein foods are good for you.

8. Cows are happy. Nothing that is happy is violent. So no cows are violent.

9. Some cats like fur. All cats are dogs. So some dogs like fur.

* 10. No dogs are cats. No dogs are rats. So no cats are rats.

11. Cats are nice. After all, cats can dance, and anything that can dance must be nice.

12. People are strange, so people are bound to be violent, because anything violent is strange.

13. Can anybody doubt that whales are evil? They're big, aren't they? And aren't big things evil? Right?

* 14. Somebody must like cheese. After all, some people buy it, and anyone who buys cheese likes cheese.

11.12 Generalizing Carroll Diagrams to More Terms

We have used Carroll diagrams involving three terms to represent sets and statements and to check arguments for validity.[2] What about set expressions involving four or more terms? The answer is that indeed we can modify Carroll diagrams to handle such expressions.

For one term, a Carroll diagram has two regions.

$$1 = A$$
$$2 = \overline{A}$$

For two terms, a Carroll diagram has four regions.

$$1 = A \cap B$$
$$2 = A \cap \overline{B}$$
$$3 = \overline{A} \cap B$$
$$4 = \overline{A} \cap \overline{B}$$

For three terms, a Carroll diagram has eight regions.

$$1 = (A \cap B) \cap C$$
$$2 = (A \cap \overline{B}) \cap C$$
$$3 = (\overline{A} \cap B) \cap C$$
$$4 = (\overline{A} \cap \overline{B}) \cap C$$
$$5 = (A \cap B) \cap \overline{C}$$
$$6 = (A \cap \overline{B}) \cap \overline{C}$$
$$7 = (\overline{A} \cap B) \cap \overline{C}$$
$$8 = (\overline{A} \cap \overline{B}) \cap \overline{C}$$

Not unexpectedly, if n is the number of terms, a Carroll diagram will have 2^n regions. Thus for four terms, a Carroll diagram (page 395) has 16 regions.

Beyond four terms, things get rather sticky. But in practice, one rarely needs to go beyond three terms.

[2]This section may be omitted without loss of continuity.

11.13 / Other Types of Logic Graphs

$1 = (A \cap B) \cap (C \cap D)$ $9 = (\overline{A} \cap \overline{B}) \cap (C \cap \overline{D})$
$2 = (\underline{A} \cap \overline{B}) \cap (C \cap D)$ $10 = (A \cap B) \cap (C \cap \overline{D})$
$3 = (\overline{A} \cap B) \cap (C \cap D)$ $11 = (\overline{A} \cap B) \cap (\overline{C} \cap D)$
$4 = (\overline{A} \cap \overline{B}) \cap (C \cap D)$ $12 = (A \cap B) \cap (\overline{C} \cap D)$
$5 = (A \cap \underline{B}) \cap (C \cap \overline{D})$ $13 = (A \cap \underline{B}) \cap (\overline{C} \cap \overline{D})$
$6 = (A \cap \overline{B}) \cap (C \cap \overline{D})$ $14 = (\underline{A} \cap \overline{B}) \cap (\overline{C} \cap \overline{D})$
$7 = (\underline{A} \cap \overline{B}) \cap (\overline{C} \cap D)$ $15 = (\overline{A} \cap \overline{B}) \cap (\overline{C} \cap \overline{D})$
$8 = (A \cap B) \cap (\overline{C} \cap D)$ $16 = (\overline{A} \cap B) \cap (\overline{C} \cap \overline{D})$

EXERCISE 11.12

To generalize to five terms, Carroll modified his four-term figure by adding a slash (|) in each of the 16 regions, as follows. Thus each region was partitioned into two subregions. In the following diagram, number the 32 regions and describe them set theoretically.

11.13   Other Types of Logic Graphs

We have seen that logic graphs have two fundamental uses in set theory: representing sets and representing statements about sets.[3] Representing sets is important for a number of tasks, perhaps the most important being that of simplification of set expressions. Representing set statements is, as we have seen throughout all of the preceding sections, important in assessing quantificational arguments. There

[3]This section may be omitted without loss of continuity.

are many types of logic graphs besides Carroll diagrams: Venn diagrams, Euler diagrams, Veitch diagrams, and Karnaugh maps. Each has its strong points and its weak points.

Especially useful in simplifying set descriptions are Veitch diagrams and Karnaugh maps. Let us work through an example to illustrate these two devices. Consider the set:

$$[(\overline{A \cap B}) \cap C] \cup [(A \cap B) \cap C] \cup [(A \cap B) \cap \overline{C}]$$

It is customary when using Veitch and Karnaugh diagrams to use + for union and simply place letters next to each other for intersection. Thus the previous set is written more economically as:

$$\overline{A}BC + ABC + AB\overline{C}$$

A Veitch diagram[4] for three variables is not unlike a Carroll diagram, but with a difference regarding the representation of the third set.

  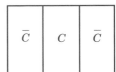

Superimposed, we get:

		A		$\overline{A}$	
B	$AB\overline{C}$	ABC	$\overline{A}BC$	$\overline{A}B\overline{C}$
$\overline{B}$	$A\overline{B}\overline{C}$	$A\overline{B}C$	$\overline{A}\overline{B}C$	$\overline{A}\overline{B}\overline{C}$
	$\overline{C}$	C		$\overline{C}$	

Now, in a Veitch diagram, simplification is carried out by coupling adjacent regions.

Adjacent cells couple horizontally and vertically; the C cells couple to each other horizontally as if the edges touched (visualize the diagram wrapped around a cylinder).

---

[4]See E.W. Veitch, "A Chart Method for Simplifying Truth Functions" in *Proceedings of the Association for Computing Machinery,* Pittsburgh, PA (May 1952), pp. 127–133.

11.13 / Other Types of Logic Graphs

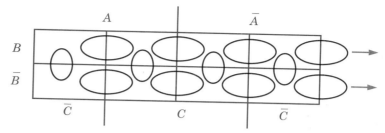

Filling in the Veitch diagram with ones in the indicated regions for $\overline{A}BC + ABC + AB\overline{C}$, we get:

[Veitch diagram with 1s in three cells in the B row: under A$\overline{C}$, AC, and $\overline{A}$C, with an oval enclosing them]

We can identify two couples that cover completely the numbered regions, and thus read off a simpler expression:

$AB + BC$

Karnaugh maps are similar to Veitch diagrams, but are generalizable more easily to more than three sets. In a Karnaugh map, we indicate a set by 1 and its complement by 0, as in the diagram that follows.

C \ AB	00	01	10	11
0			1	1
1				1

Again, we circle adjacent regions and describe them as simply as possible.

$AC + AB$

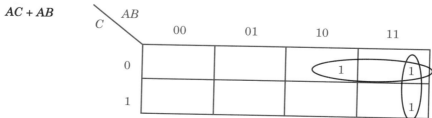

Why bother? Why bother devising diagrams that enable us to simplify set expressions (and, it turns out, truth-functional expressions)? The answer is that the design of logic circuits—such as the microprocessor chips used in personal computers—crucially involves simplification techniques.

Turning now to the evaluation of arguments, both Venn and Euler diagrams are common. Venn and Euler diagrams employ circles, but employ them in slightly different ways. We can best see how these devices function by seeing how each represents the four categorical propositions.

All S is P

Carroll diagram:

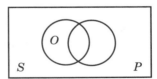

Venn diagram:

Euler diagram:

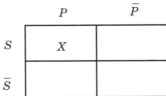

Some S is P

Carroll diagram:

Venn diagram:

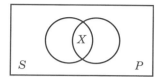

Euler diagram:

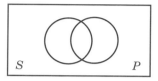

No S is P

Carroll diagram:

	P	$\bar{P}$
S	O	
$\bar{S}$		

Venn diagram:

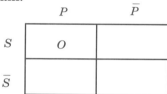

Euler diagram:

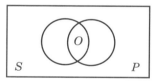

Some S is not P

Carroll diagram:

	P	$\bar{P}$
S		X
$\bar{S}$		

Venn diagram:

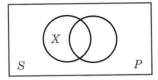

Euler diagram:

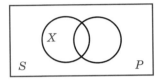

Euler and Venn diagrams are used to test syllograms for validity in much the same way Carroll diagrams are. Why did we choose Carroll diagrams rather than Venn or Euler diagrams? For practical reasons. First, Carroll diagrams are easier to draw, at least easier than Euler diagrams. More importantly, Carroll diagrams are close to Veitch and Karnaugh maps, which are of importance in computer science.

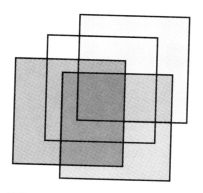

12

The Logic of Properties

12.1 Referring vs. Characterizing Expressions

In Chapters 9 and 10 we developed an artificial language, propositional logic (PL), of sufficient power to analyze truth-functionally compound statements and assess arguments whose validity depends on the truth-functional connectives. But consider this argument:

1. All men are evil.
2. Frank is a man.
∴ Frank is evil.

Such an argument is valid, yet if we symbolize it in PL terms, we get:

1. *A*
2. *M*
∴ *E*

which is invalid. We need to extend PL to include a stronger tool, which we shall call *quantificational logic* (QL). As before, we will not lay out the language all at once. Instead, we will develop and learn to use QL in a spiral fashion. The key insight required to understand the structure of QL is an understanding of the difference between refer-

401

ring and characterizing expressions, or more exactly, subjects and predicates. Think about these sample statements:

Benazhir is pretty.
Fred is dead.
José and Kim are cousins.
Barstow lies between Los Angeles and Las Vegas.
Barstow has only one good restaurant.

Such statements are common in ordinary language. They involve the speaker picking out or 'referring to' some individuals or things, and then describing or 'characterizing' them. The expressions used to refer to things we call *referring expressions*, and the things referred to we call *subjects*. Referring expressions include *proper names*, such as 'Joni Mitchell,' 'Sam Spade,' or 'Nelson Mandela.' Second, referring expressions include *pronouns*, such as 'he,' 'she,' 'they,' or 'it.' Third, referring expressions include *definite descriptions*, such as 'the happiest man in the world,' 'the loneliest teacher in Topeka,' and 'the ugliest dog in Manhattan.' Finally, referring expressions include indefinite descriptions such as 'a lonely bull' or 'a friendly ferret.' The expressions used to characterize subjects we call *characterizing expressions*, and the characteristics ascribed by characterizing expressions we call *predicates*. Characterizing expressions include properties, that is, predicates attributed to single objects:

_____ is happy.
_____ is friendly.
_____ is ugly.

Characterizing expressions also include relations, i.e., predicates attributed to two or more subjects:

_____ is the brother of _____.
_____ is the wife of _____.
_____ lies between _____ and _____.
_____ is a function of _____, _____, and _____.

The difference between referring expressions and subjects is like the difference between sentences and statements: the former are linguistic expressions of the latter, and thus the former vary from language to language. Similarly, characterizing expressions are lin-

Figure 12.1

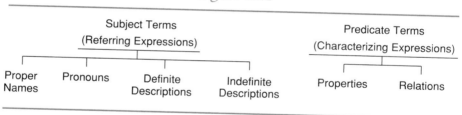

guistic expressions of predicates. In this chapter, we will focus only on referring expressions and properties, deferring relations until Chapter 13. Figure 12.1 summarizes the points just discussed.

12.2 Particular Statements

Particular statements are statements that attribute predicates to named subjects. Examples:

> Fred is ugly.
> Kim Soon is married.
> Gloria is president.

The first thing we need to have in the vocabulary of QL is proper names. We will use small letters a through t as proper names. The same subject, say, John Wayne, can be referred to by different proper names in different languages: 'John Wayne' in English, 'Juan Wayne' in Spanish, or j in QL. Besides proper names, we will need words in QL to stand for properties. These examples attribute the same property to different individuals:

> Fred is dead.
> Gary is dead.
> Samira is dead.
> Jonathan Livingston Seagull is dead.

These sentences have the form:

> _____ is dead.

where the place marker is filled in with a name. We will use the lowercase letters $u, v, w, x, y,$ and z as place markers.

Another word for names is *individual constants,* and another for the place markers is *individual variables*. These are good phrases to use because they suggest an analogy with algebra. Symbols like 1, 2, 3, and so on (the Arabic numerals) are *numerical constants*; whereas the symbols x, y, and z are used in algebra as *numerical variables*. And—stretching the analogy further—just as the different numerals 5 (an Arabic numeral) and V (a Roman numeral) refer to the same number, so the two individual constants f and 'Fred' refer to the same individual. The capital letters A, B, C, and so on, are called *predicate constants*. We will call phrases in which a predicate constant is followed by an individual variable *statement functions*. Statement functions have no truth value and are not statements, just as $(x + 1)$ names no definite number. Here are some statement functions:

Hx
Mx
Ty

We can use these symbols to abbreviate simple properties:

Hx = x is happy.
Ix = x is intelligent.
Sx = x is a singer.
Lx = x is lonely.

We need not limit ourselves to simple properties, however. We can use the truth-functional connectives introduced in PL to express compound properties as well. This is why we speak of QL as an 'extension' of PL: the vocabulary of PL is built into the vocabulary of QL. Thus the compound property of being tall and beautiful is expressed:

$Tx \& Bx$

Other examples of compound properties:

$Hx \vee Mx$
$Tx \& Hx$
$Tx \rightarrow (Bx \vee Mx)$

Having symbols for names and properties, we can now give more refined symbolizations of singular statements than were possible using PL.

12.2 / Particular Statements

If Nancy goes to the dance, then so will Juan.

will be symbolized thus (with Dx: x goes to the dance):

$Dn \to Dj$

The sentence:

If either Nancy or Juan goes to the dance, then both Betsy and Theo will go.

becomes

$(Dn \text{ v } Dj) \to (Db \text{ \& } Dt)$

Brackets belong to the vocabulary of QL and function as before. And all the heuristic tips for translating truth-functionally compound particular statements apply as before. (A quick review of Chapter 9 would be helpful.) Key steps now are to fill in omitted words and replace pronouns by the appropriate proper names, and to carry out our symbolization in stages.

Example: Symbolize:

Neither Franco nor Elaine will go to the dance.

Step 1: Let

$Dx = x$ goes to the dance.
f = Franco.
e = Elaine.

Step 2:

Neither (Franco goes to the dance) nor (Elaine goes to the dance).

Step 3:

—(Franco goes to the dance) & —(Elaine goes to the dance).

Step 4:

$-Df \text{ \& } -De$

Example:

If the Rams winning is a sufficient condition for the Steelers to go to the Super Bowl, then it is a necessary one as well.

Step 1: Let

$Wx = x$ wins.
$Gx = x$ goes to the Super Bowl.
$r =$ The Rams.
$s =$ The Steelers.

Step 2:

(The Rams winning is a sufficient condition for the Steelers to go to the Super Bowl) $\rightarrow$ (The Rams winning is a necessary condition for the Steelers to go to the Super Bowl).

Step 3:

$(Wr \rightarrow Gs) \rightarrow (Gs \rightarrow Wr)$

EXERCISES 12.2

Translate the following particular statements using the following symbolization guide. (An asterisk indicates problems that are answered in the back of the book.)

$a =$ Angie.
$b =$ Bill.
$Hx = x$ is happy.
$Rx = x$ is rich.
$Fx = x$ is friendly.

1. Angie is rich.
* 2. Bill isn't rich.
3. Either Bill is rich, or he isn't.
4. If Angie is happy, then she's friendly.
5. Only if Angie is rich will she be happy.

* 6. Either Bill or Angie is rich.

7. If and only if Bill is rich is he happy.

8. A necessary condition for Bill to be happy is that Angie be rich.

9. Only if Angie is either rich or happy will she be friendly.

* 10. If Angie is rich, then if Bill is happy she will be happy.

11. Unless Bill and Angie are rich, they are friendly.

12. Although Bill and Angie are not rich, they are friendly.

13. If not both Angie and Bill are rich, then both are not friendly.

* 14. Unless Bill is rich, Angie is neither rich nor happy.

15. While Bill is rich, Angie is both rich and happy.

16. Only if Angie is happy will her being rich be a sufficient condition for her being friendly.

17. Bill's being neither rich nor happy is a necessary condition for his being unfriendly.

* 18. Bill's being neither rich nor happy is not a sufficient condition for his being unfriendly.

19. If either Angie or Bill is rich, then either both are friendly or else neither is.

20. Only if neither is rich will both Angie and Bill be not unfriendly.

12.3 The Two Quantifiers

The vocabulary of QL so far consists of individual constants, individual variables, and predicate constants, together with the PL connectives and brackets. These suffice to symbolize particular statements. But how are we going to symbolize a sentence like:

Some women are brave.

or

All frogs are happy.

Such sentences are general (as opposed to particular). We need some extra symbolism to do the job of expressing words like 'some' or 'all,' which represent *quantities*. The extra symbols are called *quantifiers*.

Here is how they work. Look at the first example. We can paraphrase it as follows:

> For some x, x is a woman and x is brave.

which can be partially symbolized:

> For some x, Mx & Bx

We can complete the job by picking some special symbol to represent the existential quantifier "for some x." We will use $(\exists x)$ to mean "for some x" or "there exists an x." So:

> Some women are brave.

is completely symbolized as:

> $(\exists x)(Mx$ & $Bx)$

A statement with only an existential quantifier is called an *existential general statement*. We will use $(\forall x)$ to stand for the universal quantifier "for all x," "for every x," or "for any x." We can then symbolize:

> All frogs are happy.

by first paraphrasing it:

> For any x, if x is a frog, then x is happy.

and then symbolizing it as:

> $(\forall x)(Fx \rightarrow Hx)$

A statement with only a universal quantifier is called a *universal general statement*.

The vocabulary of QL is laid out in Table 12.1.

Table 12.1

Symbols	Intended use
u, v, w, x, z	Individual variables
$a, b, c, \ldots, t$	Names (individual constants)
$(\forall x), (\exists x)$	Universal and existential quantifiers
$A, B, C, \ldots$	Predicate constants
v, −, &, →, ↔	Connectives
(,), { , }, [,]	Brackets for punctuation

Quantifiers are always used with a *domain of discourse* in mind. The size of this domain varies with the topic. For example, "for all x" could mean "for any thing" in the widest sense of that term, the domain of discourse consisting of actual entity (chairs, atoms, numbers, people, and so on). Another term for the domain of discourse is 'the universe of discourse,' and we will use those terms interchangeably. QL statements will be more or less simple, depending upon the universe of discourse about which we choose to speak. Thus if our sentence is:

> Everybody at the party had a good time.

and our universe of discourse is the whole universe, then it would be paraphrased as:

> For all x, if x is a person and x was at the party, then x had a good time.

which would be symbolized as:

> $(\forall x)\{(Hx \ \& \ Px) \rightarrow Gx\}$

where

> $Hx = x$ is a person.
> $Px = x$ was at the party.
> $Gx = x$ had a good time.

But if we choose the universe of discourse to include just the human race, our example is symbolized as:

> $(\forall x)(Px \rightarrow Gx)$

And finally, if our universe of discourse is just the set of people at the party, the symbolization would be:

> $(\forall x)Gx$

12.4 Translation from English to Quantificational Logic

As before with PL, translation from ordinary English into our artificial language is not always easy. But, as before, we can give a few specific hints.

Hint No. 1: Watch out for the universe of discourse (UD).

12 / *The Logic of Properties*

Sometimes a universe of discourse will be obvious; it might even be mentioned explicitly. But other times no universe of discourse will be mentioned. Unless it is clearly otherwise, assume that the universe of discourse is the whole universe.

Hint No. 2: Always try to rephrase the statement at hand to get either "for all x . . . " or "for some x. . . . "

Examples:

Nothing is beautiful.

becomes:

For any x, x is not beautiful.

becomes:

$(\forall x){-}Bx$.

Next:

All men are dirty.

becomes:

For any x, if x is a man, then x is dirty.

becomes:

$(\forall x)(Mx \to Dx)$.

Hint No. 3: There are four forms of statements that are very common.

Traditional logic held that statements could be put in four standard forms. These forms, **categorical propositions**, were the basis of traditional logic. They are so common that we should note their structures. They are:

All S are P.

symbolized as:

$(\forall x)(Sx \to Px)$

No S are P.

symbolized as:

$(\forall x)(Sx \rightarrow Px)$

or else:

$-(\exists x)(Sx \ \& \ Px)$

Some S are P.

symbolized as:

$(\exists x)(Sx \ \& \ Px)$

Some S are not P.

symbolized:

$(\exists x)(Sx \ \& -Px)$

(Note that S and P stand for any predicates.)

Hint No. 4: Remember that properties can be truth-functionally combined.

One weakness of syllogistic logic is in requiring the user to put a statement such as "all college professors are either old or dull" into one of the four categorical forms. This required the student to choose a letter to stand for "the class of things that are either old or dull." This is awkward, and does not reveal the inner structure of the statement. A better translation is now possible:

$(\forall x)[Px \rightarrow (Ox \ \text{v} \ Dx)]$

Hint No. 5: It is almost never correct to have an existential quantifier in front of a statement function whose main connective is an arrow.

Consider:

Some men are nice.

You might be tempted to translate this as:

$(\exists x)(Mx \rightarrow Nx)$

However, this would not be a good translation, because such a claim would be true if even one thing x satisfies $Mx \rightarrow Nx$. But because of the nature of the arrow, that would be true for any x that was *not* a man, that is, for which Mx was false.

Thus any statement of the form $(\exists x)(Ax \rightarrow Bx)$ is bound to be very weak. There is a better way to translate our example:

Some men are nice.

is:

For some x, x is a man and x is nice.

which becomes:

$(\exists x)(Mx \,\&\, Nx)$

A similar point holds with respect to having a universal quantifier in front of a statement function whose main connective is an ampersand, since the resulting statement would be extremely strong. For example, you might think to symbolize:

All people are friendly.

as:

$(\forall x)(Px \,\&\, Fx)$

But that would be wrong: $(\forall x)(Px \,\&\, Fx)$, in fact, says that everything in the universe is a friendly person—which is hardly true of tables and chairs. The correct symbolization is:

$(\forall x)(Px \rightarrow Fx)$

Hint No. 6: Watch out for inverted word order.

Sentences such as "Men are all sex maniacs" are rhetorically effective, but logically deceiving. This is best rephrased as "All men are sex maniacs" and translated accordingly:

$(\forall x)(Mx \rightarrow Sx)$

Hint No. 7: Watch out for omitted quantifiers.

Some sentences omit the quantifier altogether. "Cats are sly" means "All cats are sly" and is accordingly translated:

$(\forall x)(Cx \rightarrow Sx)$

12.4 / Translation from English to Quantificational Logic

On the other hand, the sentence "Children are present" means "Some children are present" and is:

$(\exists x)(Cx \ \& \ Px)$

Hint No. 8: Watch out for omitted copulas.

Sentences often omit the copula (some tense of the verb *to be*) as for example:

All people want money.

The best translation we can make using QL, as we have defined it in this chapter, would be to rephrase it as:

All people are money-wanters.

and symbolize it as:

$(\forall x)(Px \rightarrow Mx)$

Hint No. 9: Keep in mind that English has many quantifiers besides 'all' and 'some.'

Table 12.2 lists many of the words that express quantity in English. Table 12.3 will help develop insight into translating statements that use such expressions. Move from English to the pseudocode (the intermediate stage) first, and then to the QL symbolism.

Table 12.2 *Common English Quantifiers*

Any(body)	Many	A few
Anything	Most	A lot
Every(body)	Almost all	There is/are
Everything	A	An example is
Each	An	Only
Nothing		
Not all		
All but		

Table 12.3

Original Natural English	Pseudocode Paraphrase	Final QL Translation
Anybody who likes Brubeck can't be all bad.	For all x, if x is a person and x likes Brubeck, x is not all bad.	$(\forall x) [Px \& Bx] \rightarrow -Ax]$
A tree grows in Los Angeles.	Something is a tree and is located in Los Angeles.	$(\exists x)(Tx \& Lx)$
BUT COMPARE: A good scout is honest.	For any x, if x is good and a scout, x is honest.	$(\forall x) [(Gx \& Sx) \rightarrow Hx]$
Many athletes are highly paid. or A lot of athletes are highly paid. or A few athletes are highly paid. or Most athletes are highly paid.	We lose something in our translations. Some x is an athlete and is highly paid.	$(\exists x)(Ax \& Hx)$
There is a good restaurant in Barstow.	For some x, x is good and a restaurant and in Barstow.	$(\exists x) [(Gx \& Px) \& Bx]$
Only women can dance.	For any x, if x can dance, x is a woman.	$(\forall x)(Dx \rightarrow Wx)$

EXERCISES 12.4

Symbolize the following sentences, selecting letters that the capitalized words indicate. The universe of discourse is the class of people. (An asterisk indicates problems that are answered in the back of the book.)

    1. Some people LIKE sashimi.
*   2. Everybody is HAPPY.
    3. Not everybody is LONELY.

4. Some lack COURAGE.
5. People are MONEY-grubbers.
* 6. People are mostly unHAPPY.
7. A lot of RICH people are unHAPPY.
8. Some RICH people are unHAPPY.
9. Nobody likes CHEESE.
* 10. Nobody who likes CHEESE likes TEMPURA.
11. Anybody who CHOOSES a life of crime can't be FULLY rational.
12. Unless a person is HAPPY, that person can't be FRIENDLY.
13. If a person DOES a good job, that person will be REWARDED.
* 14. Many people are UNHAPPY unless they are FREE.
15. A person will be HAPPY if and only if that person is EVIL.
16. Few VOLUNTEER for hazardous duty.
17. There are FRIGHTENED and ANGRY people.
* 18. A good TEACHER is PATIENT.
19. A DOCTOR is NEAR.
20. A good TEACHER is neither EASY to anger nor HARD to talk to.
21. Some who like CHEESE and WINE also like PRUNES.
* 22. Most GOOD restaurants TAKE credit cards.
23. Being FRIENDLY is a necessary condition for being LIKED.
24. A necessary condition for a TEACHER to be POPULAR is that the teacher ASSIGN high grades.
25. If someone wants to DANCE, that person will have to GO to the nightclub.
* 26. There are no FRIENDLY SHARKS who EAT cheese.
27. People hate LOSERS, WHINERS, CREEPS, and JERKS.
28. A necessary and sufficient condition for being HATED is that a person wear TIES.
29. A few EXECUTIVES make it to the TOP without CHEATING.
* 30. Unless she LEFT, there is a COP NEARBY.

12.5 Expansions

We can better understand the meanings of the two quantifiers if we examine the relation between quantified statements and statements about individuals.

Suppose the universe of discourse consists of just one individual, named a. What does the assertion that there is at least one individual with property P amount to? Simply that a has P. But now suppose the universe had exactly two individuals, a and b. The assertion that there is at least one individual with P then amounts to saying that either a has P or b has P (or both).

Similarly, in a model (possible universe) of three individuals, an existentially quantified statement expands into a disjunction with three disjuncts. Clearly, in an n-individual model:

$$(\exists x) Px \Leftrightarrow \{Pa_1 \lor Pa_2 \lor \ldots \lor Pa_n\}$$

Turning next to universally quantified statements, the statement that all individuals have P in a model of one individual a amounts to asserting Pa. In a model consisting of two individuals a and b, asserting that all individuals have P amounts to saying that a has P and b has P. In a model of n individuals, the statement that all individuals have P is equivalent to a conjunction of n conjuncts:

$$(\forall x)Px \Leftrightarrow \{Pa_1 \& Pa_2 \& \ldots \& Pa_n\}$$

Table 12.4 summarizes these points.

In a model of one individual, $(\exists x)Px$ and $(\forall x)Px$ are equivalent. Only when we move to a two-individual model does the difference appear. Expansion of a given quantified statement is done as indicated previously.

Example: Expand $(\forall x)(Fx \lor Gx)$ in a two-individual model.

Answer: Let the individuals be named a and b. A universally quantified statement expands into a conjunction:

$$(Fa \lor Ga) \& (Fb \lor Gb)$$

Table 12.4

Quantified Statement	One-Individual Model	Two-Individual Model	Three-Individual Model	n-Individual Model
$(\forall x)Fx$	Fa_1	$Fa_1 \& Fa_2$	$Fa_1 \& Fa_2 \& Fa_3$	$Fa_1 \& \ldots \& Fa_n$
$(\exists x)Fx$	Fa_1	$Fa_1 \lor Fa_2$	$Fa_1 \lor Fa_2 \lor Fa_3$	$Fa_1 \lor \ldots \lor Fa_n$

Example: Expand $(\exists x)(Ax \,\&\, Bx)$ in a three-individual model.

Answer: Let the individuals be $\{a, b, c\}$. An existentially quantified statement expands into a disjunction, with the variable replaced by the names in order:

$(Aa \,\&\, Ba) \text{ v } (Ab \,\&\, Bb) \text{ v } (Ac \,\&\, Bc)$

EXERCISES 12.5

Expand each of the following in the models $\{a\}$, $\{a,b\}$, $\{a,b,c\}$. (An asterisk indicates problems that are answered in the back of the book.)

    1. $(\forall x)Mx$
*   2. $(\exists x)Mx$
    3. $(\forall x)(Ax \to Cx)$
    4. $(\exists x)(Ax \to Cx)$
    5. $(\forall x)(-Ax \,\&\, Bx)$
*   6. $(\forall x)(-Rx \to -Ax)$
    7. $(\exists x)-(Ax \text{ v } Lx)$
    8. $(\forall x)[-(-Ax \text{ v } -Bx) \,\&\, Cx]$
    9. $(\forall x)[-(-Ax \text{ v } -Mx) \,\&\, -Rx]$
* 10. $(\forall x)[(--Rx \text{ v } Zx) \to Qx]$
  11. $(\exists x)[(--Rx \text{ v } Zx) \to Qx]$
  12. $(\forall x)[(Tx \,\&\, -Mx) \to Rx]$
  13. $(\exists x)[(Wx - Wx) \text{ v } (Lx \to Wx)]$
* 14. $(\exists x)[(-Ax \,\&\, Bx) \,\&\, (Rx \,\&\, Cx)]$
  15. $(\forall x)[(-Rx \text{ v } -Bx) \text{ v } (Sx \text{ v } -Tx)]$

12.6 Quantifier Exchange

The statement:

    No people like cheaters.

can be symbolized as:

    $(\forall x)(Px \to -Cx)$

or else as:

$$-(\exists x)(Px \;\&\; Cx)$$

These expressions must be logically equivalent, but why? Here we can use our expansion technique to guide our intuitions. Suppose we have the denial of an existentially quantified statement:

$$-(\exists x)Px$$

We can expand this in an *n*-individual model:

$$-\{Pa_1 \text{ v } Pa_2 \text{ v } \ldots \text{ v } Pa_n\}$$

which we can apply De Morgan's laws (DM) to:

$$-Pa_1 \;\&\; -Pa_2 \;\&\; \ldots \;\&\; -Pa_n$$

But that last expression is just the expansion of:

$$(\forall x)-Px$$

in the same *n*-individual model.

Thus we have one form of the rule of *quantifier exchange* (QE):

$$-(\exists x)Px \Leftrightarrow (\forall x)-Px$$

To return to our earlier example:

$$-(\exists x)(Px \;\&\; Cx)$$

is equivalent to:

$$(\forall x)-(Px \;\&\; Cx)$$

which by DM is equivalent to:

$$(\forall x)(-Px \text{ v } -Cx)$$

which by implication (Imp) is equivalent to:

$$(\forall x)(Px \rightarrow -Cx)$$

as required.

12.6 / Quantifier Exchange

A second form of **QE** can also be discovered by expansion. Start with:

$$-(\forall x)Px$$

expand in an *n*-individual model:

$$-\{Pa_1 \ \& \ Pa_2 \ \& \ \ldots \ \& \ Pa_n\}$$

apply DM to get:

$$-Pa_1 \lor -Pa_2 \lor \ldots \lor -Pa_n$$

which is the expansion of:

$$(\exists x)-Px$$

So the second form of **QE** is:

$$-(\forall x)Px \iff (\exists x)-Px$$

In a like fashion, two other forms of **QE** can be derived. The following are the forms of **QE**.

$$-(\exists x)Px \iff (\forall x)-Px$$
$$-(\forall x)Px \iff (\exists x)-Px$$
$$-(\exists x)-Px \iff (\forall x)Px$$
$$-(\forall x)-Px \iff (\exists x)Px$$

The easiest way to remember these forms is to imagine moving the dash past a quantifier (in either direction). As the dash passes over the quantifier, the quantifier changes into the opposite quantifier. Double negation (DN) then may be applicable. **QE** is the first new inference rule of QL. Keep in mind that **QE** applies to parts of lines as well as whole lines.

Example: Apply QE to $-(\exists x)-(Ax \ \& \ Rx)$.

Answer: $-(\exists x)-(Ax \ \& \ Rx)$

becomes by QE:

$$(\forall x)--(Ax \ \& \ Rx)$$

becomes by DN:

$(\forall x)(Ax \ \& \ Rx)$

Example: Apply QE to $(\forall x)(Ax \lor Cx)$

Answer: We need to have some dashes to apply QE, so we apply DN:

$(\forall x)--(Ax \lor Cx)$

Apply QE:

$-(\exists x)-(Ax \lor Cx)$

We could just as well have applied DN differently:

$--(\forall x)(Ax \lor Cx)$

but when we apply QE we reach the same point:

$-(\exists x)-(Ax \lor Cx)$

EXERCISES 12.6

Apply QE to obtain expressions that do not have dashes in front of quantifiers, and then apply DN to reduce the number of dashes. (An asterisk indicates problems that are answered in the back of the book.)

    1. $-(\exists x)Ax$
*   2. $--(\forall x)Mx$
    3. $---(\exists x)-Qx$
    4. $-(\forall x)(Rx \lor -Rx)$
    5. $-(\exists x)--(Rx \lor -Rx)$
*   6. $-(\forall x)-(Ax \ \& \ Mx)$
    7. $--(\exists x)--(Ax \rightarrow -Rx)$
    8. $-(\forall x)---(Hx \rightarrow Tx)$
    9. $-(\exists x)-(-Hx \rightarrow Tx)$
* 10. $--(\forall x)---(Hx \lor Zx)$

12.7 Bondage and Scope

We already have one new inference rule: quantifier exchange (QE). Eventually we will have four more new rules, but before we can discuss them, a few concepts need to be addressed. The first is the concept of scope. The **scope** of a logical symbol is the range of application of that symbol. We indicate scope by means of brackets and also informal convention. Recall how the dash functions:

$$-A \rightarrow -(B \text{ v } C)$$

As indicated previously, the first dash applies to the nearest letter, because no brackets are around it, and we have an informal convention that says that in the absence of brackets, the dash only applies to the nearest letter. The second dash applies to the whole bracketed disjunction. We will adopt the same approach to quantifiers. The scope of a quantifier is indicated by brackets, and if no brackets occur, the quantifier only applies to the nearest simple property. For example, in:

$$(\forall x)Mx \rightarrow (\exists y)(My \text{ \& } Ty)$$

the scope of $(\forall x)$ extends only to Mx.

We turn next to the concept of *bondage*. How many words are in the following sentence?

The boy took the dog to the vet.

In one sense, eight—but in another sense, only six. One word, 'the,' has three occurrences. Similarly, in this QL sentence:

$$(\forall x)(Ax \rightarrow (Bx \text{ v } Cx))$$

the same variable has four occurrences. An occurrence of a variable is **bound** if and only if it is part of or falls within the scope of a quantifier that has that variable in it. An occurrence of a variable is **free** if and only if it is not bound.

Example: Which occurrences of x are free and which are bound, in:

$$(\forall x)(Mx \rightarrow Tx) \text{ \& } -(Rx \text{ v } Hx)$$
$$12\phantom{x)\text{ \&}}3\phantom{\text{ \& }-(}4\phantom{x\text{ v }}5$$

Answer: Occurrence 1 is bound because it is part of a quantifier; 2 and 3 are bound since they fall within the scope of a quantifier that is of the appropriate sort; 4 and 5 are free because they are outside the scope of the appropriate quantifier (that is, the quantifier that has the same variable).

Example: Which occurrences are free, and which are bound, in:

$$(\forall x)(\exists y)(Ax \ \& \ By) \rightarrow My$$
$$1 2 \ \ 3 4 5$$

Answer: Occurrences 1 and 2 are bound because they are parts of quantifiers; 3 is bound by the quantifier $(\forall x)$, but not by $(\exists y)$; 4 is bound by $(\exists y)$, but of course not by $(\forall x)$; finally, 5 is free because it is outside the scope of the quantifier $(\exists y)$.

Example:

$$(\forall x)(\forall y)[(Mx \ \& \ Lx) \rightarrow Qy] \rightarrow (-Rx \ v \ Qz)$$
$$1 2 3 4 5 6 7$$

Answer: Occurrences 1 and 2 are bound because they are parts of quantifiers; 4 and 5 are bound because they fall within the scope of appropriate quantifiers, but 3 is not bound by either $(\forall x)$ or $(\forall y)$; 6 and 7 are free because they are outside the scope of any quantifier.

We now have the technical terminology to define more precisely the notion of a 'propositional function': a **propositional function** (or 'statement function') is any expression with at least one free occurrence of at least one variable.

EXERCISES 12.7

For each of the following, state which occurrences are free and which are bound. (An asterisk indicates problems that are answered in the back of the book.)

1. $(\forall x)(Ax \ \& \ Bx)$
   $1 2 3$

* 2. $(\exists x)(Ax \ \& \ By)$
   $1 2 3$

3. $(\forall x)(Ay \to By)$
      1   2   3

4. $(\forall x)(\exists y)(Mx \,\&\, Gy)$
      1   2   3   4

5. $(\exists x)(\exists y)(Mx \,\&\, Gx)$
      1   2   3   4

* 6. $(\exists x)Mx \to (\exists y)Gy$
       1   2   3 4

7. $(\exists x)[Mx \to (\exists y)\, Gy]$
      1   2   3   4

8. $(\forall x)[Mz \,\&\, (\forall y)\, Gz] \text{ v } Gx$
      1   2   3   4   5

9. $(\forall x)(\forall z)[Mz \,\&\, (\forall y)\, Gz] \text{ v } Gx$
      1   2   3   4   5   6

* 10. $(\forall x)(\forall y)(\forall z)[Az \,\&\, (Ay \text{ v } Au)] \text{ v } Az$
       1   2   3   4   5   6   7

12.8 The Concept of Instantiation

We have, on the one hand, the notion of a propositional function as involving at least one free occurrence of at least one variable, and, on the other hand, the notion of a statement. How are these two notions related? Consider the simple proposition function:

    Hx

meaning "x is happy." If we fill in the blank (remember, a variable is like a blank, a place holder) with a name (say, r = Roberta), we get a statement:

    Hr

meaning "Roberta is happy." The concept of instantiation grows out of this idea of plugging in a name for a free occurrence of a variable. On the other hand, the same proposition function:

    Hx

can be turned into a statement by binding the free variable with a quantifier:

    $(\forall x)Hx$ or $(\exists x)Hx$

Thus we will adopt the following definition of *instantiation*: Pa is an *instance* of $(\exists x)Px$ or $(\forall x)Px$ if and only if a is a name, x is a variable, and every free occurrence of x in Px is replaced by a. So instantiation takes place as the following indicates.

Example: Instantiate $(\forall x)(Lx \to Tx)$ to b.

Answer:

    Step 1: Strip away quantifiers.

$(Lx \to Tx)$

    Step 2: Note the free occurrences of the quantifier variable.

$Lx \to Tx$
 1   2

    Step 3: Replace them by the constant.

$Lb \to Tb$

Example: Instantiate $(\forall y) -(Ay \vee By)$ to c.

Answer:

    Step 1: Strip away the quantifier.

$-(Ay \vee By)$

    Step 2: Identify the free occurrences.

$-(Ay \vee By)$
    1   2

    Step 3: Replace those free occurrences by c.

$-(Ac \vee Bc)$

Besides getting an instance from a quantified statement, we should be able to go in the other direction: to determine whether a given particular statement is an instance of a given quantified statement by instantiation.

Example: Is $Fa \mathbin{\&} Gb$ an instance of $(\exists x)(Fx \mathbin{\&} Gx)$?

12.8 / The Concept of Instantiation

Answer: Try working backwards. Start with the quantified statement, and see whether or not you can get the particular statement.

$(\exists x)(Fx \ \& \ Gx)$

$Fx \ \& \ Gx$

$Fa \ \& \ Ga$

But this is not $Fa \ \& \ Gb$. We cannot get $Fa \ \& \ Gb$ from $(\forall x)(Fx \ \& \ Gx)$ by instantiation, because in instantiation we must replace all free occurrences of x by the constant.

Example: Is $Ly \rightarrow Ty$ an instance of $(\forall x)(Lx \rightarrow Tx)$?

Answer: Work backwards.

$(\forall x)(Lx \rightarrow Tx)$

$Lx \rightarrow Tx$

$Ly \rightarrow Ty$

But we instantiated by plugging in another variable, and this is not allowed, according to our concept of instantiation. We can only instantiate to constants, that is, proper names.

Example: Instantiate $(\forall x)[Hx \rightarrow (\forall y) \ Ry]$ to a.

Answer: Strip away the outside quantifier:

$Hx \rightarrow (\forall y) \ Ry$

and replace the free occurrences of x by a.

$Ha \rightarrow (\forall y) \ Ry$

Note that the result would have been the same for:

$(\forall x)(Hx \rightarrow (\forall x)Rx)$

EXERCISES 12.8A

Instantiate the following quantified statements to a. (An asterisk indicates problems that are answered in the back of the book.)

1. $(\forall x)Hx$

* 2. $(\exists x)Hx$

3. $(\forall x)(Ax \,\&\, Bx)$
4. $(\exists x)(-Rx \text{ v } Cx)$
5. $(\forall y)(-Ry \text{ v } Cy)$
* 6. $(\forall y)(-Zy \rightarrow Zy)$
7. $(\exists z)(Rx \rightarrow -Qz)$
8. $(\exists w)[(Mw \,\&\, Nw) \rightarrow (Rw \text{ v } Tw)]$
9. $(\forall z)[(Lz \,\&\, Nz) \rightarrow (Qz \,\&\, Pz)]$
* 10. $(\exists u)[--Au \rightarrow (Ru \text{ v } -Zu)]$
11. $(\forall x)[Hx \,\&\, (\forall y)Ry]$
12. $(\forall y)[(\forall x)Mx \rightarrow Ky]$
13. $(\forall y)(Ay \,\&\, -My)$
* 14. $(\exists x)[(\exists y)Ry \text{ v } -Hx]$
15. $(\forall x)[Hx \,\&\, (\forall x)Hx]$

EXERCISES 12.8B

For each pair of expressions, determine if the first is an instance of the second. If not, state why not. (An asterisk indicates problems that are answered in the back of the book.)

1. $Ra, (\forall x)Rx$
* 2. $-Ra, (\forall x)Rx$
3. $--Ra, (\forall z)--Rz$
4. $-(Re \,\&\, Te), (\exists y)-(Ry \,\&\, Ty)$
5. $Ry \rightarrow Ty, (\forall x)(Rx \rightarrow Tx)$
* 6. $Ra \,\&\, Ca, (\exists x)(Rx \,\&\, Cx)$
7. $-Wc \rightarrow --Xc, (\forall w)-(Ww \rightarrow --Xw)$
8. $Qb \text{ v } -Qb, (\forall w)-(Qw \text{ v } -Qw)$
9. $-Tc \rightarrow -Tb, (\exists x)(-Tx \rightarrow -Tx)$
* 10. $-Tc \rightarrow -Tc, (\forall x)-(Tx \rightarrow Tx)$
11. $Ra \text{ v } (\forall x)Tx, (\exists y)[Ry \,\&\, (\forall x)Tx]$
12. $Ra \rightarrow La, (\forall y)[Ry \rightarrow (\forall y) Ly]$
13. $Ba \,\&\, Ra, (\forall x)[Bx \,\&\, (\forall y) Ry]$
* 14. $--Sa \,\&\, Sb, (\forall y)(Sy \,\&\, Sy)$
15. $-(Sa \,\&\, Sb), (\forall y)-(Sy \,\&\, (\forall y) Sy)$

12.9 Two New Inference Rules

In Chapter 10, we developed a handy system of inference rules to prove valid PL arguments. Can we extend this system to do the same for arguments in QL? Yes, and with surprisingly little strain. In fact, we need add only four more rules. Each of these new inference rules—quantification rules—is intuitively clear. We begin with the rule of *universal instantiation* (UI). Consider the statement form $(\forall x)Px$. It says that everything has property P, or more precisely, that every substitution instance of the (possibly compound) propositional function Px is true. That means that for any particular individual you care to name—say, n—Pn is true. For instance, if all things are green, then this book is green. We accordingly establish the rule:

UI
$(\forall x)Px$
―――――
Pn

where Pn is a substitution instance of $(\forall x)Px$. (This restriction is assumed in all the inference rules.) The meaning of UI should be clear. If at any point in a proof we have a universal statement, we can instantiate to any individual we care to. We descend from the general to the particular. Consider an example:

1. Everybody likes money.
2. Everybody likes leisure.

∴ David Bowie likes money and leisure.
    (Enthymeme: It is known that David Bowie is a person.)

We translate this argument as:

$Px = x$ is a person.
$Mx = x$ likes money.
$Lx = x$ likes leisure.
$b =$ David Bowie.

1. $(\forall x)(Px \to Mx)$
2. $(\forall x)(Px \to Lx)$
3. Pb
4. $Pb \to Mb$ $\quad\quad$ | Mb & Lb
                   $\quad\quad$ 1, UI (instantiate to b)

5. $Pb \rightarrow Lb$	1, UI (also instantiate to b)
6. Mb	3, 4, MP
7. Lb	3, 5, MP
8. $Mb \,\&\, Lb$	6, 7, Conj

Once we have instantiated and thus gotten rid of the universal quantifiers, the rest of the proof proceeds like the proofs shown in Chapter 10.

We can get a better intuitive grasp of UI if, as we did with QE, we expand the quantified statement in a finite model—say, $\{a, b\}$. The rule:

$$(\forall x)Px$$

becomes

$$\frac{Pa \,\&\, Pb}{Pa}$$

which is just a case of simplification. More generally, in a finite universe, a universal statement is equivalent to a conjunction, and UI merely says that we may simplify to any conjunct we wish. This doesn't mean that quantifiers are dispensable, of course, since not all models are finite.

The second inference rule is equally obvious. Suppose you wanted to prove that someone has scored 100 points in a pro basketball game. All you would have to do is point to Wilt Chamberlain, to name an individual who has done it. We can properly infer $(\exists x)Px$ from any instance Pn of the statement function Px. This rule is called *existential generalization* (EG):

$$\text{EG} \; \frac{Pn}{(\exists x)Px}$$

where (again) Pn is a substitution instance of $(\exists x)\, Px$.

Consider the argument:

1. Everybody likes money.
2. Carol likes parrots.

∴ Somebody likes both money and parrots.

The translation (picking up the premises left out in this enthymeme):

12.9 / Two New Inference Rules

$Bx = x$ likes parrots.
$Px = x$ is a person.
$Mx = x$ likes money.
$c =$ Carol.

1. $(\forall x)(Px \to Mx)$
2. Bc
3. Pc $\lfloor (\exists x)[Px \& (Mx \& Bx)]$

The proof is easy:

1. $(\forall x)(Px \to Mx)$
2. Bc
3. Pc
4. $Pc \to Mc$ $\lfloor (\exists x)(Px \& (Mx \& Bx))$
                  1, UI (we can instantiate to anything, but only c is under discussion)
5. Mc 3, 4, MP
6. $Mc \& Bc$ 3, 5, Conj
7. $Pc \& (Mc \& Bc)$ 3, 6, Conj
8. $(\exists x)(Px \& (Mx \& Bx))$ 7, EG

The heart of the proof, steps 4–8, is essentially a PL proof. We apply the QL rules early to strip away quantifiers, then proceed by normal PL methods, and end the proof by putting quantifiers back on. This suggests a heuristic rule: try to see how the QL argument can be reduced to a PL situation. Example:

1. $(\forall x)(Ax \to Bx)$
2. $(\forall x)(Ax \to Cx)$ $\lfloor Aa \to (Ba \& Ca)$
3. Aa Assume for CP to show $Ba \& Ca$
4. $Aa \to Ba$ 1, UI
5. $Aa \to Ca$ 2, UI
6. Ba 3, 4, MP
7. Ca 3, 5, MP
8. $Ba \& Ca$ 6, 7, Conj
9. $Aa \to (Ba \& Ca)$ 3–7, CP

Let us conclude this section with a technical detail. We have stated the rules of QE, UI, and EG using the predicate constant P, the individual variable x, and the individual constant n. But our rules are intended to apply to any predicate (not just P), to any individual variable (not just x), and to any appropriate individual constant (not just n). For this reason, technically speaking, we should state our rules using metalinguistic variables, which range over the constants and variables of QL. For example, we could (as some texts do) use the Greek letter Φ (phi) to range over predicates, the Greek letter α (alpha) to range over individual variables, and the Greek letter υ (upsilon) to range over individual constants. We would then, for example, state the rule of UI as:

$$\frac{(\forall \alpha) \Phi \alpha}{\Phi \upsilon}$$

where $\Phi \upsilon$ is an instance of $(\forall \alpha)\Phi \alpha$.

However, most readers find the Greek symbols confusing. We will thus allow a slight and tolerable ambiguity in our use of P, x, and n. When they occur in the statement of an inference rule, they should be treated as metalinguistic variables: P referring to any predicate, x any variable, and n any name. Otherwise, they are just symbols of QL, with the same use as other similar symbols in its vocabulary.

EXERCISES 12.9

The following valid arguments have already been symbolized. Provide the proofs. (An asterisk indicates problems that are answered in the back of the book.)

1. 1. $(\forall x)(Hx \to Tx)$
   2. $(\forall x)(Tx \to Jx)$
   3. Hj $\lfloor Jj$

* 2. 1. $(\forall x)\{Mx \to (Tx \,\&\, Qx)\}$
   2. Ma $\lfloor (\exists x)(Qx \lor Lx)$

3. 1. $(\forall x)(Ax \to Bx)$
   2. $--(An \,\&\, Cn)$ $\lfloor (\exists x)(Bx \,\&\, Cx)$

4. 1. $(\forall x)(Ax \to Bx)$
   2. $(\forall x)(Bx \to Cx)$
   3. $(\forall x)(Cx \to Cx)$ $\lfloor An \to Dn$

5. 1. $(\forall x)(Qx \to Rx)$
   2. $(\forall x)(-Qx \to Zx)$

       3. —Zr $\lfloor Rr$

* 6. 1. $(\forall x)(Bx \leftrightarrow Tx)$
    2. $(\forall x)(Tx \leftrightarrow Zx)$
    3. Ba $\lfloor (\exists x)(Bx)$

7. 1. $(\forall y)(-Ty \leftrightarrow -Qy)$
   2. $(\forall z)(-Qz \leftrightarrow Rz)$ $\lfloor -(-Ta\ \&\ -Ra)$

8. 1. $(\forall z)(Lz \leftrightarrow Zz)$
   2. $(\forall y)(Zy \leftrightarrow Ty)$ $\lfloor -Lc\ \text{v}\ Tc$

9. 1. $(\forall x)(Mx \leftrightarrow Nx)$
   2. $(\forall x)(Mx \leftrightarrow -Nx)$
   3. Mt $\lfloor (\exists x)\ Zx$

* 10. 1. $(\forall x)(Mx \leftrightarrow Nx)$
    2. $(\forall x)(Mx \leftrightarrow -Nx)$
    3. Mt $\lfloor (\exists y)\ My$

12.10   The Rule of Universal Generalization

The rule of *universal generalization* (UG) underlies the practice of proving a universal claim by arguing from an arbitrarily chosen particular case. Recall high school geometry: to prove that all triangles have a given property P, you would say, "Let A be an arbitrary triangle. I will assume only that it is a triangle, and show that it must have P." The rule of UG can be roughly stated as:

$$\text{UG} \quad \frac{Pn}{(\forall x)Px}$$

where again Pn is a substitution instance of $(\forall x)Px$.

Consider this simple proof using UG.

    1. $(\forall x)(Ax \to Bx)$
    2. $(\forall x)(Bx \to Cx)$ $\lfloor (\forall x)(Ax \to Cx)$
    3. Aa Assume to show Ca
    4. $Aa \to Ba$ 1, UI
    5. $Ba \to Ca$ 2, UI
    6. Ba 3, 4, MP
    7. Ca 5, 6, MP
    8. $Aa \to Ca$ 3–7, CP
    9. $(\forall x)(Ax \to Cx)$ 8, UG

The relation between line 9 and line 8 is instantiation, yet in moving from 8 to 9, we do not instantiate, but do the reverse. Unlike the rules of UI and EG, which were stated without restrictions, we must add restrictions to UG to prevent invalid inferences. First, we insist that n (the name to which we instantiate) not occur in any assumption within whose scope Pn lies. We need this restriction to block inferences of this sort:

(UD = animals)
All men love money. So since Jason is a man, all animals (dogs, cats, bats, etc.) love money.

Symbolically:

1. $(\forall x)(Mx \to Lx)$ $\quad\quad\quad Mj \to (\forall x)Lx$

Without the aforementioned restriction, we could construct this bogus proof:

1. $(\forall x)(Mx \to Lx)$ $\quad\quad\quad Mj \to (\forall x)Lx$
2. Mj $\quad\quad\quad$ Assume
3. $Mj \to Lj$ $\quad\quad\quad$ 1, UI
4. Lj $\quad\quad\quad$ 2, 3, MP
5. $(\forall x)Lx$ $\quad\quad\quad$ 4, UG (wrong!)
6. $Mj \to (\forall x)Lx$ $\quad\quad\quad$ 2–5, CP

Next, we need a restriction that will ensure that UG is used only on an arbitrary individual. We accomplish this by requiring that n not occur in any of the premises. Otherwise, we could prove this invalid argument:

Jason can't dance. So nobody can dance.

symbolized:

1. $-Dj$ $\quad\quad\quad (\forall x)-Dx$

by this bogus proof:

1. $-Dj$
2. $(\forall x)-Dx$ $\quad\quad\quad$ 1, UG

12.10 / *The Rule of Universal Generalization*

Finally, we insist that when UG is used, all the occurrences of n be generalized upon. Otherwise, we could prove this invalid argument:

All Hungarians love stuffed cabbage. So anybody loves stuffed cabbage if Jason is Hungarian.

symbolized:

1. $(\forall x)(Hx \rightarrow Cx)$ $/(\forall x)(Hj \rightarrow Cx)$

by the bogus proof:

2. $Hj \rightarrow Cj$ 1, UI
3. $(\forall x)(Hj \rightarrow Cx)$ 2, UG (wrong)

Let us restate the rule of UG more precisely: (1) n is a name that does not occur in Px, any of the premises, nor in any assumption within whose scope Pn lies. (2) n does not occur in $(\forall x)Px$.

We can now work a few problems.

1. $(\forall x)(Ax \lor Dx)$ $/(\forall x)(-Dx \rightarrow Ax)$
2. $Aa \lor Da$ 1, UI
3. $Da \lor Aa$ 2, Comm
4. $--Da \lor Aa$ 3, DN
5. $-Da \rightarrow Aa$ 4, Imp
6. $(\forall x)(-Dx \rightarrow Ax)$ 5, UG

Note the common pattern: remove quantifiers (here, by UI); do a PL-type proof (here, lines 3–6); put quantifiers back on (here, by UG).

Example: Prove:

1. $(\forall x)(Ax \rightarrow Tx)$
2. $(\forall x)(-Ax \rightarrow -Qx)$ $/(\forall x)(Qx \rightarrow Tx)$

3.	Qa	To show Ta.

(It would be wrong to assume $(\forall x)\, Qx$, since the most that we could show would be something of the form $(\forall x)\, Qx \to$ whatever. It would also be wrong to assume Qx, since we can only generalize on n if n is a constant, not a variable).

4.	$Aa \to Ta$	1, UI
5.	$-Aa \to -Qa$	2, UI
6.	$Qa \to Aa$	5, Contra
7.	$Qa \to Ta$	4, 6, HS
8.	Ta	7, 3, MP
9.	$Qa \to Ta$	3–8, CP
10.	$(\forall x)(Qx \to Tx)$	9, UG

(We could have proved this result more simply without using CP. How?)

EXERCISES 12.10

Construct a proof of each of the following. (An asterisk indicates problems that are answered in the back of the book.

1. 1. $(\forall x)(Ax \to Zx)$ ⊢ $(\forall x)(-Ax \vee Zx)$
* 2. 1. $(\forall x)(Ax \to Rx)$
      2. $(\forall x)(Rx \to Ax)$ ⊢ $(\forall x)(Rx \leftrightarrow Ax)$
  3. 1. $(\forall x)(Mx \to (Tx \,\&\, Rx))$ ⊢ $(\forall x)(Mx \to Tx)$
  4. 1. $(\forall x)((Ax \vee Bx) \to Cx)$ ⊢ $(\forall x)(Ax \to Cx)$
  5. 1. $(\forall x)-Lx$ ⊢ $(\forall x)(Lx \to Tx)$
* 6. 1. $(\forall x)(Ax \vee Bx)$ ⊢ $(\forall x)(-Bx \to Ax)$
  7. 1. $(\forall x)(Ax \,\&\, -Ax)$ ⊢ $(\forall x)\, Cx$
  8. 1. $(\forall x)(Ax \to Zx)$ ⊢ $(\forall y)(Ay \to Zy)$
  9. 1. $(\forall x)\, Tx$ ⊢ $(\forall x)(Rx \to Tx)$
* 10. 1. $(\forall x)(Ax \leftrightarrow Mx)$
      2. $(\forall x)(Mx \leftrightarrow Zx)$ ⊢ $(\forall x)(Ax \leftrightarrow Zx)$

12.11 The Rule of Existential Instantiation

In ordinary life, we may know that a person is responsible for a state of affairs, without knowing who that person is. For example, we may discover that a string of murders has been committed, all done by one person, yet we do not know who. In such cases, we commonly invent a name to refer to this unknown individual: "the Boston Strangler," "the Hillside Strangler," "Mr. X," or "Jane Doe." Using such names facilitates discussion—which is why journalists devise them. The rule of *existential instantiation* (EI) codifies this common-sense technique. If we know that some individual x has some property P, we can choose a name to instantiate to, in order to facilitate the proof. Example:

> Some people like blintzes. But anybody who likes blintzes likes wine.
> Thus some people like wine.

Symbolization (UD = people):

1. $(\exists x)\ Cx$
2. $(\forall x)(Cx \rightarrow Wx)$ $\quad\vdash (\exists x)\ Wx$

The proof exemplifies our general pattern: remove quantifiers, do a PL-type proof, and put quantifiers back on.

3.	Ca	1, EI
4.	$Ca \rightarrow Wa$	2, UI
5.	Wa	3, 4, MP
6.	$(\exists x)\ Wx$	5, UG

But, as with the rule of UG, we need some restrictions on the rule to prevent fallacies. And the restrictions codify common sense. Suppose the police reasoned this way: "Someone has been stealing jewels in neighborhood X. Let's call this person, oh, say, Gary Jason." They then proceed to look my name up in the phone book, get my address, and drive out to place me under arrest. My reply to them (supposing they gave me a chance to reply) would be that in picking a name to refer to an unknown individual, they must pick a name that does not carry along with it any identifying information.

We might formulate our rule of EI accordingly:

$$\frac{(\forall x)Px}{Pn}$$

where n does not occur in $(\exists x)Px$ or any other preceding line.

And this formulation would block the construction of bogus proofs for invalid arguments such as:

> There is someone such that if Dave goes to the dance, that someone will dance with Dave. So if Dave goes to the dance, he will dance with himself.

This argument can be symbolized:

UD = people.
Mx = x goes to the dance.
Dx = x dances with Dave.

1. $(\exists x)(Md \: \& \: Dx)$ | $Md \: \& \: Dd$

Without the restriction placed upon EI, we could construct a bogus proof:

1. $(\exists x)(Md \to Dx)$ | $Md \to Dd$
2. $Md \to Dd$ 1, EI (wrong)

But the restriction blocks the move. However, something still is not quite right. Think about the argument:

> Someone is worth a billion dollars.
> So Jason is worth a billion dollars.

This is clearly invalid. But we can 'prove' it using the rule as stated (symbolized with UD = people):

1. $(\exists x)Wx$ | Wj
2. Wj 1, EI

Step 2 follows from EI as we have it stated: j (the instantiating constant) does not occur in line 1, and line 1 is the only preceding line.

12.11 / The Rule of Existential Instantiation

What has gone wrong is that, in instantiating existentially, we are instantiating to a 'sort-of' name (or a 'quasi-name'), and we have no assurance the quasi-name will not be used as a true name later in the proof. So besides having a restriction to the effect that n not occur before using EI, we need the restriction that n not be used after it has played its proper role. The concept we are groping for here is the idea of some kind of temporary assumption that has limited scope. We already have a device for handling such a concept—the bent arrow used in Sections 10.7 and 10.8 in the statement of CP and Reductio. We can put EI thus:

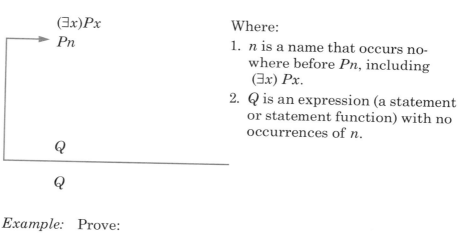

Where:

1. n is a name that occurs nowhere before Pn, including $(\exists x)\, Px$.
2. Q is an expression (a statement or statement function) with no occurrences of n.

Example: Prove:

1. $(\exists x)(Ax \,\&\, Bx)$ $\underline{|\,(\exists x)Ax}$

Proof:

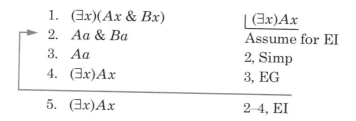

5. $(\exists x)Ax$ 2–4, EI

Note the difference between this rule of EI and the rule of CP: the line immediately before the bottom of the arrow (line 4 in this case) is absolutely identical to the line immediately following. (Is that ever true in CP?)

12 / The Logic of Properties

Example: Prove:

1.	$(\exists x)Rx$	
2.	$(\forall x)(Rx \rightarrow Hx)$	$(\exists x)Hx$
3.	Ra	Assume for EI
4.	$Ra \rightarrow Ha$	2, UI
5.	Ha	3, 4, MP
6.	$(\exists x)Hx$	5, EG
7.	$(\exists x)Hx$	3–6, EI

This example brings out an important heuristic rule. Since one of the restrictions built into EI is that the instantiated variable not occur in any previous line, it would therefore be foolish to have done the UI step first. Once you UI to a name, you cannot EI to the same name. So: EI before UI.

EXERCISES 12.11

Prove the following arguments valid. (An asterisk indicates problems that are answered in the back of the book.)

1. 1. $(\exists x)(Hx \,\&\, Gx)$ / $(\exists x)Gx$
*2. 1. $(\exists x)(Hx \,\&\, Rx)$ / $(\exists x)(Rx \vee Lx)$
3. 1. $(\exists x)(Hx \,\&\, Mx)$ / $(\exists x)(Tx \rightarrow Mx)$
4. 1. $(\exists x)Qx$
   2. $(\forall x)(Qx \leftrightarrow Mx)$ / $(\exists x)Mx$
5. 1. $(\exists x)Sx$
   2. $(\exists x)-Sx$
   3. $(\forall x)(Tx \rightarrow Sx)$ / $(\exists x)-Tx$
*6. 1. $(\exists x)(Bx \,\&\, Cx)$
   2. $(\forall x)-(Rx \,\&\, Cx)$ / $(\exists x)(-Rx)$
7. 1. $(\exists x)(Mx \,\&\, Nx)$
   2. $(\forall x)(Nx \rightarrow Tx)$
   3. $(\forall x)(Tx \leftrightarrow Zx)$ / $(\exists x)Zx$
8. 1. $(\exists x)(Cx \,\&\, Dx)$
   2. $(\forall x)(Dx \leftrightarrow Fx)$
   3. $(\forall x)(Hx \leftrightarrow Fx)$ / $(\exists x)Hx$
9. 1. $(\forall y)(Iy \leftrightarrow Ky)$
   2. $(\forall y)(Ky \leftrightarrow Ly)$
   3. $(\exists y)(Jy \,\&\, Iy)$ / $(\exists x)(Jx \,\&\, Lx)$
*10. 1. $(\exists x)(Ax \,\&\, -Ax)$ / $(\exists y)My$

12.12 Proofs Employing All the Inference Rules

We have covered five new rules specific to QL: QE, UI, EG, UG, and EI. The other rules (from PL) still apply. One key thing is that QE is a rule of equivalence, so like the other rules of equivalence, it applies to parts of lines as well as whole lines. On the other hand, UI, UG, EI, and EG only apply to complete lines. Thus the following is incorrect:

1. $-(\forall x)Hx$
2. $-Ha$ 1, UI

This suggests a good heuristic rule: apply QE to any negated quantifier before you apply the other quantifier inference rules.

Example:

1. $-(\exists x)Qx$
2. $-(\exists x)Rx$ $\vert (\forall y)(-Qy \;\&\; -Ry)$

Proof:

3. $(\forall x)-Qx$ 1, QE
4. $(\forall x)-Rx$ 2, QE
5. $-Qa$ 3, UI
6. $-Ra$ 4, UI
7. $-Qa \;\&\; -Ra$ 5, 6, Conj
8. $(\forall y)(-Qy \;\&\; -Ry)$ 7, UG

Be aware of the possibility, however, of using indirect proof on negated quantifiers, or even upon nonnegated quantifiers.

Example: Prove:

1. $(\forall x)Gx$
2. $(\forall x)(Gx \rightarrow Ax)$ $\vert -(\exists x)-Ax$

Direct proof:

3. Ga 1, UI
4. $Ga \rightarrow Aa$ 2, UI

5. Aa 3, 4, MP
6. $(\forall x)Ax$ 5, UG
7. $(\forall x)--Ax$ 6, DN
8. $-(\exists x)-Ax$ 7, QE

Indirect proof:

3. $(\exists x)-Ax$ Assume for contradiction
4. $-Aa$ Assume for EI
5. $Ga \to Aa$ 2, UI (EI before UI!)
6. Ga 1, UI
7. Aa 5, 6, MP

We cannot conjoin Aa and $-Aa$ for our contradiction, for they are still within the scope of the EI assumption (that is, they still involve the quasi-name a). But we can use a standard trick:

8. $Aa \lor Ob$ 7, Add (we picked the sentence Ob "out of the blue")
9. $Aa \lor -Ob$ 7, Add
10. Ob 8, 4, DS
11. $-Ob$ 9, 4, DS
12. $Ob \,\&\, -Ob$ 10, 11, Conj
13. $Ob \,\&\, -Ob$ 4–12, EI (copy over the last boxed sentence for EI)
14. $-(\exists x)-Ax$ 3–13 Reductio

Variables are just place markers, and thus x is no more meaningful or privileged than y or z.

Example: Prove:

1. $(\forall y)(Ly \leftrightarrow Ty)$
2. $-(\exists z)(Lz)$ $\lfloor (\exists w)Tw$
3. $(\forall z)-Lz$ 2, QE
4. $-Lr$ 3, UI (again, there is nothing privileged about the letter a)
5. $Lr \leftrightarrow Tr$ 1, UI
6. $(Lr \to Tr) \,\&\, (Tr \to Lr)$ 5, Equiv

7. $Tr \to Lr$ 6, Simp
8. $-Tr$ 4, 7, MT
9. $(\forall w)-Tw$ 8, UG
10. $-(\exists w)\,Tw$ 9, QE

It is worth repeating that quite often a QL proof can be done in three phases: first, remove quantifiers by UI and EI (EI before UI, of course); second, do a PL-style proof, using all the old strategies; third, put quantifiers back on by UG and EG. Indeed, you can often see how to work the inner PL proof by covering up the quantifiers and variables in the QL argument. For example:

1. $(\forall x)Mx$
2. $(\forall y)(My \to Ry)$ $\underline{|(\forall z)Rz}$

When you ignore the quantifiers and variables the result is:

1. M
2. $M \to R$ $\underline{|R}$

which suggests that MP will be used somewhere along the line.

Example:

1. $(\forall x)(My \leftrightarrow Qx)$
2. $(\forall y)(Qx \leftrightarrow Sx)$ $\underline{|(\forall x)(Sx \leftrightarrow Mx)}$

The strategy is fairly clear: unpack the biconditionals and show first that $S \to M$, and then that $M \to S$, and we will be done with Equiv.

3. $Ma \leftrightarrow Qa$ 1, UI
4. $Qa \leftrightarrow Sa$ 2, UI
5. $(Ma \to Qa)\,\&\,(Qa \to Ma)$ 3, Equiv
6. $(Qa \to Sa)\,\&\,(Sa \to Qa)$ 4, Equiv
7. $Ma \to Qa$ 5, Simp
8. $Qa \to Sa$ 6, Simp
9. $Ma \to Sa$ 7, 8, HS
10. $(Qa \to Ma)\,\&\,(Ma \to Qa)$ 5, Comm
11. $Qa \to Ma$ 10, Simp

12.	$(Sa \to Qa) \,\&\, (Qa \to Sa)$	6, Comm
13.	$Sa \to Qa$	12, Simp
14.	$Sa \to Ma$	11, 13, HS
15.	$(Sa \to Ma) \,\&\, (Ma \to Sa)$	14, 9, Conj
16.	$Sa \leftrightarrow Ma$	15, Equiv
17.	$(\forall x)(Sx \leftrightarrow Mx)$	16, UG

(We could also have used CP twice in the preceding proof. How?)

Not all QL proofs can be done in this three-phase fashion, however. Consider:

1.	$(\forall x)Rx \to (\forall y)(My \,\&\, Ty)$	
2.	$(\forall z)(Rx \,\&\, Mx)$	$\lfloor (\forall w)(Mw \,\&\, Tw)$

Evidently, we need to turn line 2 into the antecedent of line 1 and detach the consequent. Remember, EI, UI, EG, and UG only apply to whole lines, not parts of lines.

3.	$Ra \,\&\, Ma$	2, UI
4.	Ra	3, Simp
5.	$(\forall x)Rx$	4, UG
6.	$(\forall y)(My \,\&\, Ty)$	1, 5, MP

Now we need to turn line 6 into the conclusion by a variable change, accomplished by a standard trick:

7.	$Mb \,\&\, Tb$	6, UI
8.	$(\forall w)(Mw \,\&\, Tw)$	7, UG

EXERCISES 12.12A

Prove each of the following valid. (An asterisk indicates problems that are answered in the back of the book.)

1. 1. $(\forall x)(Bx \to -Cx)$
   2. $(\exists x)(Dx \,\&\, Bx)$ $\lfloor (\exists x)(Dx \,\&\, -Cx)$
2. 1. $(\forall x)Ax$ $\lfloor (\exists x)Ax$
3. 1. $-(\forall x)Mx$ $\lfloor (\exists x)-Mx$

12.12 / Proofs Employing All the Inference Rules

*   4. 1. $(\forall x)(Ex \to -Fx)$
       2. $(\forall x)(Gx \to Fx)$ $\lfloor (\forall x)(Gx \to -Ex)$

   5. 1. $(\forall x)(Ax \to Cx)$
      2. $(\forall y)(Cy \to Ey)$
      3. $(\forall y)(Ey \to My)$ $\lfloor (\forall x)(Ax \to Mx)$

   6. 1. $(\exists y)(Yy \,\&\, Zy)$
      2. $(\forall y)(Zy \to Ty)$ $\lfloor (\exists x)(Yx \,\&\, Tx)$

   7. 1. $(\forall x)(Hx \to Tx)$
      2. Hr $\lfloor Tr \vee Rr$

   8. 1. $(\forall x)(Ax \to Bx)$
      2. An $\lfloor (\exists x)Bx$

   9. 1. $(\exists x)(Px \,\&\, -Qx)$
      2. $(\forall y)(Py \to Ry)$ $\lfloor (\exists x)(Rx \,\&\, -Qx)$

*   10. 1. $(\exists x)Mx$
      2. $\{(Mx \vee Lx) \to Nx\}$ $\lfloor (\exists x)Nx$

  11. 1. $(\exists x)(Mx \,\&\, Rx)$
      2. $((\forall x)-Lx \to -Rx)$ $\lfloor (\exists x)(Mx \,\&\, Lx)$

  12. 1. $(\forall x)\{(Tx \to Tx) \to Tx\}$ $\lfloor (\exists y)Ty$

  13. 1. $(\forall x)(Mx \to Tx)$
      2. $-(\exists y)Ty$ $\lfloor -(\exists y)My$

*   14. 1. $(\forall w)(-Tw \leftrightarrow -Cw)$
      2. $(\forall w)(-Rw \leftrightarrow Cw)$ $\lfloor (\forall w)(Rx \leftrightarrow -Tx)$

  15. 1. $(\forall x)\{Px \to (Cx \,\&\, Dx)\}$
      2. $(\forall x)\{(Cx \vee -Dx) \to Ex\}$ $\lfloor (\forall x)(Px \to Ex)$

  16. 1. $(\forall x)\{Px \to (Rx \to Ax)\}$
      2. $(\forall x)\{(Rx \to Ax) \to Sx\}$ $\lfloor (\forall x)(Px \to Sx)$

  17. 1. $(\exists x)Rx$
      2. $(\forall x)\{Rx \to (Qx \,\&\, Fx)\}$ $\lfloor (\exists x)(Fx \,\&\, Rx)$

*   18. 1. $(\forall x)(Bx \to -Cx)$
      2. $(\exists y)(Cy \,\&\, Dy)$ $\lfloor (\exists x)(Dx \,\&\, -Bx)$

  19. 1. $(\exists x)(Ax \vee Bx)$
      2. $(\forall y)(Ay \to By)$ $\lfloor (\exists x)Bx$

  20. 1. $(\exists x)-(Rx \,\&\, Tx)$
      2. $(\forall z)Tz$ $\lfloor -(\forall x)Rx$

  21. 1. $(\exists x)-(Ax \,\&\, Bx)$
      2. $-(\exists y)-Ay$ $\lfloor (\exists x)-Bx$

*   22. 1. $-(\exists y)-Ay$
      2. $-(\exists z)-Bz$ $\lfloor (\forall x)(Ax \,\&\, Bx)$

  23. 1. $(\exists x)(Mx \,\&\, Hx)$
      2. $(\forall x)(Mx \to Ex)$ $\lfloor --(\exists x)(Hx \,\&\, Ex)$

  24. 1. $(\forall x)\{Hx \to (Bx \to -Cx)\}$

        2. $(\forall x)(Hx \to Bx)$
        3. $(\forall y)(Hy \ \& \ Ry)$ $\lfloor \neg(\exists x)Cx$
25. 1. $(\forall x)\{(Ax \lor Wx) \to Vx\}$
    2. $(\exists x)(Ax \ \& \ Fx)$
    3. $(\exists x)(Ax \ \& \ Tx)$
    4. $(\forall x)(Tx \leftrightarrow Vx)$ $\lfloor (\exists x)(Fx \ \& \ Tx)$

EXERCISES 12.12B

Symbolize and prove each of the following, selecting letters that the capitalized words indicate. UD = animals. (An asterisk indicates problems that are answered in the back of the book.)

     1. ALL CATS are SLY. Some cats are WILY. So some wily things are sly.

*   2. No CATS are SLY. Some cats LIKE cheese. So some things that like cheese are not sly.

     3. Anybody who LIKES me is SMART. Anyone who is smart will VOTE for me. So anyone who likes me will vote for me.

     4. PEOPLE like WINNERS. But to like a winner is to want to BE like him. So people want to be like winners.

     5. DOGS are all MEAN. Some dogs are even SMART. So some mean things are smart.

*   6. If a PERSON likes CHEESE, he likes WINE. If a person likes wine, he must like TOKAY. So if a person doesn't like Tokay, he must not like cheese.

     7. Some PEOPLE hate DOGS. A necessary condition for hating dogs is that a person be INSECURE. Thus some people are insecure.

     8. A sufficient condition for liking CATS is that a PERSON be STUPID. But liking cats is a sufficient condition for being EVIL. Hence being stupid is a sufficient condition for a PERSON being evil.

     9. Only if a PERSON is SMART will he graduate COLLEGE. Some have graduated, so some are smart.

*   10. If and only if a PERSON hates TUNA will she hate SUSHI. Some people don't hate tuna. So some people don't hate sushi.

12.13 Proving Invalidity

Recall that in Chapter 10 we learned how to construct PL proofs, and then learned the short truth-table method of proving invalidity. We saw that both techniques were needed to analyze cases where we did not know whether the argument was valid or not. After all, a proof

12.13 / Proving Invalidity

technique is not necessarily a discovery procedure—which means that just because we cannot figure out how to prove a given argument valid, we cannot conclude necessarily that the argument is invalid.

We can similarly devise a test for invalidity of QL to supplement our proof apparatus. This technique involves expanding the premises and conclusion of a given argument, which turns the QL argument into a PL one, and then applying the short truth-table technique to the result. Example: Prove the following argument invalid.

1. $(\exists x)Ax$
2. $(\exists x)Bx$ $\underline{|\ (\exists x)(Ax\ \&\ Bx)}$

It should be obvious that this argument is invalid. Knowing that some thing or things have property A, and some thing or things have B, does not entitle us to conclude that the same things have both. Some things are dogs, some things are frogs, but does it follow that some things are both frogs and dogs?

Step 1: Expand. We can expand in a one-individual model (a one-model), a two-model, or a three-model, and so on. It is rarely profitable to expand in a one-model. For one thing, in a model with only one individual, existentially quantified statement and the corresponding universally quantified statement are equivalent. If we expand our sample argument in a one-model $\{a\}$, we get:

1. Aa
2. Ba $\underline{|Aa\ \&\ Ba}$

which is valid (by conjunction).

So we expand in a two-model. We get (assuming the names are a and b):

1. $Aa\ \vee\ Ab$
2. $Ba\ \vee\ Bb$ $\underline{|(Aa\ \&\ Ba)\ \vee\ (Ab\ \&\ Bb)}$

Step 2: Try to assign Ts and Fs to the simple statements to make the premises true (T) and the conclusion false (F). Set up a table:

Aa	Ab	Ba	Bb

To make the conclusion F, we must make $(Aa \mathbin{\&} Ba)$ and $(Ab \mathbin{\&} Bb)$ both F. We do not want to make all the basic statements F, since we must make the premises T. So try making one of the Ax's T, and one of the Bx's also T:

Aa	Ab	Ba	Bb
F	F	T	T

This works, as you can verify for yourself.

Example: Prove the following argument invalid by expansion in the two-model $\{a, b\}$.

1. $(\forall x)(Ax \to Bx)$
2. $(\exists x)Bx$ $\quad\quad\quad\quad$ | $(\exists x)Ax$

Answer:

Step 1: Expand.

1. $(Aa \to Ba) \mathbin{\&} (Ab \to Bb)$
2. $Ba \vee Bb$ $\quad\quad\quad\quad$ | $Aa \vee Ab$

Step 2: Set up the table and find an assignment of truth values that makes the premises T and the conclusion F.

Aa	Ab	Ba	Bb

We must make Aa and Ab F to make the conclusion false. This automatically makes premise one T. All we have to do is make one or both of Ba and Bb T.

Aa	Ab	Ba	Bb
F	F	T	T

12.13 / Proving Invalidity

There is no guarantee that if an argument is invalid, expansion in a two-model will reveal the invalidity. We may have to move to a three-, four-, or five-model. For example, the argument:

1. $(\exists x)(Rx \mathbin{\&} Gx)$
2. $(\exists x)(Rx \mathbin{\&} Tx)$
3. $(\exists x)(Gx \mathbin{\&} Tx)$ $\quad\quad$ $(\exists x)[Rx \mathbin{\&} (Gx \mathbin{\&} Tx)]$

is valid in a two-model, but invalid in a three-model. This technique then becomes messy indeed. And there is a theoretical limitation to this technique: there are arguments valid in all finite models, but not valid in infinite ones.

EXERCISES 12.13

Prove each of the following arguments invalid by expansion in the model $\{a, b\}$. (An asterisk indicates problems that are answered in the back of the book.)

1. $(\exists x)(Ax \vee Bx)$
   $(\exists x)(Ax \vee Cx)$ $\quad\quad$ $(\exists x)(Bx \vee Cx)$

* 2. $(\exists x)(-Ax \mathbin{\&} Bx)$
   $(\exists x) Cx$ $\quad\quad$ $(\exists x)(Cx \mathbin{\&} Bx)$

3. $(\forall x)(Ax \leftrightarrow Bx)$ $\quad\quad$ $(\exists x)Ax$

4. $(\forall x)(Ax \rightarrow Zx)$
   $(\forall x)(-Zx \rightarrow Tx)$ $\quad\quad$ $(\forall x)(Ax \rightarrow Tx)$

5. $(\exists x)(Tx \mathbin{\&} Zx)$
   $(\exists x)(-Zx \rightarrow Mx)$ $\quad\quad$ $(\exists x)(Tx \mathbin{\&} Mx)$

* 6. $(\forall x)[Tx \rightarrow (Lx \vee Qx)]$
   $(\forall x)[(Lx \mathbin{\&} Qx) \rightarrow Zx]$ $\quad\quad$ $(\forall x)(Tx \rightarrow Lx)$

7. $(\forall x)[Tx \leftrightarrow (Lx \vee Qx)]$
   $(Ax)[(Lx \mathbin{\&} Qx) \rightarrow Zx]$ $\quad\quad$ $(\forall x)Zx$

8. $(\exists x)(Mx \mathbin{\&} Tx)$
   $(\forall x)[Mx \mathbin{\&} (Tx \rightarrow Zx)]$ $\quad\quad$ $(\forall x)Zx$

9. $-(\forall x)Rx$
   $-(\forall x)Tx$ $\quad\quad$ $-(\forall x)(Rx \vee Tx)$

* 10. $(\forall x)(-Tx \leftrightarrow Rx)$
   $-(\exists x)-(Rx \leftrightarrow Lx)$ $\quad\quad$ $(\exists x)-(-Tx \leftrightarrow Lx)$

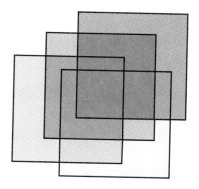

13

The Logic of Relations

13.1 Relations and Singular Statements

In Chapter 12, we developed a logic of properties, called quantificational logic (QL). The vocabulary for QL includes individual constants, which function as proper names; quantifiers; individual variables; brackets and connectives [from propositional logic (PL)]; and property constants. We said that properties are one-place predicates, and we only mentioned relations (two- or more-place predicates). We now want to extend QL to handle relations.

We will use variables and capital letters to create statement functions. Examples:

Pxy = x is the pal of y.
Rxy = x is to the right of y.
$Bxyz$ = x is between y and z.

The number of variables following a capital letter determines the number of places in that predicate. Thus, Px should not be confounded with Pxy, nor the latter with $Pxyz$. The first symbolizes some property (perhaps "x is a pig"); the second symbolizes a two-place relation (perhaps "x is a parent of y"); the third a three-place relation (perhaps "x puts y on z").

449

The order in which the variables occur in a statement function is crucial. Rxy means "x is to the right of y," while Ryx means "y is to the right of x." This is basic in translation.

We can use the truth-functional connectives to make compound statement functions out of simple ones. Examples:

$Rxy \to -Lxy$	= If x is to the right of y, then x is not to the left of y.
$Hxy \ \& \ Mxy$	= x is hotter and more massive than y.
$Sxy \leftrightarrow Mxy$	= x is the son of y if and only if y is the mother of x.
$Bxyz \ \text{v} \ Byxz$	= Either x is between y and z or y is between x and z.
$-Fxy \to Exy$	= If it is not the case that x is a friend of y, then x is an enemy of y.

In the last chapter, one way in which we can turn statement functions into statements is by plugging in the individual constants (names) for the free occurrences of variables. Examples:

In QL we can have:

Px	= x is a person.
(Instance) Pa	= Alfredo is a person.

We can go on to have:

Pxy	= x is proud of y.
(Instance) Pab	= Alfredo is proud of Ben Hassan.
Lxy	= x loves y.
(Instance) Las	= Alfredo loves Sara Lou.
(Instance) Laa	= Alfredo loves Alfredo.
$Bxyz$	= x is between y and z.
(Instance) $Bblv$	= Barstow is between Los Angeles and Las Vegas.
$Ewxyz$	= w is equidistant from x, y, and z.
(Instance) $Etcad$	= Topeka is equidistant from Chicago, Atlanta, and Detroit.

As you can see, singular statements involving relations present no special problems.

EXERCISES 13.1

Translate the following, using the following symbolic guide. (An asterisk indicates problems that are answered in the back of the book.)

$Hxy = x$ hits y.
$Cx = x$ cries.
$Swxyz = x$ is seated between y and z.
$f = $ Fred.
$s = $ Suzy.
$h = $ Hassan.

1. If Fred cries, Suzy will hit him.
* 2. If Fred hits Suzy, he will cry.
3. If Hassan is seated between Suzy and Fred, he'll cry.
4. If Hassan is seated between Fred and Suzy, they'll cry.
5. If Hassan is seated between Fred and Suzy, he'll hit them both.
* 6. If and only if Hassan hits Suzy will Fred hit Hassan.
7. If Hassan and Suzy cry, then Fred will hit them both.
8. If Fred is seated between Hassan and Suzy, then Suzy is seated between Hassan and Fred.
9. If Fred cries, then if Fred is seated between Suzy and Fred, Suzy will hit Hassan and Hassan will cry.
* 10. A necessary condition for Hassan to hit Fred is that both Fred and Suzy hit Hassan.
11. If neither Suzy nor Hassan hits Fred, neither will cry.
12. Only if neither Hassan nor Fred cry will Suzy be seated between them.
13. If Fred isn't seated between himself and Suzy, he will cry.
14. Fred is seated between himself and himself.
15. Hassan and Fred are seated between Suzy and Fred.

13.2 Multiple Quantifiers

Having strengthened QL to include many-place relations, we can now expand our use of quantifiers as well. Consider the statement "Everybody loves Bo." If we take the universe of discourse to be the class of people, we can paraphrase this as:

For every x, x loves Bo.

which can be symbolized as:

$(\forall x)Lxb$

Similarly, we can symbolize the statement "Bo loves everybody" as:

$(\forall x)Lbx$

But what about the statement "Everybody loves somebody?" To symbolize such statements, we have to use *multiple quantifiers*. A multiple-quantified statement function has the form:

$Q_1 Q_2 \ldots Q_n P x_1 x_2 \ldots X n$

where each quantifier Q_1 is either $(\forall x_i)$ or $(\exists x_i)$ for some variable of x_i, and where each quantifier applies to the expression to its right. Examples:

$(\forall x)(\forall y)Fxy$
$(\exists y)(\forall z)(\exists z)(Mzy \to Tzx)$
$(\forall x)(Ay)(\exists z)(Mzy \to Tzx)$

This greatly increases our power of translation. Consider "Everybody loves somebody." We can paraphrase this as [universe of discourse (UD) = persons]:

For every x, there is some y such that x loves y.

and symbolize it as:

$(\forall x)(\exists y)Lxy$

On the other hand, consider "Somebody loves everybody." The best paraphrase of this is (again, UD = persons):

There is at least one x such that for every y, x loves y.

which is easily symbolized:

$(\exists x)(\forall y)Lxy$

Thus which quantifier comes first makes all the difference. Table 13.1 lists four possible combinations of the two quantifiers.

13.2 / Multiple Quantifiers

Table 13.1

Quantifier Combination	Meaning
$(\forall x)(\forall y)$	For every x and for every y.
$(\forall x)(\exists y)$	For every x there is a y such that.
$(\exists x)(\forall y)$	Some x is such that for every x.
$(\exists x)(\exists y)$	There is an x and a y such that.

Example:

If someone wants to see me, I will be surprised.

First, paraphrase by replacing the pronouns *me* and *I* by the name of the person who utters the sentence:

If there is an x such that x wants to see Jason, then Jason will be surprised.

This is symbolized:

$(\exists x)Wxj \rightarrow Sj$

Example:

If someone wants to see me, that person will be surprised.

Paraphrased:

If there is an x such that x wants to see Jason, then that x will be surprised.

But this has the force of a universal:

If any x wants to see Jason, then that x will be surprised.

Symbolized:

$(\forall x)(Wxj \rightarrow Sx)$

Remember our tip: avoid the existential quantifier applying to a statement function whose main connective is an arrow.

Example:

> Everybody loves somebody sometime.

Paraphrase:

> For every x, if x is a person, then there is a y and a z such that y is a person and z is a time and x loves y at z.

Symbolized:

> $(\forall x)[Px \rightarrow (\exists y)(\exists z)(Py \ \& \ Tz \ \& \ Lxyz)]$

Example:

> Anybody who loves everybody will be loved by somebody.

Paraphrase:

> For any x, if x loves everybody, then somebody loves x.

Symbolized:

(UD = persons):

> $(\forall x)[(\forall y)Lxy \rightarrow (\exists y)Lyx]$

All of the examples so far have involved statements that have a leading quantifier whose scope applies to the whole statement. But not all multiple quantified statements are of that sort.

Example:

> If everyone loves salami, then if everyone who loves salami loves crackers, then everyone loves crackers.

Here we have separate statements connected by truth-functional connectives.

> (Everyone loves salami) → [(Everyone who loves salami loves crackers) → (Everyone loves crackers)]

Symbolized (UD = unrestricted):

13.2 / Multiple Quantifiers

$Px = x$ is a person.
$Lxy = x$ loves y.
s = salami.
c = crackers.

$(\forall x)(Px \to Lxs) \to [(\forall x)[((Px \& Lxs) \to Lxc] \to (\forall x)(Px \to Lxc)]$

Example:

If horses are animals, then a leg of a horse is a leg of an animal.

Partially paraphrased:

If for every x, if x is a horse, then x is an animal, then for any y, if y is a horse's leg, then y is an animal's leg.

Partially translated:

$\{(\forall x)(Hx \to Ax)\} \to \{(\forall y)[y \text{ is a horse's leg} \to y \text{ is an animal's leg}]\}$

To see how to translate "y is a horse's leg," paraphrase it first:

There is some x such that x is a horse and y is the leg of x.

which is symbolized:

$(\exists x)(Hx \& Lyx)$

So our final translation becomes:

$(\forall x)(Hx \to Ax) \to (\forall y)[(\exists x)(Hx \& Lyx) \to (\exists x)(Ax \& Lyx)]$

EXERCISES 13.2A

Symbolize the following (UD = animals), selecting letters that the capitalized words indicate. (An asterisk indicates problems that are answered in the back of the book.)

1. DOGS LIKE Lassie (l = Lassie).
* 2. Lassie HATES CATS.
3. DOGS HATE CATS.
4. DOGS HATE themselves.

5. DOGS HATE each other.
* 6. Some DOGS HATE Lassie.
7. Some CATS are LIKED by Lassie.
8. Some DOGS LIKE some CATS.
9. CATS and DOGS HATE RATS.
* 10. Any CAT LIKES a FRIENDLY DOG.
11. Any CAT who LIKES a DOG is FRIENDLY.
12. If DOGS HATE RATS, then rats hate dogs.
13. Only if DOGS LIKE CATS will cats like rats.
* 14. PEOPLE LIKE RATS only if RATS hate cats.
15. Nobody can TRUST anybody.
16. Nobody can TRUST themselves.
17. Any woman who TRUSTS herself can trust anyone.
* 18. Not everybody can trust themselves.
19. PEOPLE LIKE Lassie if and only if Lassie is a DOG.
20. Unless DOGS LIKE CATS, CATS don't like DOGS.
21. Lassie is a DOG that everyone LIKES.
* 22. Some DOGS are liked by everyone.
23. If Lassie is a DOG that everyone LIKES, then there is a dog that everyone likes.
24. Only if Lassie is a CAT is there a cat that everyone LIKES.
25. If there is a DOG that everyone LIKES, then if all CATS are dogs there is a cat everyone likes.

EXERCISES 13.2B

Symbolize the following. State what the universe of discourse (UD) is, and provide a dictionary of the predicate and individual constraints you have chosen. All sentences are from Dr. Isadore Rosenfeld, Second Opinion. *(An asterisk indicates problems that are answered in the back of the book.)*

Example: There are no discussions of rare or exotic diseases in this book.
Answer: Let UD = unrestricted.
b = this book.
Dxy = x is discussed in y.
Rx = x is rare.
Ex = x is exotic.
Dx = x is a disease.

Paraphrase:
It is not the case that there exists an x such that x is a disease and x is rare or exotic, and that x is discussed in this book.

Symbolization: $-(\exists x)[Dx \,\&\, (Rx \lor Ex) \,\&\, Dxb]$

1. The diet doesn't work for everybody.
* 2. Each of these views on medicine is perfectly legitimate.
3. If you [that is, anybody] are feeling poorly and one consultant can't help you, don't stop there.
4. Every physician has a hard time keeping up with all there is to know.
5. When you are in a medical situation never hesitate to ask for the extra input of a second opinion.
* 6. There are many doctors who welcome requests for consultation.
7. Some doctors are even angry when a patient is candid.
8. If you can't choose between two divergent recommendations, and you don't know instinctively which is better for you, you will need to find a final arbitrator.
9. There are patients who take several "one-a-day" vitamins every morning.
* 10. Hyperthyroid individuals are apt to be less alert than those around them.
11. If you suddenly and for no good reason develop "heart failure," make sure to ask whether your thyroid function has been checked.
12. If you are suddenly "nervous" in response to a crisis, and remain that way even after it is over, see to it that your thyroid function is evaluated.
13. Any iodine introduced in whatever form ends up in the thyroid.
* 14. Undergoing a sophisticated, expensive new test is no guarantee that the result will be accurate.
15. Should the surgeon cut a nerve that supplies the vocal cords, you will be hoarse forever.

13.3 Expansions and Instantiation

Having now expanded QL to include relations and multiple quantifiers, we need to examine instantiation and expansions.

To begin with instantiation, we can easily modify the step-by-step approach given in Chapter 12.

Step 1: Strip away the left-most quantifier.

Step 2: Identify the free occurrences of the relevant variable.

Step 3: Replace those occurrences by the instantiating constant.

Step 4: Go to the left-most quantifier, and repeat.

Example: Instantiate the following quantified statement first to a, then to b:

$$(\forall x)(\exists y)(Lxy \rightarrow -Hxy)$$

Step 1: Strip away the outside quantifier:

$$(\exists y)(Lxy \rightarrow -Hxy)$$

Step 2: Identify the free occurrence of the appropriate variable:

$$(\exists y)(\underset{1}{Lxy} \rightarrow -\underset{2}{Hxy})$$

While both occurrences of x fall within the scope of $(\exists y)$, neither is bound by it.

Step 3: Replace:

$$(\forall y)(Lay \rightarrow -Hay)$$

Step 4: Repeat—strip away the quantifier:

$$(Lay \rightarrow -Hay)$$

Step 5: Repeat—spot free occurrences:

$$(\underset{1}{Lay} \rightarrow -\underset{2}{Hay})$$

Step 6: Replace:

$$Lab \rightarrow -Hab$$

Example: Instantiate the following to a, b, b:

$$(\forall x)(\forall y)(\exists z)(Lxyz \rightarrow Mzyx)$$

13.3 / Expansions and Instantiation

Step 1: $(\forall y)(\exists z)(Lxyz \to Mzyx)$
Step 2: $(\forall y)(\exists z)(Lxzy \to Mzyx)$
$\quad\quad\quad\quad\quad\quad\quad\quad\quad\ 1 \quad\quad\quad\ \ 2$
Step 3: $(\forall y)(\exists z)(Layz \to Mxya)$
Step 4: $(\exists z)(Layz \to Mzya)$
Step 5: $(\exists z)(Layz \to Mzya)$
$\quad\quad\quad\quad\quad\quad\ 1 \quad\quad\quad\ \ 2$
Step 6: $(\exists z)(Labz \to Mzba)$
Step 7: $(Labz \to Mzba)$
$\quad\quad\quad\quad\ 1 \quad\quad\ \ 2$
Step 8: $Labb \to Mbba$

Nothing we have said so far forbids our instantiating different quantifiers to the same constant.

One use of instantiation is to expand quantified statements into logically equivalent singular statements in a given finite model, and then apply our short truth-table technique to prove invalidity. The trick is to expand one quantifier at a time, starting with the left-most.

Example: Expand the following sentence in the two-model $\{b, c\}$.

$(\forall x)(\exists y)Axy$

Step 1: Expand around the left quantifier. Remember that the universal quantifier unpacks into a conjunction.

$(\exists y)Aby$ & $(\exists y)Acy$

Step 2: Expand each conjunct.

$[Abb \lor Abc]$ & $[Acb \lor Acc]$

Example: Expand the following sentence in the three-model $\{a, b, c\}$.

$(\exists x)(\forall y)(\exists z) - Axyz$

Step 1: Expand the $(\exists x)$ by stepwise instantiation:

$[(\forall z)(\exists z) - Aayz] \lor [(\forall y)(\exists x) - Abyz] \lor [(\forall y)(\exists z) - Acyz]$

Step 2: Expand each $(\forall y)$:

$[(\exists z)-Aaaz$ & $(\exists z)-Aabz$ & $(\exists z)-Aacz]$ v $[(\exists z)-Abaz$ & $(\exists z)-Abbz$ & $(\exists z)-Abcz]$ v $[(\exists x)-Acaz$ & $(\exists z)-Acbz$ & $(\exists z)-Accz]$

Step 3: Expand each $(\exists z)$:

$\{[-Aaaa$ v $-Aaab$ v $-Aaac]$ & $[-Aaba$ v $-Aabb$ v $-Aabc]$ & $[-Aaca$ v $-Aacb$ v $-Aacc]\}$ v $\{[-Abaa$ v $-Abab$ v $-Abac]$ & $[-Abba$ v $-Abbb$ v $-Abbc]$ & $[-Abca$ v $-Abcb$ v $-Abcc]\}$ v $\{[-Acaa$ v $-Acab$ v $-Acac]$ & $[-Acba$ v $-Acbb$ v $-Acbc]$ & $[-Acca$ v $-Accb$ v $-Accc]\}$

Proving invalidity proceeds just as in Section 12.13. We expand and then seek an assignment of truth values to the components that makes the premises true (T) and the conclusions false (F).

EXERCISES 13.3

Expand each of the following, first in the two-model $\{a, b\}$, then in the three-model $\{a, b, c\}$. (An asterisk indicates problems that are answered in the back of the book.)

1. $(\forall x)(\forall y)Mxy$
* 2. $(\exists y)(\exists x)Mxy$
3. $(\exists x)(\exists y)Mxy$
4. $(\forall x)(\exists y)--Mxy$
5. $(\exists x)(\forall y)(Mxy \rightarrow -Nxy)$
* 6. $(\forall x)(\forall y)(Nxxx \rightarrow -Nyy)$
7. $(\forall x)(\exists y)(-Nxy$ & $-Nyx)$
8. $(\forall x)(\forall y)(\forall z)-Mxyz$
9. $(\exists x)(\exists y)(\exists z)--Mxyz$
* 10. $(\exists x)(\forall y)(\exists z)(Mxya \rightarrow Mzzy)$
11. $(\forall u)(\forall x)(\forall y)(\forall z)Auxyz$
12. $(\forall x)(\exists y)(\forall z)(\exists u)Auxzy$
13. $(\forall x)(\exists y)(\forall u)(\exists z)(Axy \rightarrow -Mxuz)$
14. $(\exists y)(\forall x)(\forall u)(\exists z)Auuxxyz$
15. $(\forall y)(\forall x)(\exists u)(\exists z)Auzzzuxxy$

13.4 The Five Quantificational Logic Rules with Relations and Multiple Quantifiers

What we have seen so far is that allowing QL to include relations (which also leads to multiple quantifiers being employed) dramatically increases the power of QL to translate intricate English sentences, and thus greatly improves the ability of QL to reveal the logical structure of statements. But how do we employ our five QL inference rules with multiple quantifiers and relations? The rules discussed in Chapter 12 are quantifier exchange (QE), universal instantiation (UI), universal generalization (UG), existential instantiation (EI), and existential generalization (EG).

The answer is to work top-down. Apply any appropriate rule first to the left-most quantifier, treating the rest of the expression as one whole statement function. Then it may be possible to apply another (or even the same) rule again.

Example: Apply QE to:

$-(\forall x)(\forall y)(\forall z) Lxyz$

Apply QE once, to get:

$(\exists x)-(\forall y)(\forall z) Lxyz$

QE can be applied again because unlike UI, UG, EI, and EG, it is a logical equivalence that can apply to parts of lines:

$(\exists x)(\exists y)-(\forall z) Lxyz$

Apply QE again:

$(\exists x)(\exists y)(\exists z)-Lxyz$

(We could reverse those steps just as easily.)

Example: Apply EI to:

$(\exists x)(\exists y)Axy$

Apply it to the outside quantifier:

$(\exists y)Aay$

(We assume a has not occurred in the proof before that line.)
Apply EI again:

Aab (b must be new)

Example: Existentially generalize:

Maa

Apply EG once:

$(\exists x)Mxx$

Or we can apply EG twice, generalizing on the occurrences of a taken singly:

$(\exists y)Mya$

then

$(\exists x)(\exists y)Myx$

And indeed this makes sense: if Alfred is mad at Alfred, then it follows that some man is mad at himself, but it also follows (more weakly) that some man is mad at some man.

Example: Universally generalize:

Laa

Here, we cannot allow:

$(\forall x)Lxa$

since a still occurs in the expression $(\forall x)Lxa$, which violates the rule. The answer is:

$(\forall x)Lxx$

Example: Apply UI to:

$(\forall x)Mxx$

We have only one choice (if <u>a</u> is the instantiating constant):

Maa

13.4 / *The Five Quantificational Logic Rules*

since

Mxa

has a variable occurring freely in it. But:

$(\forall x)(\forall y)Mxy$

can be instantiated:

$(\forall y)May$

then

Mab

or, if we choose,

Maa

If everybody loves everybody (UD = persons), then it follows that Alfred loves Bonnie, but it also follows that Alfred loves Alfred!

Example: Construct a proof of:

1. $(\forall x)(\forall y)Hxy$ $\underline{\ (\forall y)(\forall x)Hxy\ }$

Proof:

1. $(\forall x)(\forall y)Hxy$ $\underline{\ (\forall y)(\forall x)Hxy\ }$
2. $(\forall y)Hay$ 1, UI
3. Hab 2, UI
4. $(\forall x)Hxb$ 3, UG
5. $(\forall y)(\forall x)Hxy$ 4, UG

Example: Prove:

1. $(\forall x)(Mx \to Lxa)$
2. Mc $\underline{\ (\exists x)Lxa\ }$

Proof:

1. $(\forall x)(Mx \to Lxa)$
2. Mc $\underline{\ (\exists x)Lxa\ }$
3. $Mc \to Lca$ 1, UI
4. Lca 2, 3, MP
5. $(\exists x)Lxa$ 4, EG

Example: Prove:

1. $(\exists x)(\forall y)Mxy$ / $(\forall y)(\exists x)Mxy$

Proof:

1. $(\exists x)(\forall y)Mxy$ / $(\forall y)(\exists x)Mxy$
2. $(\forall y)May$ Assume for EI
3. Mab 2, UI
4. $(\exists x)Mxb$ 3, EG
5. $(\exists x)Mxb$ 2–4, EI
6. $(\forall y)(\exists x)Mxy$ 5, UG

Notice that the converse argument is not valid:

1. $(\forall y)(\exists x)Mxy$ / $(\exists x)(\forall y)Mxy$
2. $(\exists x)Mxa$ 1, UI
3. Mba Assume for EI
4. $(\forall y)Mby$ 3, UG (wrong! *a* occurs in an assumption within whose scope *Mba* lies)
5. $(\exists x)(\forall y)Mxy$ 4, EG
6. $(\exists x)(\forall y)Mxy$ 3–5, EI

Pay attention to scope, and remember that EI, UI, EG, and UG apply only to whole lines (not parts of lines). Consider:

1. $(\forall x)((\exists y)Lxy \to (\exists y)Qxy)$
2. $(\forall x)Lxx$ / $(\exists x)(\exists y)Qxy$

The wrong way to proceed would be to instantiate line 1 twice:

3. $(\exists y)Lay \to (\exists y)Qay$ 1, UI (OK so far)
4. $Lab \to Qab$ Assume for EI (wrong! EI cannot be applied to parts of lines)
5. Lab 2, UI (wrong! Must replace all free occurrences of the variable)
6. Qab 4, 5, MP
7. $(\exists y)Qay$ 6, EG
8. $(\exists x)(\exists y)Qxy$ 7, EG
9. $(\exists x)(\exists y)Qxy$ 4–8, EI

13.4 / The Five Quantificational Logic Rules

The correct proof is:

3.	$(\exists y)Lay \rightarrow (\exists y)Qay$	1, UI
4.	Laa	2, UI
5.	$(\exists y)Lay$	4, EG (yes, perfectly legal to EG on one of the constants)
6.	$(\exists y)Qay$	5, 3, MP
7.	$(\exists x)(\exists y)Qxy$	6, EG

A common strategy in proving arguments with universal conclusions is to apply conditional proof (CP) and then generalize. Example:

1.	$(\forall x)(\forall y)(Fxy \rightarrow Gxy)$	$(\forall x)(Fxx \rightarrow Gxx)$
2.	Faa	Assume to show Gaa
3.	$(\forall y)(Fay \rightarrow Gay)$	1, UI
4.	$Faa \rightarrow Gaa$	3, UI
5.	Gaa	4, 2, MP
6.	$Faa \rightarrow Gaa$	2–5, CP
7.	$(\forall x)(Fxx \rightarrow Gxx)$	6, UG

EXERCISES 13.4A

Prove the following valid. (An asterisk indicates problems that are answered in the back of the book.)

1. 1. $(\forall x)Lxxx$
   $Laaa$

* 2. 1. $(\forall x)(\forall y)Lxy$
   $Lab \lor Laa$

3. 1. $(\forall x)(\forall z)Lxy$
   $(\forall x)(\forall z)Lxz$

4. 1. $(\forall x)(\forall y)(Lxy \rightarrow Lyx)$
   2. Lab
   $Laa \lor Lba$

5. 1. $(\forall x)(\forall y)(Mxy \rightarrow -Nyx)$
   2. $(\forall x)(\forall y)(Lxy \rightarrow Nyx)$
   $(\forall x)(\forall y)(Mxy \rightarrow -Lxy)$

* 6. 1. $(\forall x)(\forall z)(Rzx \leftrightarrow Mxz)$
   2. $(\forall x)(\forall z)(Mxz \leftrightarrow Lxz)$
   $(\forall x)(\forall z)(Rzx \leftrightarrow Lxz)$

7. 1. $(\forall x)(\forall y)(\forall z)Rxyz$
   $(\forall x)(\forall y)(\forall z)(Rxyz \lor -Rxyz)$

8. 1. $-(\exists x)(\exists y)(\exists z)Rxyz$
   $(\forall x)(\forall y)(\forall x)(Rxyz \lor - Rxyz)$

9. 1. $(\forall x)(\forall y)(\forall z)-Mxyz$
   $(\forall x)(\forall y)(\forall z)(Mxyz \rightarrow Rxyz)$

* 10. 1. $(\forall x)(\forall y)(\forall z)-Mxyz$ $\mid (\forall x)(\forall y)(\forall z)(Mxyz \to Rxy)$
11. 1. $(\forall x)(Qx \to Rx)$
    2. $(\forall w)(Sw \to Tw)$ $\mid (\forall x)(Rx \to Sx) \to (\forall y)(Qy \to Ty)$
12. 1. $(\exists x)(\forall y)(Jx \leftrightarrow Ky)$ $\mid (\forall y)(\exists x)(Jx \leftrightarrow Ky)$

EXERCISES 13.4B

Prove each of the following. The proofs are trickier, in that you must pay attention to scope. (An asterisk indicates problems that are answered in the back of the book.)

    1. 1. $(\exists y)Ay \lor (\exists y)By$
       2. $(\forall y)-Ay$ $\mid (\exists y)By$

* 2. 1. $(\forall x)[Ax \to (\forall y)(My \to Ny)]$ $\mid (\forall x)Ax \to (\forall y)(My \to Ny)$

    3. 1. $(\exists x)Lx \to (\exists y)My$ $\mid (\exists x)[Lx \to (\exists y)My]$

    4. 1. $(\exists x)Nx \to (\forall y)Oy$ $\mid (\forall x)[Lz \to (\forall y)Oy]$

    5. 1. $(\exists x)Ax \to (\forall y)(Yy \to Zy)$ $\mid (\exists x)(Ax \& Yx) \to (\exists y)(Ay \& Zy)$

* 6. 1. $(\exists y)By \to -(\exists x)Hy$ $\mid (\forall y)[(\exists x) Bx \to -Hy]$

    7. 1. $(\exists x)Bx \to (\forall y)(Cy \to Dy)$
       2. $(\exists x)Ex \to (\exists y)Cy$ $\mid (\exists x)(Bx \& Ex) \to (\exists y)Dy$

    8. 1. $(\forall y)(Ay \to By)$
       2. $(\forall y)(By \to Cy)$ $\mid (\exists x)Ax \to (\forall x)Cx$

    9. 1. $(\exists x)Bx \lor (\exists y)Cy$
       2. $(\exists x)(Bx \to Cx)$ $\mid (\exists y)Cy$

* 10. 1. $(\forall x)(\exists y)(Ax \lor By)$ $\mid (\forall x)Ax \lor (\exists y)By$

    11. 1. $(\forall x)(\exists y)(Ax \& By)$ $\mid (\forall x)Ax \& (\exists y)By$

    12. 1. $(\forall x)\{Bx \to [(\forall y)(Cy \to Dy) \to Ex]\}$
       2. $(\forall x)\{Ex \to [(\forall y)(Cy \to Fy) \to Gx]\}$ $\mid (\forall y)[Cy \to (Dy \& Fy)] \to (\forall x)(Bx \to Gx)$

    13. 1. $(\forall x)[Mx \to ((\exists y)Ny \to Ox)]$
       2. $(\forall x)[Ox \to ((\exists y)Py \to Rx)]$ $\mid (\exists x)(Nx \& Sx) \to [(\forall y)(Sy \to Py) \to (\forall w)(Mw \to Rw)]$

* 14. 1. $(\exists x)[Gx \& (\forall y)Mxy]$ $\mid (\exists x)(Gx \& Mxb)$

    15. 1. $(\forall x)(\forall y)Lxy$ $\mid (\forall y)(\forall x)Lxy$

    16. 1. $(\exists x)(Mx \& (\forall y)Rxy)$
       2. $(\forall x)\{Mx \to [(\forall y)Rxy \to Nx]\}$ $\mid (\exists x)Nx$

17. 1. $(\exists x)(\forall y)[(\exists z)Byz \to Byx]$
     2. $(\forall y)(\exists z)Byz$ $\underline{\,(\exists x)(\forall y)Byx\,}$

* 18. 1. $(\forall x)[Fx \to (\forall y)(Gy \to Hxy)]$
     2. $(\exists x)[Fx \,\&\, (\exists y){-}Hxy]$ $\underline{\,(\exists x){-}Gx\,}$

19. 1. $(\forall x)(Aax \to Bxb)$
     2. $(\exists x)Bxb \to (\exists y)Bby$ $\underline{\,(\exists x)Aax \to (\exists y)Bby\,}$

20. 1. $(\exists x)[Ax \,\&\, (\forall y)(By \to Cxy)]$ $\underline{\,(\forall x)(Ax \to Bx) \to (\exists y)(By \,\&\, Cyy)\,}$

21. 1. $(\forall x)[Bx \to (\forall y)(Cy \to Dxy)]$
     2. $(\forall x)[Ex \to (\forall y)(Dxy \to Fy)]$ $\underline{\,(\forall x)(Bx \,\&\, Ex) \to (\forall y)(Cy \to Fy)\,}$

13.5 Useful Theorems

We are now in a position to state and prove some useful theorems. In what follows, let us use Fx and Gx to stand for any proposition functions that have at least one free occurrence of x. We will use P to stand for any proposition or proposition function that does not have any free occurrences of x. Remember that Fx can be any proposition function with a free occurrence of x; for instance:

$(\forall y)Fxy$

$(\exists x)Fx \to Hx$

$(\forall y)(\forall z)(Mxyz \to Txy)$

and so on. With that notation, we can formulate useful theorems.

T1: $(\forall x)(Fx \,\mathrm{v}\, {-}Fx)$
T2: $[(\forall x)Fx \,\&\, P] \Leftrightarrow (\forall x)(Fx \,\&\, P)$
T3: $[(\forall x)Fx \,\mathrm{v}\, P] \Leftrightarrow (\forall x)(Fx \,\mathrm{v}\, P)$
T4: $[P \to (\forall x)Fx] \Leftrightarrow (\forall x)(P \to Fx)$
T5: $[(\exists x)Fx \to P] \Leftrightarrow (\forall x)(Fx \to P)$
T6: $[(\exists x)Fx \,\&\, P] \Leftrightarrow (\exists x)(Fx \,\&\, P)$
T7: $[(\exists x)Fx \,\mathrm{v}\, P] \Leftrightarrow (\exists x)(Fx \,\mathrm{v}\, P)$
T8: $[(\exists x)Fx \,\mathrm{v}\, (\exists x)Gx] \Leftrightarrow (\exists x)(Fx \,\mathrm{v}\, Gx)$
T9: $(\exists x)(Fx \,\&\, Gx) \Rightarrow [(\exists x)Fx \,\&\, (\exists x)Gx]$

(The converse of T9 does not hold.)

We will proceed with a few of the proofs, leaving the remainder as exercises.

Proof of T1: no premises are here. We will prove the theorem by contradiction.

1.	$-(\forall x)(Fx \vee -Fx)$	Assume for contradiction
2.	$(\exists x)-(Fx \vee -Fx)$	1, QE
3.	$-(Fa \vee -Fa)$	Assume for EI
4.	$-Fa \& --Fa$	3, DM
5.	$-Fa$	4, Simp

We now employ our standard trick for obtaining a contradiction not involving the EI constant.

6.	$-Fa \vee (P \& -P)$	5, Add
7.	$--Fa \& -Fa$	4, Comm
8.	$--Fa$	7, Simp
9.	$P \& -P$	6,8, DS
10.	$P \& -P$	3–9, EI
11.	$(\forall x)(Fx \vee -Fx)$	1–10, Reductio

Proof of T2: we have to prove both

$$[(\forall x)Fx \& P] \Rightarrow [(\forall x)Fx \& P]$$

and the converse:

$$(\forall x)(Fx \& P) \Leftarrow [(\forall x)Fx \& P]$$

Proof of the first logical implication:

1.	$(\forall x)Fx \& P$	$(\forall x)(Fx \& P)$
2.	$(\forall x)Fx$	1, Simp
3.	$P \& (\forall x)Fx$	1, Comm
4.	P	3, Simp
5.	Fa	2, UI

where a is any constant that does not occur in P. We must require this, since UG does not allow us to generalize unless the instantiated constant does not occur in any premise.

13.5 / Useful Theorems

6. $Fa \& P$ 4, 5, Conj
7. $(\forall x)(Fx \& P)$ 6, UG

Now the converse:

1. $(\forall x)(Fx \& P)$ | $(\forall x)Fx \& P$
2. $Fa \& P$ 1, UI (remember: P has no free occurrences of x)
3. Fa 2, Simp
4. $(\forall x)Fx$ 3, UG
5. $P \& Fa$ 2, Comm
6. P 5, Simp
7. $(\forall x)Fx \& P$ 6, 4, Conj

Proof of T9:

1. $(\exists x)(Fx \& Gx)$ | $(\exists x)Fx \& (\exists x)Gx$
2. $Fa \& Ga$ Assume for EI
3. Fa 2, Simp
4. $(\exists x)Fx$ 3, EG
5. $Ga \& Fa$ 2, Comm
6. Ga 5, Simp
7. $(\exists x)Gx$ 6, EG
8. $(\exists x)Fx \& (\exists x)Gx$ 4, 7, Conj
9. $(\exists x)Fx \& (\exists x)Gx$ 2–8, EI

To see why the converse does not hold, let us try to prove it.

1. $(\exists x)Fx \& (\exists x)Gx$ | $(\exists x)(Fx \& Gx)$
2. $(\exists x)Fx$ 1, Simp
3. Fa Assume for EI
4. $(\exists x)Gx \& (\exists x)Fx$ 1, Comm
5. $(\exists x)Gx$ 4, Simp
6. Ga Assume for EI (wrong! a has been used in an assumption within whose scope 6 lies)
7. $Fa \& Ga$ 3, 6, Conj
8. $(\exists x)(Fx \& Gx)$ 7, EG
9. $(\exists x)(Fx \& Gx)$ 6–8, EI
10. $(\exists x)(Fx \& Gx)$ 3–9, EI

The converse of T9 does not hold because the fact that something has property F and something has property G does not entitle us to conclude that the same thing has both properties.

EXERCISES 13.5

1. Prove T3.
* 2. Prove T4.
3. Prove T5.
4. Prove T6.
5. Prove T7.
* 6. Prove T8.

13.6 Proofs of Invalidity

As before, we can prove arguments invalid by expanding them in finite models and then using the short truth-table method to find a falsifying case.

Example: Prove:

1. $(\forall x)(\exists y)Lxyy$
∴ $(\forall x)(\forall y)Lxyy$

invalid by expanding in the two-model $\{a, c\}$.

Answer: Do the expansion in steps:

Premise
$(\forall x)(\exists y)Lxyy$
$(\exists y)Layy \;\&\; (\exists y)Lxyy$
$[Laaa \text{ v } Lacc] \;\&\; [Lcaa \text{ v } Lccc]$

Conclusion
$(\forall x)(\forall y)Lxyy$
$(\forall y)Layy \;\&\; (\forall y)Lcyy$
$[Laaa \;\&\; Lacc] \;\&\; [Lcaa \;\&\; Lccc]$

The premise is T and the conclusion is F under this assignment:

Laaa	Lacc	Lcaa	Lccc
T	F	T	T

(Under which other assignments is the premise T and conclusion F?)

EXERCISES 13.6

Expand each of the following arguments in the two-model {a, b} and prove each invalid.

     1. $(\forall x)Axxx$ $(\forall x)\!-\!Axx$
* 2. $(\forall x)Lxxx$ $(\forall x)(\forall y)Lxy$
     3. $(\exists x)\!-\!Lxx$ $(\forall x)(\forall y)\!-\!Lxy$
     4. $(\exists x)(\exists y)Lxy$ $(\exists x)Lxxx$
     5. $(\exists x)(\exists y)(Lxy \text{ v } Lxx)$ $(\exists y)Lyy$
* 6. $(\exists x)(\exists y)(\exists z)Bxyz$ $(\exists x)Bxxx$
     7. $(\exists x)(\forall y)Lxy$ $(\forall x)(\exists y)Lxy$
     8. $(\forall x)(\exists y)Lxy$ $(\exists x)(\forall y)Lxy$
     9. $(\exists x)(\forall y)(\forall z)Lxyz$ $(\forall x)(\forall y)(\forall z)Lxyz$
* 10. $(\forall x)(\forall y)(\exists z)Lxyz$ $(\forall x)(\forall y)(\forall z)Lxyz$

13.7 Properties of Dyadic Relations

We call predicates of one place 'properties' and those of more than one place 'relations.' We can further refine our terminology by calling predicates of two places *dyadic* relations, and those of more than two places *polyadic* relations. (Properties are sometimes called **monadic** relations.) In this section, we will review three important properties of dyadic relations: reflexivity, symmetry, and transitivity.

We say that a dyadic relation is *reflexive* when that relation holds between any object and itself. Symbolically:

R is reflexive = def $(\forall x)Rxx$

The relation of 'being the mother of' is not reflexive: not everybody is his or her own mother (indeed, nobody is). But the relation of identity is reflexive: everything is identical to itself.

We say a dyadic relation is **symmetric** when, if the relation holds between x and y, it also holds between y and x. Symbolically:

$$R \text{ is symmetric} = \text{def } (\forall x)(\forall y)(Rxy \rightarrow Ryx)$$

For instance, the relation of 'being the brother of' is symmetric: if x is the brother of y, then y is the brother of x.

Finally, we say a dyadic relation is **transitive** if whenever the relation holds between x and y, and between y and z, it holds between x and z. Symbolically:

$$R \text{ is transitive} = \text{def } (\forall x)(\forall y)(\forall z)[((Rxy) \& (Ryz)) \rightarrow Rxz]$$

For instance, the relation of 'being bigger than' is transitive: if x is bigger than y, and y is bigger than z, then x is bigger than z. On the other hand, the relation of 'being the mother of' is not transitive: if x is the mother of y, and y is the mother of z, then x is not the mother of z, but the grandmother.

EXERCISES 13.7

For each of the following dyadic relations, determine whether it is reflexive, transitive, or symmetric. (An asterisk indicates problems that are answered in the back of the book.)

1. Being a sister of
* 2. Being in front of
3. Being heavier than
4. Being a relative of
5. Being to the left of (spatially)

13.8 Identity and Definite Description

One of the most valuable dyadic (two-place) relations is *identity*. We often express identity by the word 'is,' in the sense of "is identical to" or "is the same as." The word 'is' is ambiguous, however. The word 'is' can express identity, as in the following examples.

13.8 / *Identity and Definite Description*

    Tully is Cicero.
    Mark Twain is Samuel Clemens.
    George Eliot is Mary Ann Evans.
    The president is the commander-in-chief.
    Lewis Carroll is Charles L. Dodgson.

But 'is' also can be used to express existence, as in the following examples.

    There is a king of Sweden.
    A talking frog is admired by all as the best emcee in the business.
    A planet is nearby.

Moreover, 'is' can express predication:

    Snow is white.
    Grass is green.
    Billie Mae is black.

In order to symbolize identity statements, we will use the equality sign (=):

$$a = b$$

will symbolize

    a is b

in the sense of "a is the same as b." Thus:

    Mark Twain is Samuel Clemens.

is translated

$$m = s$$

    Identity is reflexive, symmetric, and transitive. That is, for any x, x is identical to x; if x is identical to y, then y is identical to x, and if x is identical to y, while y is identical to z, then x is identical to z.

We must understand what a statement about identity asserts. Consider:

Mark Twain is Samuel Clemens.

That statement asserts that the same individual is named or referred to by two names. The reason such a statement is informative is that we might not know that "Mark Twain" was the pen name Samuel Clemens used.

Using the identity relation symbol, we can translate even more accurately statements expressed in ordinary language.

Example: Translate:

There are at least three happy pigs.

Paraphrase: (*UD* = pigs)

For some *x* and some *y* and some *z*, *x* is happy and *y* is happy and *z* is happy, and *x* is distinct from *y* and *x* is distinct from *z* and *y* is distinct from *z*.

Symbolization:

$(\exists x)(\exists y)(\exists z)[Hx \& Hy \& Hz \& -(x = y) \& -(x = z) \& -(y = z)]$

Example:

There is at most one happy pig.

Paraphrase: (*UD* = pigs)

For any *x* and any *y*, if *x* is happy and *y* is happy, then *x* must be *y*.

Symbolization:

$(\forall x)(\forall y)[(Hx \& Hy) \rightarrow (x = y)]$

In saying that there is *at most* one happy pig, we are not saying there is such a pig.

13.8 / Identity and Definite Description

Example:

There are at most two happy pigs.

Paraphrase: (*UD* = pigs)

For all x and y and z, if x is happy and y is happy and z is happy, then either x is y or x is z or y is z (that is, given any three happy pigs, at least two must be the same).

Symbolization:

$(\forall x)(\forall y)(\forall z)\{(Hx \,\&\, Hy \,\&\, Hx) \to [(x = y) \text{ v } (x = z) \text{ v } (y = z)]\}$

Example:

There is exactly one happy pig.

Paraphrase: (*UD* = pigs)

There is at least one happy pig, and at most one happy pig.

Symbolization:

$(\exists x)Hx \,\&\, (\forall x)(\forall y)[(Hx \,\&\, Hy) \to (x = y)]$

Example:

Porky is the only happy pig.

Paraphrase: (*UD* = pigs)

Porky is a happy pig, and for any x, if x is a happy pig, then x is no other than Porky.

Symbolization:

$Hp \,\&\, (\forall x)(Hx \to (x = p))$

Example:

There are exactly two happy pigs.

Paraphrase (*UD* = pigs):

> There is an *x* and a *y* such that *x* is happy and *y* is happy and *x* is distinct from *y*, and for all *z*, if *z* is happy, then either *z* is *x* or *z* is *y*.

Symbolization:

> (∃*x*)(∃*y*){(H*x* & H*y*) & −(*z*=*y*) & (∀*z*)[H*z* → ((*z*=*x*) v (*z*=*y*))]}

We can generalize the previous approach to statements involving 'exactly three,' 'exactly four,' and so on, but the symbolizations become difficult.

Early in Chapter 12 (Section 12.1) we noted that definite descriptions are referring expressions. Recall that definite descriptions are expressions of the form "the such-and-such." With the identity symbol we can now symbolize statements involving definite descriptions, following an analysis first laid down by Bertrand Russell ("On Denoting," *Mind* 14:479–493, 1905). Consider:

> The present king of France is bald.

Such a statement can be paraphrased:

> There is at least one present king of France; and there is at most one king of France; and that individual is bald.

We can then symbolize (*UD* = persons):

> K*xy* = *x* is the king of *y*.
>   *f* = France.
>   B*x* = *x* is bald.
>
> (∃*x*){K*xf* & (∀*y*)(K*yf* → (*y* = *x*)) & B*x*}

The statement can be F in three different ways: (1) if there is no present king of France, (2) if there is more than one king of France, or (3) if there is one and only one king of France, but he is not bald. In general, then, we will symbolize any statement of the form:

> The *x* that has *F* has *G*.

as

> (∃*x*){F*x* & (∀*y*)(F*y* → (*y* = *x*)) & G*x*}

13.8 / Identity and Definite Description

We can simplify significantly symbolization of statements involving definite descriptions by introducing a symbol $\daleth$ as an abbreviation:

$$F(\daleth x\, Gx)$$

to mean:

$$(\exists x)[Gx \,\&\, (\forall y)(Gx \to (y = x)) \,\&\, Fx]$$

For example:

The king of France is bald.

can now be symbolized:

$$B\,\daleth x\,(Kxf)$$

Similarly:

Juanita is the teacher of this class.

can be symbolized ($Txy = x$ is the teacher of y, j = Juanita, c = this class):

$$j = (\daleth x)\, Txc$$

(It helps to read $\daleth x\, Gx$ as "the x which is G," whatever G may be.)

EXERCISES 13.8

Symbolize the following using the following dictionary (UD = persons). (An asterisk indicates problems that are answered in the back of the book.)

a = Amy
c = Charlene
$Ax = x$ is arrogant
$Sxy = x$ is the sister of y
b = Babylon
h = Hollywood
$Qxy = x$ is the queen of y

1. Amy is Charlene.
* 2. Hollywood is Babylon.
3. Amy is the queen of Babylon.
4. If Amy is the queen of Hollywood, then she is arrogant.
5. Charlene and Amy are sisters.
* 6. There is at most one queen of Hollywood.
7. If there is a queen of Hollywood, then there is a queen of Babylon.
8. There are at most two queens of Babylon.

9. There is exactly one queen of Babylon.
* 10. Amy has exactly one sister.
11. Charlene is the sister of Amy.
12. If there is exactly one queen of Hollywood, then she is arrogant.
13. The queen of Hollywood is arrogant.
* 14. If Amy is a sister of Charlene, then she is the only sister of Charlene.
15. There is exactly one queen of Babylon, and she is Charlene.
16. Charlene is the queen of Babylon.
17. Any queen of Babylon is a queen of Hollywood, and is arrogant.
* 18. Amy is no other than Charlene's only sister.
19. The queen of Hollywood is the queen of Babylon.
20. Amy, Charlene, the queen of Babylon, and the queen of Hollywood are all one and the same.

13.9 Inference Rules for Identity

To enable us to prove valid arguments involving identity statements, we need two new inference rules: *Leibnitz's Law* (LL) and *identity introduction* (II)

LL states that if two individuals are identical, then any property one has, the other has. Symbolically:

LL
$n = m$
$\underline{Fn}$ (n, m any constants)
Fm

II allows the introduction of the logical truth that anything is identical to itself:

II
$a = a$

With these, we can now construct proofs of a wider class of arguments than before.

Example: Translate and prove:

1. Sally is the teacher of this class.
2. Sally is happy.

∴ The teacher of this class is happy.

13.9 / Inference Rules for Identity

Symbolize:

Txy = x is a teacher of y.
c = this class.
s = Sally.
Hx = x is happy.

1. Tsc & $(\forall y)(Tyc \to (y=s))$
2. Hs

$\quad\quad\quad\quad\quad\quad\quad\quad\quad\quad\quad\quad$ | $(\exists x)\{[Txc$ & $(\forall y)(Tyc \to (x=y))]$ & $Hx\}$

Proof:

3. Tsc $\quad\quad\quad\quad\quad\quad\quad\quad\quad\quad$ 1, Simp
4. $(\forall y)(Tyc \to (y=s))$ & Tsc $\quad$ 1, Comm
5. $(\forall y)(Tyc \to (y=s))$ $\quad\quad\quad$ 4, Simp
6. $Tac \to (a=s)$ $\quad\quad\quad\quad\quad$ 5, UI
7. $a=a$ $\quad\quad\quad\quad\quad\quad\quad\quad\quad$ II
8. Tac $\quad\quad\quad\quad\quad\quad\quad\quad\quad$ Assume
9. $a=s$ $\quad\quad\quad\quad\quad\quad\quad\quad\quad$ 6, 8 MP
10. $s=a$ $\quad\quad\quad\quad\quad\quad\quad\quad\quad$ 7, 9 LL
11. $Tac \to (s=a)$ $\quad\quad\quad\quad\quad$ 8–10, CP
12. $(\forall y)(Tyc \to (s=y))$ $\quad\quad\quad$ 11, UG
13. Tsc & $(\forall y)(Tyc \to (s=y))$ $\quad$ 12, 3, Conj
14. $[Tsc$ & $(\forall y)(Tyc \to (s=y))]$ & Hs $\quad$ 2, 13, Conj
15. $(\exists x)\{[Txc$ & $(\forall y)(Tyc \to (x=y))]$ & $Hx\}$ $\quad$ 14, EG

Example: Translate and prove:

1. Fred loves Mary.

∴ Someone loves Mary, and he is Fred.

Symbolize:

Lxy: x loves y
F: Fred
M: Mary

Proof:

1. Lfm $\quad\quad\quad\quad\quad\quad$ | $(\exists x)(Lxm$ & $(x=f))$
2. $f=f$ $\quad\quad\quad\quad\quad\quad$ II
3. Lfm & $(f=f)$ $\quad\quad$ 2, 1, Conj
4. $(\exists x)(Lxm$ & $(x=f))$ $\quad$ 3, EG

EXERCISES 13.9

Prove the following. (An asterisk indicates problems that are answered in the back of the book.)

1. 1. $a = b$
   2. Hab / Hbb

* 2. 1. $(\forall x)Mxx$
   2. $a = b$ / Mab

3. 1. $(\forall x)(\forall y)(Lxy \to Lyx)$
   2. $m = r$
   3. Lra / Lam

4. 1. $(Rst \ \& \ Rmt) \ \& \ Rat$ / $(\exists x)(Rxt \ \& \ (x = m))$

5. 1. Tg
   2. $(\forall x)[(Qx \ \& \ Wbx) \to (g = x)]$ / $(\forall x)[(Qx \ \& \ Wbx) \to Tx]$

* 6. 1. $a = b$ / $b = a$

7. 1. $Ma \ \& -Mb$ / $-(a = b)$

8. 1. $a = b$
   2. $b = c$ / $a = c$

9. 1. $(\forall x)(Ax \ v \ Bx)$
   2. $Am \ \& \ Bc$ / $-(c = m)$

* 10. 1. $(\forall x)(Hxc \to Mxc)$
   2. $(\exists x) Hxc$
   3. $c = a$ / $(\exists x) Mxa$

11. 1. $a = b$
    2. $b = c$
    3. $(\forall x)Lxa$ / $(\forall x)Lxc$

12. 1. $-(a = c)$
    2. $a = b$ / $-(b = c)$

13. 1. $a = b$
    2. $b = r$
    3. $Sr \ \& -St$ / $-(t = a)$

* 14. 1. $Pb \to Sa$
    2. $c = a$
    3. $-Sc$
    4. $b = r$ / $-Pr$

15. 1. $(\forall x)(Mxb \to Rxc)$
    2. $c = b$
    3. $(\exists x)Mxc$ / $(\exists x)Rxb$

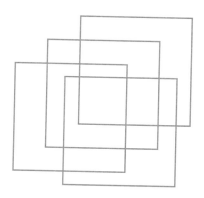

Appendix

Mill's Methods

A.1 The Relation of Cause and Effect

In this appendix, we will take a first look at scientific method, focusing on the work of the eminent logician John Stuart Mill (1806–1873). Mill elaborated five rules or 'methods,' which we will examine. We state those methods informally and then look at some actual cases from scientific research. Next, we give a formal account of those methods, employing the truth-functional techniques developed in Chapter 9. We then conclude by discussing some limitations of Mill's account.

Scientists are interested in finding the causes of diverse phenomena. What causes earthquakes, cancer, unemployment? Mill's methods are rules for determining the cause of a given phenomenon under investigation. But what precisely do we mean by the term 'cause'? We need to clarify that ambiguous term.

Aristotle, the Greek logician who lived in the fourth century B.C., delineated four types of causes: efficient, material, formal, and final. The **efficient cause** of an event or thing is the event that either starts it or stops it. For example, the efficient cause of an explosion is the ignition of the gunpowder. The notion of 'efficient cause' applies only to events that begin at a definite moment in time.

The **material cause** of a thing is the matter of which it is made. To use Aristotle's example, the material cause of a statue is the bronze of which it is made. This may strike the reader as an odd way to talk, but Aristotle had a point: we do say, for example, that the car rusted because it was made of metal (also, that the fishing pole did not rust because it was made of fiberglass).

The **formal cause** of a thing is its essence or nature. The essence of something is the set of properties that define it, that is, that necessarily characterize it. Aristotle's example is that of the octave in music: the cause of the octave is the relation of 2 to 1 that holds between the frequencies of a given note and a note that is an octave higher. This way of talking strikes the modern reader as odd, because we normally think of cause as efficient cause. But Aristotle had a good point: we do say, for example, that dogs mark territory with urine because they are territorial by nature.

The **final cause** of a thing is the purpose for which that thing is done or made. Thus the final cause of Lucia's choice of major may be her desire to get a high-paying job. Explanations that refer to purposes or intentions (final causes) are called 'purposive' or 'teleological' explanations.

In what follows, we will use 'cause' to mean only efficient cause. Most modern texts conform to this usage.

Another ambiguity inherent in the word 'cause' centers around the temporal closeness of the cause and the effect. Consider the following sequence of events. A car is parked on a hill. The parking brake cable snaps, the brake gives way, the car rolls down the hill and hits a pig, which subsequently dies. What is the cause of the pig's death? In one sense, being hit by the car was the cause of the pig's death. But in another sense, the snapping of the cable was the cause. When a finite sequence of events is linked by cause-and-effect relations, we call the sequence a finite **causal chain**. The **remote** cause of an event in a causal chain is the first event in the chain; the **proximate** cause is the event immediately prior to the event.

Obviously, causal chains can be infinitely long, in which case what event we take to be 'the' remote cause depends upon our point of view. In the previous case of the unfortunate pig, we can trace the causal chain back indefinitely far: pig crushed; prior to that, car hits pig; prior to that, car rolls down hill; prior to that, the brake fails; prior to that, the metal-fatigued cable snaps; prior to that, the car owner puts tension on the cable; and so on, back to the building of the car, and even earlier to the mining of the ore from which the car's metal was extracted. What we pick as the remote cause depends upon the context of discussion: a lawyer might focus upon the failure of the parking brake, whereas a mechanical engineer might focus upon the metal

Appendix

fatigue of the cable, and the pig's owner might emphasize our having decided to park on that hill in the country.

In what follows, we will have in mind proximate rather than remote causes.

Another area of ambiguity in the notion of causality concerns whether we are speaking of necessary or sufficient causes. The distinction here is the same as the distinction between necessary and sufficient conditions we drew in Chapter 9. A necessary cause or condition is one in whose absence the effect cannot occur; a sufficient cause or condition is one whose presence by itself leads to the effect. For instance, a match being in the presence of oxygen is a necessary cause of its lighting, but not a sufficient one—strike a wet match even in the open air and it will not light. Many different sufficient causes may apply to a given effect. Swallowing poison is sufficient to cause death, but so is a bullet through the brain, a bullet through the heart, and so on. A given effect may have a number of necessary causes. To light, a match needs to be struck *and* be dry *and* be in the presence of oxygen.

Whether a necessary or a sufficient cause is sought depends upon the context of research. If we are interested in producing the effect, we look for a sufficient cause. If we are interested in preventing the effect, we look for a necessary cause (to try to prevent it).

In what follows, we will take the cause of an effect to be the set of conditions (one or more) that are singly necessary and jointly sufficient to bring about the effect.

In summary, the notion of causation is ambiguous. But our discussion of Mill's methods will be clearer if we fix upon the notion of cause as the efficient, proximate set of conditions that are individually necessary and together sufficient for the occurrence of the effect.

EXERCISES A.1

I. *Each of the following problems describes a state of affairs and four possible causes of it. Identify the causes as efficient, material, formal, or final. Explain your answers. (An asterisk indicates problems that are answered in the back of the book.)*

1. The Statue of Liberty exists because:
   (a) It was cast by workers;
   (b) The people of France wanted to present a gift to the American people;

(c) The bronze of which it is made exists;
(d) It symbolizes liberty.
2. The house burned down because:
   (a) It was set on fire by a can of gas;
   (b) The arsonist wanted to collect insurance on the building;
   (c) It was made of wood, and wood burns.
3. Cats chase mice because:
   (a) Cats are hunters by nature;
   (b) Something inside a cat's brain causes it to chase a mouse when it sees one;
   (c) Cats are carnivorous.
4. There is a car parked outside because:
   (a) It is a machine with the power to move about at the direction of the driver;
   (b) The FBI is observing the occupants of the house;
   (c) Somebody applied the brake, which brought the car to a halt.

II. Each of the following problems describes a state of affairs and two contrasting explanations of it. Discuss the different concepts of causality implicit in each explanation.

5. The man was killed because:
   (a) The bullet went through his heart;
   (b) He knew too much (that is, he was a potential witness).
6. The woman had a heart attack because:
   (a) Insufficient blood reached the heart (she had hardening of the arteries);
   (b) She got little exercise and ate high-cholesterol food.
7. The man got the flu because:
   (a) He was exposed to the virus;
   (b) He was exposed to the virus and his immune system could not control the virus.

A.2 Mill's Methods—An Informal Account

Scientists search for causes (causal explanations) of phenomena. Mill formulated five methods for determining causes: the method of agreement, the method of difference, the joint method of agreement and difference, the method of concomitant variation, and the method of residues. Mill intended these methods as tools for both discovering and proving causal laws. In our terms, Mill offered these techniques as both heuristic and logical rules.

The *method of agreement* is if two or more instances of the phenomenon under investigation have only one circumstance in common, the circumstance in which alone all the instances agree is the cause (or effect) of the given phenomenon. For example, if we investigate baldness and find that potassium shortage is the only circumstance that all bald persons have in common, then we may conclude that potassium shortage causes baldness (or that baldness causes potassium shortage).

The *method of difference* states that if an instance in which the phenomenon under investigation occurs and an instance in which it does not occur have every circumstance in common except one, that one occurring in the former, then the circumstance in which alone the two instances differ is the cause (or effect, or part of the cause) of the phenomenon. For example, if someone drinks a lot of alcohol and gets liver disease, and his identical twin, who eats the same diet and has all other habits identical to the first brother but does not drink and does not get liver disease, then we may conclude that drinking causes liver disease (or else that liver disease causes one to drink).

The *joint method of agreement and difference* states that if two or more instances in which the phenomenon occurs have only one circumstance in common, while two or more instances in which it does not occur have nothing in common except the absence of that circumstance, then the circumstance in which alone the two sets of instances differ is the cause (or effect, or part of the cause) of the phenomenon. For example, if we examine those who live past 75 years of age and those who do not, and find that the one factor that the former have in common that all the latter lack is regular exercise, then we may conclude that regular exercise is the cause of longevity.

The *method of concomitant variation* states that whatever phenomenon varies in any manner, whenever another phenomenon varies in some other particular manner, is the cause (or effect) of that phenomenon. For example, if the testosterone level in the blood varies directly with the degree of hair loss in men, then we may conclude that high testosterone level causes hair loss.

The *method of residues* states that if we subtract from any phenomenon such part as is known by previous inductions to be the effect of specific antecedents, then the residue of the phenomenon is the effect of the remaining antecedents. For example, suppose some poor woman is losing her teeth and eyesight and has liver disease. We discover that she never eats carrots, never drinks milk, but drinks lots of alcohol. If we know from prior research that alcohol destroys the liver and that lack of carrots causes eyesight loss, then we may conclude that lack of milk causes tooth loss.

Applying Mill's methods comes down to setting up experiments that seek to isolate factors that may or may not be causally linked to a given phenomenon. Biological and medical science in particular involve experiments and research accurately describable in Mill's terms.

Consider as an example a description of the evidence for viewing the cause of depression as low levels of the neurotransmitter norepinephrine:[1]

> Some other lines of evidence also underscore the possible role of norepinephrine. For example, electroconvulsive shock therapy, which despite its bad name is the most effective treatment for severe depression (as well as a surprisingly safe one), has been shown to elevate norepinephrine levels in laboratory animals. Salts of the metal lithium, a remarkably simple and effective drug treatment for the manic phase of manic-depressive psychosis, also may produce their effect by altering norepinephrine metabolism. Finally, compounds that deplete norepinephrine in the brain, or that interfere with its synthesis, or that poison neurons that make norepinephrine, have been shown to cause obvious symptoms of depression in monkeys, rats, and other laboratory animals.

Here, no instances of reversal of depression are examined, and one common factor is discerned: elevation of norepinephrine level. This is an application of the method of agreement.

Social and medical scientists often conduct studies of twins raised apart. For instance, to discover whether alcoholism is genetic in origin, researchers will examine twins who were separated at birth and raised in markedly different environments. Presumably, all that the twins have in common is their genetic structure (environmental factors being ruled out), so by applying the method of agreement, genetic predisposition must be the cause (or part of the cause) of alcoholism.

The joint method is the logic behind control group experiments. A control group, say, of smokers is given a placebo (a sugar pill), and another group of smokers is given a newly developed anti-smoking drug. If none of the control group loses the craving for cigarettes, and all of the others (who have been given the real drug) do lose their craving, then we conclude that the drug causes the drop in desire for cigarettes.

---

[1] Taken from Melvin Konner, *The Tangled Wing: Biological Constraints on the Human Spirit* (New York: Harper & Row, 1982).

Appendix

The method of concomitant variation is a common mode of reasoning in the social sciences. A political scientist may notice a strong correlation between, say, the rate of unemployment and the crime rate, and then infer that unemployment causes (or is at least part of the cause) of crime. The limitations of such inferences will be discussed in Section A.6 in this appendix.

EXERCISES A.2

For each of the following passages, determine which of the five Mill's methods is being employed. All passages are from Melvin Konner, The Tangled Wing: Biological Constraints on the Human Spirit *(New York: Harper & Row, 1982).*

1. Those results are essentially as follows: In every sample of infants, in widely separated social classes and cultures around the world, the percentage of infants who fret, cry, or show other signs of distress when the mother leaves or when a strange person appears rises markedly after the age of six months. In each sample, some or even many infants do not show obvious fear at all, and some respond positively to strangers.

But the growth change is universal in this sense: Before the age of six months, and especially between birth and four months of age, signs of either fear are for all intents and purposes nonexistent, whereas after seven or eight months of age they are quite common, and in many samples predominate. In infants from the professional class as well as the working class in the United States, in infants in an Israeli kibbutz where mother-infant contact is by our standards quite limited, in rural and urban Guatemala, and among the !Kung San of Botswana, who have the closest, most intimate, and most indulgent mother-infant relationship ever systematically described, the rise of separation fear and stranger fear occurs during the same age period. There are large cross-cultural variations in the percentage of infants who cry at any given age, in the age at which the percentage reaches its peak, and in the steepness and duration of the subsequent decline with further growth. But the variations in the shape and timing of the rising portion of the curve between the ages of six and fifteen months are minor or nonexistent; cultural training does not much accelerate or decelerate the change.

It is thus most likely that the change is a result of growth and that, like the rise of social smiling during the first few months of postnatal life, it is primarily the result of maturation in the nervous system—changes most of which are not in principle different from those that go on prenatally.

2. In every culture there is at least some homicide, in the context of war or ritual or in the context of daily life, and in every culture men are mainly responsible for it. Among the !Kung San of Botswana, noted for their pacifism

as well as for equality between the sexes, the perpetrators in twenty-two documented homicides were all men. Fights over adultery or presumed adultery were involved in several cases, and a majority of the others were retaliations for previous homicides. In a sample of 122 distinct societies in the ethnographic range, distributed around the world, weapon making was done by men in all of them. There are of course exceptions, certainly at the individual level and, in rare cases—such as modern Israel or nineteenth-century Dahomey—transient partial exceptions at the group level. What we are dealing with, to be sure, is a difference in degree, but one so large that it may as well be qualitative. Men are more violent than women.

3. One of the most impressive experiments of the kind produced "pseudohermaphrodite" monkeys by administering male gonadal hormones to female fetuses before birth. As they grew, these females showed neither the characteristic low female level of aggressive play nor the characteristic high male level, but something precisely in between.

4. In 1874, Carl Wernicke, a then obscure twenty-six-year-old neurologist, described a new aphasic syndrome characterized by fluent and rapid but largely meaningless speech, with most of the content words missing, and by almost total loss of comprehension of speech, despite normal hearing. These patients often had lesions in a quite different region—now called Wernicke's area—behind and adjacent to the area of the cerebral cortex involved in the first-level interpretation of auditory patterns. (Wernicke's area would be approximately above the ear, although this association with hearing is purely coincidental.) In his impressive paper on the subject, Wernicke went on to advance a theory of the brain mediation of language, taking into account his own findings as well as Broca's. According to this theory, the area he identified, adjacent to the primary higher processing center for hearing, was responsible for the analysis of sound patterns at the level of speech comprehension.

5. Nineteen of the subjects appeared at birth to be ambiguously female, and were viewed and reared as completely normal females by their parents and other relatives. At puberty they first failed to develop breasts and then underwent a completely masculine pubertal transformation, including growth of a phallus, descent of the testes (which had previously been in the abdominal cavity), deepening of the voice, and development of muscular masculine physique. Physically and psychologically they became men, with normal or occasionally hypernormal sexual desire for women and with a complete range of sexual functions except for infertility due to abnormal ejaculation (through an opening at the base of the penis). After many years of experience with such individuals, the villagers identified them as a separate group, called *guevedoce*, "penis at twelve," or *machihembra*, "man-woman." The physiological analysis undertaken by Imperato-McGinley and her colleagues revealed that these individuals are genetically male—they have one X and one Y chromosome—but lack a single enzyme of male sex-hormone synthesis, due to a defective gene. The enzyme, 5a-reductase,

changes testosterone into another male sex hormone, dihydrotestosterone. Although they lack dihydrotestosterone almost completely, they have normal levels of testosterone itself. Evidently these two hormones are respectively responsible for the promotion of male external sex characteristics at birth (dihydrotestosterone) and at puberty (testosterone). Despite the presence of testosterone, the lack of "dihydro" makes for a female-looking newborn and prepubertal child. The presence of testosterone makes for a more or less normal masculine puberty.

6. Testosterone promotes aggression, certainly in males and possibly in females, in a much more specific way. Indeed, generalized stress is likely to decrease the level of testosterone. Yet, in members of various species, especially in males, testosterone injections can increase aggressiveness in various situations and male castration can decrease it. Naturally occurring variations in testosterone level can accompany fighting behavior, and fighting can in turn affect that level. For example, in an experiment in which two groups of rhesus monkeys were made to fight, the losers experienced a large decrease after the fight (actually in two stages, the second perhaps corresponding to final acceptance of the loss), while the winners did not.

A.3 A More Formal Approach to Mill's Methods

We can gain a deeper understanding of Mill's methods if we use the resources of symbolic logic to study their underlying structure. First, we will focus upon the direct method of agreement. Mill's rule is that if researchers are interested in discovering the cause of a phenomenon, they should look at cases in which E is present (or occurs) and seek a common factor C. That common factor would be the cause of E.

Two points need to be made. The first is that the number of possible causal factors the researcher could examine is infinite. In practice, however, the researcher starts with a list of candidates, for possible causes to investigate. If one of the factors on the candidate list is shared by all the E cases, then presumably that factor is the cause of E. But if none of the candidates turn out to be shared by all the E cases, then presumably the list will have to be expanded.

The second point is that the direct method of agreement discovers only necessary conditions. With the definition we are using, the cause C of an effect E is both a necessary and sufficient condition for the occurrence of E: when C is present, E is present; when C is absent, E is absent as well. But the direct method only involves cases in which C is present, thus it does not allow the researcher to conclude anything about whether E would not occur in the absence of C. However, the direct method does help us eliminate possible causes, since if a factor is not a necessary condition for E, it cannot be the cause of E.

Let us put these points in propositional logic (PL) terms. Recall from Chapter 9 that "C is a necessary and sufficient condition for E" is symbolized as $C \leftrightarrow E$. Clearly, $C \leftrightarrow E$ can be shown to be false (F) if E is shown to be true (T) and C F. Put another way, the following valid argument form functions as an *elimination rule* (let us call it Elimination Rule 1, or ER1 for short):

1. $E \ \& -Ci$

$\therefore -(Ci \leftrightarrow E)$

The researcher begins with a list of possible causes $\{C_1, C_2, \ldots, C_n\}$, that is, a set of possible causal hypotheses $\{C_1 \leftrightarrow E, C_2 \leftrightarrow E, \ldots, C_n \leftrightarrow E\}$, and eliminates possible candidates by examining E cases and employing ER1.

Consider an example. A deadly disease called "chokosis" has broken out, and we suspect the following factors are causes of the disease: zinc deficiency, presence of a strain of bacteria, exposure to asbestos, and excessive consumption of corn whiskey. We examine some chokosis patients and discover the following: Mr. A is badly deficient in zinc, but has no trace of the bacteria, has never been around asbestos, but is quite fond of corn whiskey. Dr. B is also a fan of corn whiskey, does have the bacteria in her system, but is not zinc deficient, nor has she ever been exposed to asbestos. Ms. C has all the zinc a person can possibly have, but also is loaded with bacteria. She has been exposed to asbestos and admits to nipping corn whiskey now and then.

How can we use ER1 to find the cause of chokosis? By means of a table, akin to the truth tables we used in Chapter 9. The shell of the table is given as follows:

The cases examined	Properties examined	Phenomenon being investigated	Hypotheses examined

That is, an elimination table consists of the names of the individuals examined in our example.

Cases
Mr. A
Dr. B
Ms. C

Appendix

Further, such a table includes the list of candidates for a necessary condition in our example: zinc deficiency (Z), bacterial infection (B), asbestos (A), and corn whiskey (W). It also includes the factor for whose cause we are searching—chokosis (C)—and whether each case examined has it.

Cases	Z	B	A	W	C
a					
b					
c					

We next fill in the results of the investigation, indicating whether a given individual has a given factor by T or F.

Cases	Z	B	A	W	C
a	T	F	F	T	T
b	F	T	F	T	T
c	F	T	T	T	T

Finally, list the hypotheses involved, and eliminate from contention any factor that is not present in every case. We could do this by inspecting the table, but a more mechanical technique is to list the hypotheses and do truth-functional calculations directly underneath them. To say that Z is a possible cause for C is to say that $Z \leftrightarrow C$ is always T, and so we just do the table:

							Hypotheses		
						(1)	(2)	(3)	(4)
Cases	Z	B	A	W	C	$Z \leftrightarrow C$	$B \leftrightarrow C$	$A \leftrightarrow C$	$W \leftrightarrow C$
a	T	F	F	T	T	~~T~~ T ~~T~~	~~T~~ (F) ~~T~~	~~T~~ (F) ~~T~~	~~T~~ (T) ~~T~~
b	F	T	F	T	T	~~T~~ (F) ~~T~~	Hypothesis eliminated	Hypothesis eliminated	~~T~~ (T) ~~T~~
c	F	T	T	T	T	Hypothesis eliminated			~~T~~ (T) ~~T~~

So if any of the candidates are the cause, it must be W.

The logic underlying the direct method of agreement should now be clear. The researcher begins with a wide candidate list of causal hypotheses and examines a large enough number of cases to eliminate all but one. The researcher might not get down to one candidate, in which case more E cases must be examined. The researcher may well eliminate all the candidates, in which case more causal hypotheses must be sought.

We can extend the elimination table technique to handle compound factors. After all, we want to allow the possibility that the cause of a given phenomenon might be a conjunction of conditions, each one of which is necessary and which are jointly sufficient for the occurrence of the phenomenon. Compound factors can be expressed with the usual truth-functional connectives. Thus, if P refers to potassium shortage and B to the presence of bacteria, $-B$ would be the absence of the bacteria, $P \vee B$ the factor of either potassium shortage or bacterial presence, and $P \,\&\, B$ for potassium shortage and bacterial presence.

As an example, suppose a researcher suspects that one of $\{B, C, -B, -C, -B \,\&\, C, B \vee -C\}$ is the cause of D. She checks four D individuals ($a, b, c, d,$) and finds that a has B and C, b has B but not C, c has B but not C, and d has neither B nor C. What should she conclude?

Construct the table:

Cases	B	C	D	(1) $B \leftrightarrow D$	(2) $C \leftrightarrow D$	(3) $-B \leftrightarrow C$	(4) $-D \leftrightarrow C$	(5) $(-B \,\&\, C) \leftrightarrow D$	(6) $(B \vee -C) \leftrightarrow D$
a	T	T	T	T T T	T T T	T T (F) T	T T (F) T	T T T (F) T	T T (T) T
b	T	F	T	T T T	T (F) T	Eliminated	Eliminated	Eliminated	T T T T (T) T
c	T	F	T	T T T	Eliminated				T T T T (T) T
d	F	F	T	T (F) T					T T T T (T) T
				Eliminated					

Only hypothesis (6) survives. Thus she should conclude that of the group of possible causes of D, only $(B \vee -C)$ is a live option.

EXERCISES A.3

1. Complete the elimination table for the list $\{A, L, -A, L \,\&\, A, -L \vee A\}$ of possible causes for M, using the direct method of agreement.

Cases	A	L	M
a	T	T	T
b	T	F	T
c	F	F	T

Appendix

2. Complete the elimination table for the list $\{R, S, T, S \vee -R, -R \& -T, -S \vee -T\}$ as possible necessary conditions for C.

Cases	R	S	T	C
a	T	T	F	T
b	F	F	T	T
c	F	T	F	T
d	F	F	F	T

3. You suspect one of $\{M, R, S, M \& -R, M \vee -S\}$ is a necessary condition for the occurrence of O. You examine cases a, b, c, d, each of which has O, and find that a has M but lacks R and S; b has M and R but lacks S; c lacks S and M but has R; d lacks all three. Set up the elimination table and draw the appropriate conclusion.

A.4 The Inverse Method of Agreement

In Section A.3 we saw that Mill's first method can be viewed as the application of an elimination rule (ER1) to a list of causal hypotheses. We prove that a factor C is not a necessary condition for E by finding cases where E is present and C absent.

By a similar process of thought, we can prove that a factor C is not a sufficient condition for E and hence cannot be the cause of E if we can point to cases where C is present and E is not. Symbolically:

ER2
1. $C_i \& -E$

$\therefore -(C_i \leftrightarrow E)$

This is the inverse method of agreement: we begin (as before) with a list of candidates for the cause of E, and then examine cases where E is absent. Any property that is present when E is absent cannot be the cause of E. So for example, zinc deficiency cannot be the cause of male baldness if some zinc-deficient men have full heads of hair.

The elimination table for the inverse method looks exactly as it does for the direct method, but in all the cases E is F.

An example: Suppose we are trying to find the cause of depression. We suspect drinking alcohol, playing volleyball, reading romantic novels, and excessive coffee consumption as possible causes. We examine only people who are never depressed. Ms. A plays volleyball and drinks coffee but not alcohol. Ms. A hates romantic novels. Ms. B never drinks alcohol or coffee, or plays volleyball, but she loves romantic novels. Ms. C lives for volleyball, loves coffee, but hates romantic novels and alcoholic beverages. We construct the table.

Factors
    A = Alcohol consumption
    V = Volleyball playing
    R = Romantic novel reading
    C = Coffee consumption

Cases	A V R C	(1) D	(2) $A \leftrightarrow D$	(3) $V \leftrightarrow D$	(4) $R \leftrightarrow D$	(5) $C \leftrightarrow D$
a	F T F T	F	↟ T ↟	↟ Ⓕ ↟ Eliminated	↟ T ↟	↟ Ⓕ ↟ Eliminated
b	F F T F	F	↟ T ↟		T Ⓕ ↟ Eliminated	
c	F T F T	F	↟ T ↟			

We conclude that if any of the list is the cause of D, it must be A.

As in Section A.3, we can consider also compound properties. Suppose we suspect one of the candidates $\{M, -M, Z, M \vee -Z, M \& Z\}$ is the cause of Q. We again examine cases of not Q and find:

Cases	M Z	Q	(1) $M \leftrightarrow Q$	(2) $-M \leftrightarrow Q$	(3) $Z \leftrightarrow Q$	(4) $(M \vee -Z) \leftrightarrow Q$	(5) $(M \& Z) \leftrightarrow Q$
a	F T	F	↟ T ↟	↟↟ Ⓕ ↟ Eliminated	↟ Ⓕ ↟ Eliminated	↟ ↟↟T T ↟	↟ ↟↟T T ↟
b	T F	F	↟ Ⓕ ↟			↟ ↟↟↟ Ⓕ ↟	↟ ↟ ↟ T F
c	T T	F	Eliminated				↟ ↟ ↟ Ⓕ ↟ Eliminated

Here, we must conclude that none of the candidates on the original list can be the cause of Q. We would have to expand out list.

Allowing conjunction of simple properties is valuable. For example, several factors must be present to cause longevity: we must eat well *and* exercise *and* have the right genetic inheritance *and* so on.

Appendix

EXERCISES A.4

1. Complete the following table for the list $\{E, F, -E, -E \,\&\, F\}$ of candidates for the causes of R. What can be concluded?

Cases	E	F	R
a	T	F	F
b	T	F	F
c	F	T	F

2. Complete the following table for the list $\{A, B, C, -B, -C, A \vee -B, B \,\&\, -C\}$ of candidates as possible causes of H. What can be concluded?

Cases	A	B	C	H
a	T	T	T	F
b	F	F	T	F
c	F	T	F	F
d	T	T	F	F
e	F	F	F	F

3. Complete the following table for the list $\{L, M, N, -N, -L, L \,\&\, M, N \vee L\}$ of candidates as possible causes of P. What can be concluded?

Cases	L	M	N	P
a	T	T	F	T
b	F	F	T	T
c	F	T	F	T
d	F	F	F	T

A.5 The Combined Method of Agreement

We began by viewing the cause of a phenomenon E as a necessary and sufficient condition (simple or compound) for the occurrence of that phenomenon. Strictly speaking, the direct method of agreement involves looking for cases where E is present and eliminating candidates for necessary conditions by the rule:

Appendix

1. $-Ci \ \& \ E$

∴ $-(E \leftrightarrow Ci)$

Also strictly speaking, the inverse method involves looking at cases where E is absent and eliminating candidates for sufficient conditions by the rule:

ER1:
1. $Ci \ \& -E$

∴ $-(Ci \leftrightarrow E)$

ER2:
1. $Ci \ \& -E$

∴ $-(Ci \leftrightarrow E)$

We set up our tables using the elimination rules: taking the position that if something is not a necessary condition, it cannot be the cause, and if something is not a sufficient condition, it cannot be the cause.

We can now easily combine the methods, formulating the general rule:

1. $(-Ci \ \& \ E) \ \text{v} \ (Ci \ \& -E)$

∴ $-(Ci \leftrightarrow E)$

That is, we look at cases where E is present and other cases where E is absent, ruling out candidates that are either present when E is not, or else are absent when E is not. The table is done exactly as in Sections A.3 and A.4. Thus, for example, we might want to know the cause of an outbreak of dysentery. We examine both people who have the sickness and people who do not. The factors we consider candidates are $\{S, L, M, -L, S \& -L, -M\}$. We examine Mr. A, who has the disease, and find that he has S, has L, but lacks M. Ms. B does not have the disease, has S and M, but lacks L. Mr. C has the disease and lacks S, L, and M. What can we conclude? We set up the table as before. We can conclude that if any of our list is the cause of D, $-M$ is. In the cases we have examined, only $-M$ is present when D is, and absent when D is.

Cases	S	L	M	D	(1) $S \leftrightarrow D$	(2) $L \leftrightarrow D$	(3) $M \leftrightarrow D$	(4) $-L \leftrightarrow D$	(5) $(S \& -L) \leftrightarrow D$	(6) $-M \leftrightarrow D$
a	T	T	F	T	T T T	T T T	T Ⓕ T	T T Ⓕ T	T T T T Ⓕ T	T T T
b	T	F	T	F	T Ⓕ T	T T T	Eliminated	Eliminated	Eliminated	T T T
c	F	F	F	T	Eliminated	T Ⓕ T				T Ⓣ T
						Eliminated				

Appendix

EXERCISES A.5

Complete the tables and draw the appropriate conclusion.

1. $\{A, B, -A, -B, -A \vee -B, A \& -B, -A \& B\}$: candidates for cause of C.

Cases	A	B	C
a	T	T	T
b	T	F	F
c	F	F	F
d	T	T	F

2. $\{Z, X, R, -X, -R, R \vee X, R \& Z, -R \& -Z\}$: candidates for the cause of L.

Cases	Z	X	R	L
a	T	F	F	T
b	F	F	F	F
c	T	T	T	F
d	T	F	T	F

3. $\{E, F, -E, E \& F, E \& G, -E \vee -F\}$: candidates for the cause of Q.

Cases	E	F	G	Q
a	T	F	F	F
b	F	T	F	T
c	T	F	T	F

A.6 The Limitations of Mill's Methods

Mill claimed that his methods were both heuristic rules for discovery and rules for proof or causal hypotheses. To assess his claim, let us reconsider the direct method of agreement, our remarks applying to the other methods as well.[2]

---

[2]For a full elaboration of these criticisms, see Morris Cohen and Ernest Nagel, *An Introduction to Logic and Scientific Method* (New York: Harcourt, Brace and World, 1934), ch. 13.

We ask first if Mill's methods are effective as methods of discovery. The answer seems to be no. Take the method of agreement. To discover the cause of baldness, we are told that we need only find the factor that all bald men have in common. But the number of potential factors is infinite. Do we have to check every imaginable property? Color of eyes? Toenail length? Favorite music? Amount of sardines consumed? Distance between the man's scalp and Venus?

The point is that to even begin searching for a cause of a given phenomenon, we must first have an idea, an *hypothesis*, about what properties are relevant. Indeed, in our discussion we were careful to stipulate that our tables were applied to preexisting lists of candidates. But where does this list come from in practice? Mill gives us no hint.

Moreover, what counts as a *factor* is not clear. The way we describe things will, in part, dictate what factors we isolate as being in common. Suppose, for example, we examine three bald men and analyze the factors thus:

Factor 1	Factor 2	Factor 3
Mr. *A*: Brown male	Eats greasy food.	Drinks heavily.
Mr. *B*: Black male	Eats greasy food.	Doesn't drink.
Mr. *C*: White male	Eats greasy food.	Drinks heavily.

The method of agreement tells us that factor 2 (eating greasy food) is the cause of baldness. But now analyze the factors thus:

Factor 1	Factor 2	Factor 3
Mr. *A*: Male	Eats greasy food while being brown.	(As before)
Mr. *B*: Male	Eats greasy food while being black.	(As before)
Mr. *C*: Male	Eats greasy food while being white.	(As before)

The method implies that factor 1 is the cause!

To apply the methods, we must first have a hypothesis about how factors are to be analyzed. Thus, as methods of discovery, these methods seem inadequate as they stand.

How about Mill's canons as methods of proof? Here, things are more promising. Certainly the canons are not methods of absolute proof. We begin with a list of factors, and we may eliminate all but one. But we cannot ever know with certainty that the cause was among the list of candidates, because we can never be sure that the next case we examine will not eliminate whatever candidates have so far survived. In that sense, Mill's rules are *inductive*.

However, we can view the methods as providing knowledge that something is not a cause of the phenomenon under investigation. Mill's methods are deductively valid only as methods of elimination.

Within limits, Mill's methods are accurate descriptions of much scientific research. For this reason, they are well worth our studying.

Glossary

Accent Changing the meaning of a word or phrase by stressing or omitting part of it.

Accident The fallacy of applying a general rule in an atypical way or to an atypical case.

Action-oriented conversation A conversation directed primarily at persuading people to act.

Aesthetics-oriented conversation A conversation directed primarily at entertainment or literary display.

Ambiguous Having more than one meaning.

Amphiboly A sentence that is ambiguous due to grammatical construction.

Analogy A common mode of nondeductive reasoning that is a comparison of two things.

Announcement A statement about where a person or group stands on some issue.

Appeal to fear Using threats or scare tactics instead of evidence to get a point accepted.

Appeal to ignorance The fallacy of arguing that something must be true because no one can prove it false (or alternatively, that something must be false because no one can prove it true).

Appeal to pity Evoking the feeling of pity instead of giving evidence to get a point accepted.

Appeal to the crowd Arousing feelings of group identity instead of giving evidence to get a point accepted.

Argument A set of one or more statements, called the *premises*, taken as evidence for another statement, called the *conclusion*.

Argument by analogy When an analogical premise is used in an argument.

Argument form An array of variables, connectives, and brackets, but no constants, which has the property that when statements replace the variables, an argument is the result.

Argumentum ad baculum Appeal to fear.

Argumentum ad hominem Attacking the person.

Argumentum ad ignorantiam Appeal to ignorance.

Argumentum ad misericordiam Appeal to pity.

Argumentum ad populum Appeal to the crowd.

Argumentum ad verecundiam Bad appeal to authority.

Artificial language A language designed for a specific purpose.

Attacking the person The fallacy of criticizing a person who puts forward a proposal or claim rather than giving evidence to logically refute the person's point of view.

Bad appeal to authority Using the testimony of an expert as evidence for a point, when that expert is not identified, not competent in the area under discussion, not quoted in full, not current, or not unbiased.

Balance-indicator A word that signals that the elements on both sides—be they premises, conclusions, or arguments—are of the same kind.

Begging the question Assuming during the course of your argument the very thing you are supposed to prove.

Biased description A situation in which loaded language is used to slant evidence.

Biconditional A compound with two components. A *biconditional* is true if and only if both components have the same truth value.

Bound An occurrence of a variable is *bound* if and only if it is part of or falls within the scope of a quantifier that has that variable in it.

Carroll diagram A graphical device for representing sets and statements about sets.

Categorical proposition A statement of the form "All/some S is/are P," that is, [quantifier] [subject term] [copula][predicate term].

Categorical syllogism A two-premise argument made up of categorical propositions.

Causal chain A finite sequence of events linked by cause-and-effect relations.

Glossary

Cognitive science A science that experimentally investigates the nature of knowledge and reasoning.

Coining definition One that aims to assign a meaning to a new word.

Collective use (of a general term) Using that term to refer to the members of the group collectively, that is, as a whole.

Complement The set of those members of the universal set that are not members of the original set.

Composition Either assuming that what is true of the parts is true of the whole, or else using a general term distributively in the premises and collectively in the conclusion.

Conclusion-indicator A word that signals that the clause following it expresses the conclusion.

Conditional A statement that if some component statement A (called the *antecedent*) is true, so will a second statement B (called the *consequent*).

Conjunction A compound with two components. A *conjunction* is true if and only if both components are true.

Contingent A statement is *contingent* if it is possible for it to be false and possible for it to be true.

Contradiction A statement that cannot possibly be true.

Contradictories Two statements are *contradictories* if they cannot both be true and cannot both be false.

Contraries Two statements are *contraries* if they can both be false, but cannot both be true.

Corrective answer (to a question) A statement that indicates that one of the presuppositions to a question is false.

Counterexample A realistic case that fits the premises of a given argument, but for which the conclusion does not hold.

Deductive logic The study of arguments to determine which are valid and why.

Deductively valid An argument is *deductively valid* if and only if it is impossible for the premises to be true and the conclusion to be false.

Definiendum The word being defined in a definition.

Definiens The phrase that defines the word being defined in a definition.

Definition Giving the meaning of a word.

Deontically sufficient An argument is *deontically sufficient* if and only if it is reasonable to adopt the conclusion given that the premises are true.

Describe To give a set of statements to characterize a situation.

Detensifier An adverb of degree used to downtone (to diminish the power of) a predicate (the word or words that express what is affirmed or denied of the subject).

Dialogue An array of questions and statements, structured in rounds, in an information-oriented conversation.

Direct answer (to a question) A statement that completely answers the question, but gives no more information than is needed.

Discourse In the broadest sense of the term, any oral or written conversation between people.

Disjunction A compound with two components. A *disjunction* is true if and only if at least one of the components is true.

Distributive use (of a general term) Using that term to refer to members of the group as individuals.

Division Either assuming that what is true of the whole is true of the parts, or else using a general term collectively in the premises and distributively in the conclusion.

Efficient cause (of an event) The event that either starts it or stops it.

Emotion-oriented conversation A conversation aimed at expressing and inducing emotions.

Empty set The set containing no members.

Enthymeme An argument with unstated premises or conclusion.

Equivocation Shifting from one meaning (of an ambiguous word or phrase) to another during the course of an argument.

Exclusive sense A condition that requires a selection of one thing or another, but not both.

Exemplify To give an illustration of a claim.

Fallacy An illogical argument, that is, one that is neither valid nor strong.

False analogy A comparison that overlooks significant differences between the things compared.

False cause Arguing that A causes B merely on the basis that A and B are linked in time.

False dilemma A dilemma in which either significant options are overlooked or consequences of stated options are misstated.

Final cause (of a thing) The purpose for which that thing is done or made.

Formal cause (of a thing) Its essence or nature.

Free An occurrence of a variable is *free* if and only if it is not bound.

General term A term that refers to a group.

Genetic fallacy A case of attacking the person in which an idea is dismissed on the basis of the alleged defects of the group who originated that idea.

Hasty generalization Generalizing on the basis of atypical or too few cases.

Hedging The fallacy of changing a statement during an argument by understating it.

Heuristics The study of how to discover solutions to problems.

Ignoratio elenchi Ignoring the issue.

Ignoring the issue Arguing about something other than the point at hand.

Inclusive sense A disjunction that allows a choice of one thing or the other, or both.

Inductive logic The study of arguments to determine their degree of strength and why.

Inductively strong An argument is *inductively strong* if and only if it is not impossible, but it is unlikely, that the conclusion would be false given that the premises are true.

Inference The process of reasoning from starting points to some final conclusion.

Inference rule A simple, valid argument form.

Information-oriented conversation A conversation directed primarily at exchanging or acquiring information.

Intersection The set composed of those objects common to both sets.

Laden A word is *laden* if it has emotional or theoretical connotations.

Linked support Two or more premises that work together to support a conclusion.

Loaded language Using laden words to bias the statement of evidence in favor of a predetermined conclusion.

Loaded question A question with a false or debatable presupposition.

Logic The normative study of the evidential relations between premises and conclusions of arguments.

Logical equivalence Two statements are *logically equivalent* if it is impossible for them to differ in truth value.

Logic graphs Graphical devices for representing sets.

Material cause (of a thing) The matter of which it is made.

Mathematics The study of number, form, arrangement, and rules.

Material equivalence The notion of having the same truth value.

Metalanguage A language used to carry out the investigation of a language.

Metalogic The study that assesses logical systems.

Modality A claim about the degree of confidence the speaker has in a given statement.

Natural language A language that evolves naturally in a social context.

Negation A compound with one component. A negation is true if and only if the component is false.

Nonsequiturs Types of arguments in which absolutely no evidence whatsoever is given for the conclusion.

Object language A language under study.

Particular affirmative A statement of the form "Some S are P."

Particular negative A statement of the form "Some S are not P."

Petitio principii "Arguing in a circle"; see *Begging the question*.

Pooh-poohing Dismissing rather than arguing against your opponent's point.

Positive assertion The emphatic assertion of a claim without any warrant or proof.

Precising definition One that aims to reduce the vagueness of a term.

Premise cluster A group of mutually linked premises.

Premise-indicator A word that signals that the clause following it expresses a premise.

Presupposition (of a question) A statement that has to be true if that question is to have any true answer.

Propositional function Any expression with at least one free occurrence of at least one variable; also called "statement function."

Proximate cause The event immediately prior to the event in a causal chain.

Question A request for information.

Reflexive A two-place relation is *reflexive* if that relation holds between any object and itself.

Reformative definition A definition that is intended to establish a new meaning for an existing term.

Refutation by logical analogy To give an argument of the same logical form as the one you are trying to refute, and that clearly has true premises and a false conclusion.

Remote cause The first event in a causal chain.

Repeated assertion Repeating a claim without offering any proof of it.

Rephrase To restate a point using different words.

Reportive definition A definition that gives the meaning speakers actually assign to the word defined.

Retroductively strong An argument is *retroductively strong* if and only if the conclusion is plausible, given that the premises are true.

Rhetoric The study of persuasive communication.

Scope (of a symbol) The range of application of that symbol.

Sentence A grammatical sequence of words of a language.

Set Any collection of things.

Set equality Any statement to the effect that any two sets are identical.

Set inclusion A statement to the effect that one set is included in another set.

Set membership A statement to the effect that a given individual is a member of a given class.

Set theory That branch of mathematics that studies sets.

Shifting the burden of proof Trying to make the other person prove what you should prove.

Single support A premise taken singly that supports a conclusion.

Sophistry Illogical but persuasive argumentation.

Sound An argument is *sound* if and only if it is valid and all the premises are true.

Special pleading Biasing the presentation of evidence for a point, often involving not mentioning the negative evidence against a proposition being advocated.

Statement The meaning of any sentence used to convey information.

Subalternation The universal implication of the particular statement.

Subcontraries Two statements are *subcontraries* if they cannot both be true, but can both be false.

Suggestive definition One that suggests a given meaning be assigned to the word defined.

Summary A set of statements that highlights or repeats points made earlier in a passage.

Syllogism A two-premise argument.

Symbolic logic The study of logic using artificial languages.

Symmetric A two-place relation is symmetric if whenever that relation holds between any objects x and y, it also holds between y and x.

Tautology A statement that cannot possibly be false.

Testify To ask people to accept a claim on one's own say-so.

Transitive A two-place relation is *transitive* if whenever that relation holds between x and y, and between y and z, it holds between x and z.

Truth-functional compound A compound whose truth value is determined solely by the values of the components.

Tu quoque A form of attacking the person in which the speaker is accused of hypocrisy.

Understate To use words that diminish the content or force of a claim.

Union (of two sets) The set composed of all the elements of the two component sets.

Universal affirmative A statement of the form "All S are P."

Universal negative A statement of the form "No S are P."

Universal set The set containing everything.

Vague A word is *vague* if it has imprecise meaning.

Valid argument An argument is *valid* if it is impossible for the premises to be true while the conclusion is false.

Variables Symbols that may vary in their reference.

Answers to Selected Exercises

Chapter One

1.2

(Answers to problems 1–9 courtesy of Raymond Smullyan.)

1. A remarkably large number of people arrive at the wrong answer, that the man is looking at his own picture. They put themselves in the place of the man looking at the picture and reason as follows: "Since I have no brothers or sisters, then my father's son must be me. Therefore I am looking at a picture of myself." The first statement of this reasoning is absolutely correct; if I have neither brothers nor sisters, then my father's son is indeed myself. If the second clause of the problem had been, "This man is my father's son," then the answer to the problem would have been "myself." But the problem didn't say that; it said, "This man's father is my father's son." From which it follows that this man's father is myself (since my father's son is myself). Since this man's father is myself, then I am this man's father, hence this man must be my son. Thus the correct answer to the problem is that the man is looking at a picture of his son.

(1) This man's father is my father's son.

Substituting the word 'myself' for the more cumbersome phrase 'my father's son,' we get

(2) This man's father is myself.

Now are you convinced?

2. The statement on the gold and lead caskets say the opposite, hence one of them must be true. Since, at most, one of the three statements is true, then the statement on the silver casket is false, so the portrait is actually in the silver casket. This problem could be alternatively solved by the following method: If the portrait were in the gold casket, we would have two true statements (namely, on the gold and lead caskets), which is contrary to what is given. If the portrait were in the lead casket, we would again have two true statements (this time on the lead and silver caskets). Therefore the portrait must be in the silver casket. Both methods are correct, and this illustrates the fact that in many problems there can be several correct ways of arriving at the same conclusion.

3. Bellini made the casket. If a son of Bellini had made the casket, the statement would be false, which is impossible. If Cellini or a son of Cellini had made the casket, the statement would be true, which is impossible. Therefore Bellini made it.

4. It is impossible for either a knight or a knave to say, "I'm a knave," because a knight wouldn't make the false statement that he is a knave, and a knave wouldn't make the true statement that he is a knave. Therefore A never did say that he was a knave. So B lied when he said that A said that he was a knave. Hence B is a knave. Since C said that B was lying and B was indeed lying, then C spoke the truth, hence he is a knight. Thus B is a knave and C is a knight. (It is impossible to know what A is.)

5. Given any proposition P, if any person A on the island of knights and knaves says, "If I'm a knight then P," then the speaker must be a knight and P must be true! This is quite surprising, and we can prove this in two ways: (1) suppose that A is a knight. Then the statement "If A is a knight, then P" must be a true statement (since knights always tell the truth). So A is a knight and it is true that if A is a knight, then P. From these two facts it follows that P must be true. Thus the assumption that A is a knight leads to P as a conclusion. Therefore..., we have proved that if A is a knight, then P. But this is precisely what A asserted! Therefore A must be a knight. And since we have just proved that if A is a knight, then P, then it follows that P must be true. Let us apply this principle to our puzzle. If we take P to be the proposition that B is a knight, then we see that A must be a knight and his statement is true, hence B is a knight. Thus the answer is that A and B are both knights.

6. C is either a knight or a knave. Suppose he is a knight. Then there really are at least two knaves, hence they must be A and B. Then he must be a werewolf (since he says he isn't a knave). So if C is a knight, then the werewolf is a knave (since he must be B). On the other hand, suppose C is a knave. Then it is not true that at least two of them are knaves, so there is at most one knave. This knave must be C, since A

and *B* are both knights. Since *A* is a knight and claims that *C* is a werewolf, then *C* really is a werewolf. So in this case, the werewolf is again a knave—namely, he is *C*. Therefore, regardless of whether *C* is a knight or a knave, the werewolf is a knave (though a different person in each case). So the answer to the first question is that the werewolf is a knave. Also, we have proved that the werewolf is either *B* or *C*; hence if you wish to choose someone who is definitely not a werewolf, then pick *A*.

7. The only days the Lion can say "I lied yesterday" are Mondays and Thursdays. The only days the Unicorn can say "I lied yesterday" are Thursdays and Sundays. Therefore the only day they can both say that is on Thursday.

8. I shall first show that at least one, *A* or *C*, is guilty. If *B* is innocent, then it's obvious that *A* and/or *C* is guilty—since by statement 1, no one other than *A*, *B*, and *C* is guilty. If *B* is guilty, then she must have had an accomplice (since she can't drive), so again *A* or *C* must be guilty. So *A* or *C* (or both) is guilty. If *C* is innocent, then *A* must be guilty. On the other hand, if *C* is guilty, then by statement 2, *A* is also guilty. Therefore *A* is guilty.

9. It is not possible to tell what Bal means, but we can tell that the speaker must have been human. Suppose Bal means yes. Then Bal is the truthful answer to the question whether Bal means yes. So in this case, the speaker was human. Suppose Bal means no. Then no is the truthful English answer to the question whether Bal means yes, therefore Bal is the truthful native answer to the question. So again the speaker is human. Regardless of whether Bal means yes or no, the speaker is human.

10. The third student reasons to herself as follows: "Suppose I had on a green hat. Then since the first student saw the second student raise his hand, that would mean that the first student would know that either he or I have on a red hat. But this would mean—since I am supposing that I am wearing a green hat—that he (the first student) must be wearing a red hat. But he can't tell what color hat he is wearing, so I must not be wearing green. So I am wearing a red hat."

Chapter Two

2.2A

2. You are a nice husband.

6. An important advantage of philosophy is that it helps overcome suspicion and false religion.

10. Parents are important in the education of their children, especially in the preschool years.

14. Enjoy life while you can.

2.2B

2. Exercise of any sort, be it individual or team sports, jogging, walking, or any other activity of that sort, will, it can be said with some assurance, lower your chances of heart disease.

6. Ronald Reagan, as anyone can attest who was in a position to know and not inclined to lie about the matter, held the high office of president of the United States of America, during the year of our Lord 1982.

10. In the world of music, in the domain of the performing arts, in the dimension of the aesthetic, it must be said that spontaneous live performance of that genre of music, which has over the years become known as "jazz," is preferable in the aesthetic sense to the jazz music that is recorded.

2.4A

2. 1. The Padres have weak pitching.
   2. Their best hitter is injured.
   ∴ The Padres won't win this year.

6. 1. Jazz musicians constantly perform.
   2. Jazz musicians don't get much sleep.
   3. Jazz musicians drink and take drugs.
   4. Jazz musicians eat high-cholesterol food.
   ∴ Jazz musicians can be expected to have heart disease.

10. 1. The murderer kills one person.
    2. The traitor, if he succeeds, kills a whole nation.
    ∴ Treason is worse than murder.

14. 1. If we leave the kid home, he will burn the house down.
    2. We don't want him to burn the house down.
    ∴ We should take the kid with us.

18. 1. If Mark was in the car crash, the police would have notified us by now.
    2. The police have not notified us.
    ∴ Mark was not in the car crash.

Answers to Selected Exercises　　　　　　　　　　　　　　　　　　　　**513**

2.4B

2. (This is a case in which an argument is reported rather than given.)
   1. Engineers don't allow for curvature of the earth in building canals or railroads.
   2. The opposite shore of Lake Winnebago is clearly visible from 12 miles away.

   ∴ The earth is flat.

6. 1. Dance is a performing art.
   2. At many universities, dance is housed in the fine or performing arts area.

   ∴ Dance courses should be included as partial fulfillment of the units required for the Associate of Arts in Humanities and Creative and Performing Arts.

10. 1. To lift the incomes of the poor will also lift the incomes of the rich by greater absolute amounts.

    ∴ Inequality may grow as poverty declines.

14. 1. We all have a sense of the sacredness of human identity.

    ∴ We recoil from those who would intrude upon our privacy.

18. 1. The Supreme Soviet is in session only 6 days a year.
    2. In its 43 years of existence, the Supreme Soviet, there has never been a single vote against a motion the government submitted.
    3. None of the Supreme Soviets have ever questioned any government official.

    ∴ The Supreme Soviet is merely a rubber stamp.

2.5

2. 1 = People hate and fear snitches.
   2 = People hate Richard.
   3 = Richard is a snitch.
   4 = People fear Richard (enthymeme)

   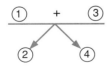

6. 1 = Rhonda's dog is a Pomeranian.
   2 = Pomeranians bark like crazy.
   3 = Rhonda's dog will bark a lot.
   4 = Old man Crandall hates noise.
   5 = Crandall will hate Rhonda's dog.

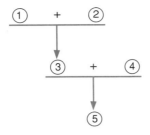

10. 1 = People are basically greedy.
    2 = Communism will fail.
    3 = Communism requires that people be unselfish.
    4 = Capitalism will win.
    5 = If communism fails, then capitalism wins.

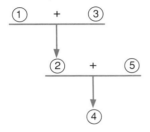

14. 1 = The use of computers in education is to be welcomed.
    2 = Computers are endlessly patient.
    3 = Computers force students to interact.
    4 = Computers remind students of video games.
    5 = We should be putting more computers into schools.
    6 = We need to start training more teachers in computer science.

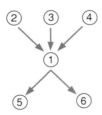

18. 1 = Computers have the power of creative thought.
    2 = Computers will eventually be able to replace highly paid workers.
    3 = We should no longer treat computers as mere inanimate objects.

Answers to Selected Exercises **515**

2.6A

2. Positive assertion.
6. 1. Health insurance is a must.
   2. The company won't volunteer anything.
   ∴ We have to put health insurance in our contract.

2.6B

2. 1. Rising living standards are intertwined with our identities.
   2. We will enter a period of slow growth.
   ∴ We will experience disappointment.
6. No argument here, just a series of assertions.
10. No argument here, just assertions.
14. 1. Any diversion causes your readers to lose interest in what you have to say.
    2. You shouldn't want people to lose interest.
    ∴ You should avoid introducing irrelevancies.
18. 1. Hitler's description of a lower-class family matches what we know of his family in respect to the drunkenness of the father.
    ∴ Hitler was probably describing his own family life.

2.7

2. 1. Buffalo Bill was a good storyteller.
   2. Indians love a good storyteller.
   ∴ The Indians he took on tour loved Buffalo Bill.

   Purpose: to explain.

2.8

2. a. What movie shall we see?
   b. Presupposes we are going to see a movie.
   c. We should see *Dogs Night Out*.
   d. I don't want to see a movie.
6. a. Why are Americans so unhappy?
   b. Presupposes that Americans are unhappy.
   c. They are unhappy because they fear unemployment.
   d. Americans are happy.
10. a. How does one light the furnace?
    b. Presupposes that the furnace can be lit.
    c. Push the pilot button for 30 seconds, then light.

d. Your gas has been cut off, so your furnace cannot be lit.
14. a. When did Mom say the train would arrive?
   b. Presupposes Mom specified a time of arrival.
   c. 10:00 p.m.
   d. Mom died 20 years ago.

2.9

2.

	Vida	Gary
Round 1	I'm going to have to work harder if I want to get to medical school.	Did you do poorly this semester?
Round 2	Yes.	Without good grades a person doesn't have a chance of getting into medical school.
Round 3	Yes.	[Pass]

4.

	Fan A	Fan B
Round 1	The Yankees will not win this year.	1. The Yankees have good hitters. 2. The Yankees have good pitchers. 3. The Yankees have the best catcher. ∴ The Yankees will win.
Round 2	[Pass]	What else is necessary for the Yankees to win?
Round 3	The Force.	What is the Force?
Round 4	The Universal Force mentioned in *Star Wars*.	You are weird.

Answers to Selected Exercises **517**

Chapter Three

3.1

2. Action (to get you to petition for graduation).

6. Aesthetics.

10. Aesthetics.

14. Action (to get you to read the book).

3.2

2. Valid.

6. Neither valid nor strong.

10. Valid. (If the dice are not loaded, it mathematically follows that the chances are high that a 7 will not be rolled. The fact that the last 12 rolls were 7s is irrelevant.)

3.4

2. 1. Men and women currently carry deep feelings of misogyny and unconscious sexism.
   2. These feelings will have an impact upon the first generation that today's parents raise.

   ∴ The changes we propose cannot be implemented in one generation.

   Modality: None given.

6. 1. If the universe is closed, we should see only a fraction of 20 to 30 parts per million of deuterium.
   2. We see 20 to 30 parts per million of deuterium.

   ∴ The universe is open.

   Modality: Not clear in this case. "Must conclude" seems to indicate certainty, but the qualifier "unless there is some other way to explain . . ." indicates probability.

3.6

2. Counterexample: Joni is not in love, but is polite.
   Logical analogy: If I were rich, I would eat every day. I am not rich. So I do not eat every day.

6. Counterexample: Kim is going to study medicine.
   Logical analogy: You are either living or dead. So you are dead.

Chapter Four

4.3

2. Problem: No evidence is given that the article in question was unfair, only assertions that it was and a threat to rescind subscription.
   Label: Appeal to fear.

6. Problem: The picture of death is used to scare the reader into buying vitamins.
   Label: Appeal to fear.

10. Problem: The advertisement only appeals to the million lasses who use the product.
    Label: Ad populum.

14. Problem: A subtle appeal to one's desire not to smell bad.
    Label: Ad populum.

4.6

2. Problem: Schirra makes a joke out of the question.
   Label: Ignoring the issue.

6. Problem: The advertisement appeals to the fact that many Americans have enjoyed the product.
   Label: Ad populum.

10. Problem: Presupposes that Hammer Mortgage Company is different.
    Label: Loaded question.

14. Problem: No, question is—how can they be priced so cheap?
    Label: Ignoring the issue.

18. Problem: No evidence is given that Dannemeyer's claim (some comparison between Horton and Mandela) is false; instead, Dannemeyer is called names.
    Label: Attacking the person.

22. Problem: The letter-writer merely insults the critics as "twisted."
    Label: Ad hominem.

26. Problem: The president pooh-poohs the charge, then proceeds to ignore the issue (which was not whether the stealth bomber program was known but whether confirming it was prudent).
    Label: Pooh-poohing, ignoring the issue.

30. Problem: Volkswagen (in a parody of the Chevrolet advertisement) appeals to patriotic sentiments.
    Label: Ad populum.

34. Problem: Slippery slope from guns to cars.
    Label: Ignoring the issue.

Answers to Selected Exercises

Chapter Five

5.4

2. What team of psychologists?
   Label: Ad verecundiam.

6. The couple overlooked the difference in social impact between men's and women's bare chests.
   Label: False analogy.

10. Doesn't take into account that you can know something about a subject without knowing everything about it.
    Label: False dilemma.

14. Even if the actress Polly Bergen is an expert on sewing, her affiliation with the company biases her testimony.
    Label: Ad verecundiam.

5.6

2. Boor generalizes on his own case.
   Label: Hasty generalization.

6. The writer argues that since abortion was legalized, child abuse rates shot up, so the one caused the other. But some other social change might have caused both.
   Label: False cause.

10. The writer cites only his own case as evidence that large cars use less fuel than small cars.
    Label: Hasty generalization.

14. What professor? Where?
    Label: Ad verecundiam.

18. Does that mean the music caused the lust? Hardly.
    Label: False cause.

22. One case cited.
    Label: Hasty generalization. No mention of unsuccessful cases.
    Label: Special pleading. Also false cause (he carried the mustard seed and life changed, so . . .).

26. The writer had a lousy time at her prom, so concludes that most people have a bad time.
    Label: Hasty generalization.

Chapter Six

6.3A

2. We should make the most of what remains of our life before we die.

6.3B

2. Loaded: 'Savage,' 'drastically' insinuates that the cuts are excessive. 'Concerted effort' insinuates that the people who cut the funding intend to destroy public programming. 'Insulting rubbish,' 'wasteland' insinuate that commercial TV produces inferior entertainment.

6. Excessively vague: 'Lonely,' 'almost as destructive,' 'active form of solitude' (rather vague if reading is active while TV is passive).

10. Loaded: 'Featherbrained,' 'wasting,' 'outrageous,' 'squander' all bias the facts in the direction of the conclusion that the study (of the mating habits of the sparrow) is worthless.

6.4

2. Detensifier: I am positive that pizza tastes rather good.
   Qualifier: I am positive that pizza tastes good to some people.
   Contradictory substitution: I am positive that pizza does not taste bad.
   Weaker modality: I am persuaded that pizza tastes good.

6. Detensifier: Clearly, Darwin helped devise modern evolutionary theory.
   Qualifier: Clearly, Darwin devised this aspect of modern evolutionary theory.
   Contradictory substitution: Clearly, Darwin was not uninvolved in the devising of modern evolutionary theory.
   Weaker modality: Probably, Darwin devised modern evolutionary theory.

10. Detensifier: It is a fact that people sort of hate the truth.
    Qualifier: It is a fact that people hate the truth about these matters.
    Contradictory substitution: It is a fact that people do not like the truth.
    Weaker modality: It is likely that people hate the truth.

6.5A

Answers can be obtained from your dictionary.

6.5B

2. "Big cars save more gas" (part of the headline) is amphibolous. In the headline, it sounds as if a big car gets better gas mileage than a small car does—which indeed would overturn the "small car myth." But in the story, the other meaning of the sentence comes out: big cars are more fuel efficient now than before.

6. Amphibolous headline misleads the reader into thinking that the story is about Communists going to war.

Answers to Selected Exercises 521

10. People desire their own happiness, not that of humankind. Composition.

14. Because the average food price will rise, doesn't mean that the price of butter and eggs will. Division.

6.6

Answers can be obtained from your dictionary.

Chapter Seven

7.1A

2. Falwell assumes that because leaving the classroom would not be traumatic for him, it would not be so for any child. Hasty generalization.

6. Who says you are? Complex question.

10. Can't one worry about both people and animals? False dilemma.

14. Personal attacks against Watt. Ad hominem.

18. Patriotic appeal. Ad populum.

22. No, it is not reasonable to say that because part of the universe can think, the whole thing can. In fact, it's stupid. Composition.

26. Griffith pooh-poohs the question (she has answers, but doesn't have the time to give them!)

30. Personal attacks rather than reasons. Ad hominem.

34. Atypical case—assassins are not easily deterred. Just because guns failed to defend Reagan that one time, doesn't mean they don't usually succeed. Accident.

38. Slippery slope from school lunches to communism. Or better—since the move is directly from lunches to communism, strawman. Ignoring the issue.

42. Assumes the debatable claim that life was simple in the past. Loaded question.

46. Small sample. Hasty generalization.

Chapter Eight

8.1

2. Reading Kepler's account of his discoveries is like watching a slow motion film of the creative act.

6. The universe is like the skin of a balloon.

10. A body near another body is like a straw caught in an eddy.

8.2A

2. Sue concentrates like a laser.

8.2B

2. Mouse
6. Tall
10. Eagle

8.3

2. The liver is like a muscle. Analogy used to argue that one ought to drink a lot. Weak argument.

8.4

2. Positive analogy: People and animals share basic biological needs and drives.
   Negative analogy: People are more intelligent, more capable of moral choice.
   Neutral analogy: People may not be able to control their impulses.
6. Positive analogy: Life and poker have winners and losers.
   Negative analogy: Poker is a game played for amusement, life is not.
   Neutral analogy: Economic life may be a zero-sum game (as is poker).
10. Positive analogy: College and prison both involve certain limitations of freedom.
    Negative analogy: Being in college is voluntary (being in prison is not).
    Neutral analogy: College and prison may both make people better.

Chapter Nine

9.5

2. $-Y \vee C$
   T **F** T T

6. $A \mathbin{\&} --X$
   T **F** F T F

10. $-(A \vee Y) \mathbin{\&} (B \vee Z)$
    F T T F **F** T T F

14. $[Z \vee (X \vee X)] \vee (Y \mathbin{\&} A)$
    F F F F **F** F F T

Answers to Selected Exercises

18. [B v (Z & X)] v −−[(A & Z) & B]
    T T F F F T F T T F F F T

22. −{[A v (B v C)] & [X v (Y v Z)]}
    T T T T T T F F F F F F

9.6

2. J v K

6. (J & B) v (−J & −B)

10. −−K & −−−T

14. (−S & −D) v (A & B)

18. −−−R & −−−G

22. L v (R & C)

9.7

2. −A → −B
   F T T F T

6. (A → Z) v (C → X)
   T F F F T F F

10. (A → A) → (X → X)
    T T T T F T F

14. (C → X) → [(X → Y) v (A → B)]
    T F F T F T F T T T

18. −X → [(Y → Z) v (Z → A)]
    T F T T F F T F T T

9.8A

2. D → R

6. S → (−R & −D)

10. C → (−S & J)

14. −C → (R → −S)

18. (S → −J) → K

9.8B

2. E → A

6. M → S

10. −(C → A)

14. −−(−−A → −M)

9.8C

2. $I \to R$
6. $-[L \to (P \vee (R \vee G))]$
10. $-(U \to S)$
14. $-D \& -S$
18. $(H \vee L) \to [((P \vee D) \vee L) \& -B]$

9.9A

2. $A \leftrightarrow B$
   T T T

6. $(X \vee A) \leftrightarrow (Y \vee B)$
   F T T   T   F T T

10. $(--A \vee -B) \leftrightarrow -(A \& B)$
   T F T T F T   F   F T T T

14. $(-X \leftrightarrow Y) \leftrightarrow (-C \vee -Z)$
   T F F T F   T   F T T T F

18. $[A \to (Y \to X)] \leftrightarrow [A \to (A \to Y)]$
   T T   F T F   F   T F   T F   F

22. $[-(A \& B) \leftrightarrow (-B \leftrightarrow C)] \leftrightarrow [(X \& A) \leftrightarrow (A \to B)]$
   F T T T   T F T F T   F   F F T   F T T T

9.9B

2. $(-R \& -H) \leftrightarrow C$
6. $P \leftrightarrow B$
10. $W \to (R \leftrightarrow B)$

9.10A

2.

A	B	$A \to B$	$-A$	$-B$
T	T	T T T	F T	F T
T	F	T F F	F T	T F
F	T	F T T	T F	F T	← Argument invalid
F	F			

6.

M	N	$M \to N$	M	$\to$	$(M$	$\&$	$N)$
T	T	T T T	T	T	T	T	T
T	F	T F F	T	F	T	F	F
F	T	F T T	F	T	F	F	T	Argument valid

Answers to Selected Exercises

10.

X	Y	Z	X → Y	Y → Z	Z → X
T	T	T	T T T	T T T	T T T
T	T	F	T T T	T F F	F T T
T	F	T	T F F	F T T	T T T
T	F	F	T F F	F T F	F T T
F	T	T	F T T	T T T	T F F ←
F	T	F			
F	F	T			
F	F	F			

Argument invalid

14.

X	M	N	−X v M	−M v N	−N → −X
T	T	T	FT T T	FT T T	FT T FT
T	T	F	FT T T	FT F F	TF F FT
T	F	T	FT F F	TF T T	FT T FT
T	F	F	FT F F	TF T F	TF F FT
F	T	T	TF T T	FT T T	FT T TF
F	T	F	TF T T	FT F F	TF T TF
F	F	T	TF T F	TF T T	FT T TF
F	F	F	TF T F	TF T F	TF T TF

Argument valid

18.

A	B	C	D	(A → B) & (C → D)	−B v −D	−A v −C
T	T	T	T	T T T T T T T	FT F FT	FT F FT
T	T	T	F	T T T F T F F	FT T TF	FT F FT
T	T	F	T	T T T T F T T	FT F FT	FT T TF
T	T	F	F	T T T T F T F	FT T TF	FT T TF
T	F	T	T	T F F F T T T	TF T FT	FT F FT
T	F	T	F	T F F F T F F	TF T TF	FT F FT
T	F	F	T	T F F F F T T	TF T FT	FT T TF
T	F	F	F	T F F F F T F	TF T TF	FT T TF
F	T	T	T	F T T T T T T	FT F FT	TF T FT
F	T	T	F	F T T F T F F	FT T TF	TF T FT
F	T	F	T	F T T T F T T	FT F FT	TF T TF
F	T	F	F	F T T T F T F	FT T TF	TF T TF
F	F	T	T	F T F F T T T	TF T FT	TF T FT
F	F	T	F	F T F F T F F	TF T TF	TF T FT
F	F	F	T	F T F F F T T	TF T FT	TF T TF
F	F	F	F	F T F F F T F	TF T TF	TF T TF

Argument valid

9.10B

2. 1. $-D \to S$
    2. $-S$
    ∴ $-D$ Invalid: $D = T$
                       $S = F$

6. 1. $-A \to --C$
    2. $C \to M$
    ∴ $-M \to A$ Valid

10. 1. $L \to (-C \:\&\: -R)$
    2. $(R \:\&\: C) \to W$
    ∴ $L \to W$ Invalid: $L = T$ $W = F$
                             $C = F$ $R = F$

9.11A

2. Contingent
6. Tautology
10. Contingent
14. Tautology

9.11B

2. Tautology
6. Not a tautology
10. Tautology
14. Not a tautology
18. Tautology

9.12

2.

X	Y	(X	&	Y)	→	(Y	&	X)
T	T	T	T	T	T	T	T	T
T	F	T	F	F	T	F	F	T
F	T	F	F	T	T	T	F	F
F	T	F	F	F	T	F	F	F

Answers to Selected Exercises 527

6.

X	Y	Z	[(X & Y) & Z] → [Z & (X & Y)]
T	T	T	T T T T T T T T T
T	T	F	T T T F F T F F T T
T	F	T	T F F F T T T F F F
T	F	F	T F F F F T F F F F
F	T	T	F F T F T T T F F T
F	T	F	F F T F T F F F F T
F	F	T	F F F T T T F F F F
F	F	F	F F F F T T F F F F

9.13A

2. $-(X \otimes X)$
    T F F F

6. $(A \to X) \& (A \leftarrow X)$
    T F F F T T F

10. $(A \downarrow A) \downarrow (X \downarrow X)$
    T F T F F T F

14. $--(A \leftrightarrow A) \mid B$
    T T T T T F T

18. $-[(A \mid B) \downarrow (C \mid X)]$
    T T F T F T T F

9.13B

2. $(D \times S) \vee P$

6. $(W \leftarrow H) \downarrow (P \leftarrow H)$

10. $(F \downarrow T) \downarrow E$

Chapter Ten

10.2A

2. (a)
○
□
―
△
No form is valid.

(b)
○ v □
−□
―
△

(c)
○
−□
―
△

(d)
○ v □
−□
―
−○

6. (a) (b) (c) (d)

(c) is valid (it is just double negation).

10.2B

2. (a)
$$\frac{A \,\&\, (B \to C)}{A}$$

(b)
$$\frac{A \,\&\, [(B \vee B) \to C]}{A}$$

(c)
$$\frac{-A \,\&\, (B \to C)}{-A}$$

(d)
$$\frac{-A \,\&\, [(B \vee B) \to C]}{-A}$$

(e)
$$\frac{-A \,\&\, [(B \vee B) \to -C]}{-A}$$

6. (a)
$$\frac{-(-A \vee -B)}{A \vee B}$$

(b)
$$\frac{-(--A\vee-A)}{-A \vee A}$$

(c)
$$\frac{-(-B \vee ---B)}{B \vee --B}$$

(d)
$$\frac{-(-(A \,\&\, B) \vee -B)}{(A \,\&\, B) \vee B}$$

(e)
$$\frac{-(-(B \leftrightarrow -A) \vee -(A \vee B))}{(B \leftrightarrow -A) \vee (A \vee B)}$$

10.3

2. Add
6. Conj
10. None

Answers to Selected Exercises

10.4

2. 1. $A \,\&\, B$
   2. $A \rightarrow C$
   3. A ⌊C
   4. C 1, Simp
             2, 3, MP

6. 1. M
   2. $(M \text{ v } X) \rightarrow (Z \,\&\, A)$
   3. $(Z \text{ v } A) \rightarrow D$ ⌊D
   4. $M \text{ v } X$ 1, Add
   5. $Z \,\&\, A$ 4, 2, MP   (Classic trick:
   6. Z 5, Simp   get A v B from $A \,\&\, B$
   7. $Z \text{ v } A$ 6, Add   by Simp then
   8. D 3, 7, MP   add B.)

10. 1. $A \,\&\, Z$
    2. $B \,\&\, T$
    3. $(A \,\&\, B) \rightarrow Q$ ⌊Q
    4. A 1, Simp
    5. B 2, Simp
    6. $A \,\&\, B$ 4, 5, Conj
    7. Q 6, 3, MP

10.5

2. Comm
6. None
10. Equiv

10.6

2. 1. $A \leftrightarrow B$
   2. $-(-A \text{ v} -D)$ ⌊B
   3. $(A \rightarrow B) \,\&\, (B \rightarrow A)$ 1, Equiv
   4. $--A \,\&\, --D$ 2, DM
   5. $--A$ 4, Simp
   6. A 5, DN
   7. $A \rightarrow B$ 3, Simp
   8. B 6, 7, MP

6. 1. $-(D \,\&\, E) \rightarrow E$
   2. $-D$
   3. $-D \text{ v} -E$ ⌊$E \text{ v } Q$
   4. $-(D \,\&\, E)$ 2, Add
   5. E 3, DM
   6. $E \text{ v } Q$ 4, 1, DM
              5, Add

530 Answers to Selected Exercises

10.
1. $-(M \text{ v} -Z)$
2. $\{-L \text{ v} (X \text{ v} Z)\} \to Q$ $\lfloor Q$
3. $-M \,\&\, --Z$ 1, DM
4. $--Z \,\&\, -M$ 3, Comm
5. $--Z$ 4, Simp
6. Z 5, DN
7. $Z \text{ v} (-L \text{ v} X)$ 6, Add
8. $(-L \text{ v} X) \text{ v} Z$ 7, Comm
9. $-L \text{ v} (X \text{ v} Z)$ 8, Assoc
10. Q 9, 2, MP

14.
1. $A \text{ v} L$
2. $\{R \text{ v} (L \text{ v} A)\} \to Z$
3. $Z \to P$
4. $--P \to Q$ $\lfloor Q$
5. $(A \text{ v} L) \text{ v} R$ 1, Add
6. $(L \text{ v} A) \text{ v} R$ 5, Comm
7. $R \text{ v} (L \text{ v} A)$ 6, Comm
8. Z 2, 7, MP
9. P 8, 3, MP
10. $--P$ 9, DN
11. Q 10, 4, MP

10.7

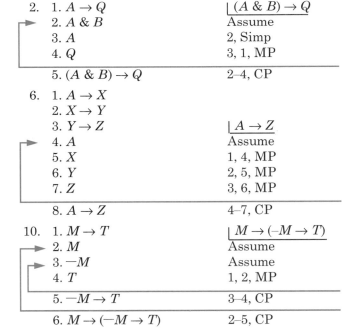

2.
1. $A \to Q$ $\lfloor (A \,\&\, B) \to Q$
2. $A \,\&\, B$ Assume
3. A 2, Simp
4. Q 3, 1, MP
5. $(A \,\&\, B) \to Q$ 2–4, CP

6.
1. $A \to X$
2. $X \to Y$
3. $Y \to Z$ $\lfloor A \to Z$
4. A Assume
5. X 1, 4, MP
6. Y 2, 5, MP
7. Z 3, 6, MP
8. $A \to Z$ 4–7, CP

10.
1. $M \to T$ $\lfloor M \to (-M \to T)$
2. M Assume
3. $-M$ Assume
4. T 1, 2, MP
5. $-M \to T$ 3–4, CP
6. $M \to (-M \to T)$ 2–5, CP

Answers to Selected Exercises

10.8

2. 1. $X \to T$
    2. $X \to -T$
    3. X $\lfloor -X$ Assume
    4. T 1, 3, MP
    5. $-T$ 2, 3, MP
    6. $T \& -T$ 4–5, Conj
    7. $-X$ 3–6, Reductio

6. 1. X
    2. $-X$
    3. $-Z$ $\lfloor Z$ Assume
    4. $X \& -X$ 1, 2, Conj
    5. Z 3–4, Reductio

10. 1. $A \to (Z \to X)$
    2. $A \to (Z \to -X)$
    3. A $\lfloor A \to Z$ Assume
    4. $Z \to X$ 1, 3, MP
    5. $Z \to -X$ 2, 3, MP
    6. Z Assume
    7. X 4, 6, MP
    8. $-X$ 5, 6, MP
    9. $X \& -X$ 7, 8, Conj
    10. $-Z$ 6–10, Reductio
    11. $A \to Z$ 3–10, CP

10.9

2. 1. $A \to B$
    2. A $\lfloor A \to (B \text{ v } C)$ Assume
    3. B 1, 2, MP
    4. $B \text{ v } C$ 3, Add
    5. $A \to (B \text{ v } C)$ 2–4, CP

6. 1. $(Q \text{ v } R) \to S$
    2. Q $\lfloor Q \to S$ Assume
    3. $Q \text{ v } R$ 2, Add
    4. S 3, 1, MP
    5. $Q \to S$ 2–4, CP

10. 1. $A \to B$
    2. $-(-A \text{ v } B)$ $\lfloor -A \text{ v } B$ Assume
    3. $--A \& -B$ 2, DM
    4. $--A$ 3, Simp
    5. A 4, DN
    6. B 5, 1, MP
    7. $-B \& --A$ 3, Comm
    8. $-B$ 7, Simp
    9. $B \& -B$ 6, 8, Conj

10. $-A \text{ v } B$ 2–9, Reductio

14. 1. $(J \text{ v } K) \to L$
     2. $L \to J$ $\lfloor L \to L$
     3. L Assume
     4. J 2, 3, MP
     5. $J \text{ v } K$ 4, Add
     6. L 5, 1, MP
     7. $L \to L$ 3–6, CP

18. 1. $-A \text{ v } B$ $\lfloor A \to B$
     2. A Assume
     3. $-B$ Assume
     4. $A \& -B$ 2, 3, Conj
     5. $--A \& -B$ 4, DN
     6. $-(-A \text{ v } B)$ 5, DM
     7. $(-A \text{ v } B) \& -(-A \text{ v } B)$ 1, 6, Conj
     8. B 3–7, Reductio
     9. $A \to B$ 2–8, CP

22. 1. $(S \text{ v } T) \to M$
     2. $(M \text{ v } T) \to \{S \to (P \to C)\}$
     3. $S \& P$ $\lfloor P \to C$
     4. S 3, Simp
     5. $S \text{ v } T$ 4, Add
     6. M 1, 5, MP
     7. $M \text{ v } T$ 6, Add
     8. $S \to (P \to C)$ 2, 7, MP
     9. $P \to C$ 8, 4, MP

26. 1. $G \to F$
     2. $F \to -P$
     3. P $\lfloor -G$
     4. $--G$ Assume
     5. G 4, DN
     6. F 1, 5, MP
     7. $-P$ 2, 6, MP
     8. $P \& -P$ 3, 7, Conj
     9. $-G$ 4–8, Reductio

10.10A

2. 1. $A \text{ v } C$
    2. $A \to B$
    3. $C \to D$ $\lfloor B \text{ v } D$

Answers to Selected Exercises

	4. $-(B \lor D)$	Assume
	5. $-B \mathbin{\&} -D$	4, DM
	6. $-B$	5, Simp
	7. $-A$	6, 2, MT
	8. $-D \mathbin{\&} -B$	5, Comm
	9. $-D$	8, Simp
	10. $-C$	9, 3, MT
	11. $-A \mathbin{\&} -C$	7, 10, Conj
	12. $-(A \lor C)$	11, DM
	13. $(A \lor C) \mathbin{\&} -(A \lor C)$	12, 1, Conj
	14. $B \lor D$	4–13, Reductio
6.	1. $A \to (B \to C)$	$(A \mathbin{\&} B) \to C$
	2. $A \mathbin{\&} B$	Assume
	3. A	2, Simp
	4. $B \to C$	1, 3, MP
	5. $B \mathbin{\&} A$	2, Comm
	6. B	5, Simp
	7. C	4, 6, MP
	8. $(A \mathbin{\&} B) \to C$	2–7, CP

10.11A

2.

A	E
T	F

6.

A	B	C
T	T	T

10.

H	J	L	X	Y	M	N
T	T	F	F	F	F	F

10.11B

2. Valid
1. $A \to (B \lor C)$
2. $C \to B$ $A \to B$

	3. A	Assume
	4. $B \lor C$	1, 3, MP
	5. $-B$	Assume
	6. C	Assume
	7. B	2, 6, MP
	8. $B \ \& -B$	5, 7, Conj
	9. $-C$	6–8, Reductio
	10. $-B \ \& -C$	5, 9, Conj
	11. $-(B \lor C)$	10, DM
	12. $(B \lor C) \ \& -(B \lor C)$	4, 11, Conj
	13. B	5–12, Reductio
	14. $A \rightarrow B$	3–13, CP

6. Valid

1. $A \rightarrow (N \ \& \ R)$ $\lfloor A \rightarrow (R \lor S)$
2. A Assume
3. $N \ \& \ R$ 1, 2, MP
4. $R \ \& \ N$ 3, Comm
5. R 4, Simp
6. $R \lor S$ 5, Add
7. $A \rightarrow (R \lor S)$ 2–6, CP

10. Invalid

Q	X	R	T	S
T	T	F	F	F

Chapter Eleven

11.3

2. {Paula, Freda, Hans, Kathy, Luisa, Lauren, Mischa}
6. {Paula, Hans, Isaac, Freda, Luisa, Lauren}
10. Σ
14. {Kathy, Mischa}
18. Σ
22. Λ

11.4A

2. $\overline{M}$
6. $\overline{S} \cup \overline{H}$
10. $(\overline{M} \cap \overline{S}) \cup (\overline{H} \cap \overline{S})$

Answers to Selected Exercises

11.4B

2. Dogs that aren't smart, together with those that are not housebroken.
6. Dogs that are not smart and also either are not housebroken or are mean.
10. Dogs that are smart, housebroken, and mean.
14. Dogs that are not smart, not housebroken, and are mean.

11.5A

2.

6.

10.

11.5B

2. $\overline{S}$

11.6A

2. $M \cup C$
6. $(M \cap C) \cup J$
10. $(\overline{M} \cap \overline{J}) \cap C$

11.6B

2. Unfriendly cats.
6. Cats that are friendly and either Siamese or tabby.

10. Friendly cats that are not either Siamese or tabby.

14. Cats that are not either friendly or tabby, together with cats that are not either friendly or Siamese.

11.6C

2.

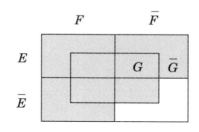

6.

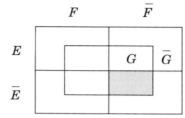

10.

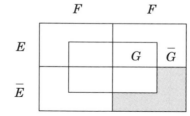

14.

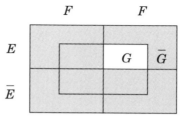

11.6D

2. $\{(\overline{R} \cap \overline{Q}) \cap S\} \cup \{(Q \cap \overline{S}) \cap \overline{R}\}$

6. $\{R \cap (Q \cap S)\} \cup \{S \cap (Q \cap R)\}$

10. $\{R \cap (S \cap Q)\} \cup \{R \cap (Q \cap S)\} \cup \{S \cap (Q \cap R)\}$

Answers to Selected Exercises

11.7

2. T

6. F

10. F

14. T

18. F

11.8

2. **A**CB
Quality = affirmative
Quantity = universal

	B	$\overline{B}$
C		O
$\overline{C}$		

6. **O**AB
Quality = negative
Quantity = particular

	B	$\overline{B}$
A		X
$\overline{A}$		

10. **A**BC
Quality = affirmative
Quantity = universal

	C	$\overline{C}$
B		O
$\overline{B}$		

11.10

2. No cats can dance.
Quality = negative
Quantity = universal

	D	$\overline{D}$
C	O	
$\overline{C}$		

538 *Answers to Selected Exercises*

6. Some burglars are clever.
   IBC
   Quality = affirmative
   Quantity = particular

	C	$\bar{C}$
B	X	
$\bar{B}$		

10. All people are loser-haters.
    APL
    Quality = affirmative
    Quantity = universal

	L	$\bar{L}$
P		O
$\bar{P}$		

14. No dogs are present.
    EDP
    Quality = negative
    Quantity = particular

	P	$\bar{P}$
D	O	
$\bar{D}$		

11.11

2. Major term = girls (*G*)
   Minor term = boys (*B*)
   Middle term = dancers (i.e., people who dance) (*D*)

   1. Some girls are dancers.
   2. Some boys are dancers.
   ∴ Some boys are girls.
   Invalid

   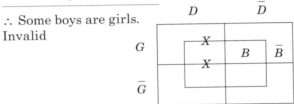

Answers to Selected Exercises **539**

6. Major term = Communists (C)
   Minor term = Democrats (D)
   Middle term = people who favor rent control (R)

   1. All Communists favor rent control.
   2. All Democrats favor rent control.

   ∴ All Democrats are Communists.
   Invalid

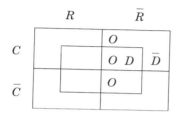

10. Major term = rats (R)
    Minor term = cats (C)
    Middle term = dogs (D)

    1. No dogs are rats.
    2. No dogs are cats.

    ∴ No cats are rats.
    Invalid

14. Major term = cheese-liking people (L)
    Minor term = people (P)
    Middle term = cheese-buying people (B)

    1. All cheese-buying people are cheese-liking people.
    2. Some people are cheese-buying people.

    ∴ Some people are cheese-liking people.
    Valid

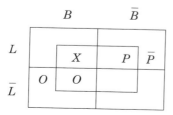

Chapter Twelve

12.2

2. $-Rb$

6. $Rb \lor Ra$

10. $Ra \rightarrow (Hb \rightarrow Ha)$

14. $(Ra \lor Rb) \rightarrow [(Fa \,\&\, Fb) \lor (-Fa \,\&\, -Fb)]$

18. $-\{(-Rb \,\&\, -Hb) \rightarrow -Fb\}$

12.4

2. $(\forall x)Hx$

6. $(\exists x)-Hx$

10. $(\forall x)(Cx \rightarrow -Px)$

14. $(\exists x)(Fx \,\&\, Ex)$

18. $(\forall x)(Tx \rightarrow Px)$

22. $(\forall x)(Gx \,\&\, Tx)$

26. $(\forall x)[(Fx \,\&\, Sx) \rightarrow Ex]$

30. $(\forall x)[Cx \,\&\, (Nx \lor Lx)]$

12.5

2. Ma
   $Ma \lor Mb$
   $Ma \lor Mb \lor Mc$

6. $-Ra \rightarrow -Aa$
   $(-Ra \rightarrow -Aa) \,\&\, (-Rb \rightarrow -Ab)$
   $(-Ra \rightarrow -Aa) \,\&\, (-Rb \rightarrow -Ab) \,\&\, (-Rc \rightarrow -Ac)$

10. $(--Ra \lor Za) \rightarrow Qa$
    $[(--Ra \lor Za) \rightarrow Qa] \,\&\, [(--Rb \lor Zb) \rightarrow Qb]$
    $[(--Ra \lor Za) \rightarrow Qa] \,\&\, [(--Rb \lor Zb) \rightarrow Qb]$
    $\,\&\, [(--Rc \lor Zc) \rightarrow Qc]$

14. $(-Aa \,\&\, Ba) \,\&\, (Ra \,\&\, Ca)$
    $[(-Aa \,\&\, Ba) \,\&\, (Ra \,\&\, Ca)] \,\&\, [(-Ab \,\&\, Bb) \,\&\, (Rb \,\&\, Cb)]$
    $[(-Aa \,\&\, Ba) \,\&\, (Ra \,\&\, Ca)] \,\&\, [(-Ab \,\&\, Bb) \,\&\, (Rb \,\&\, Cb)] \,\&\, [(-Ac \,\&\, Bc)$
    $\,\&\, (Rc \,\&\, Cc)]$

Answers to Selected Exercises

12.6

2. $(\forall x)Mx$
6. $(\exists x)(Ax \,\&\, Mx)$
10. $(\forall x)-(Hx \lor Zx)$

12.7

2. 1, 2 bound; 3 free
6. 1, 2, 3, 4 bound
10. 1, 2, 3, 4, 5 bound; 6, 7 free

12.8A

2. Ha
6. $-Za \to Za$
10. $--Aa \leftrightarrow (Ra \lor -Za)$
14. $(\exists y)Ry \lor -Ha$

12.8B

2. No. $(\forall x)Rx$ has no dash in it.
6. Yes.
10. No. Instances of $(\forall x)-(Tx \to Tx)$ have the form $-(Ta \to Ta)$, not $-Ta \to -Ta$.
14. No. a and b are different constants.

12.9

2.
1. $(\forall x)[Mx \to (Tx \,\&\, Qx)]$
2. Ma
3. $Ma \to (Ta \,\&\, Qa)$
4. $Ta \,\&\, Qa$
5. $Qa \,\&\, Ta$
6. Qa
7. $Qa \lor La$
8. $(\forall x)(Qx \lor Lx)$

$\lfloor (\exists x)(Qx \lor Lx)$
1, UI
2, 3, MP
4, Comm
5, Simp
6, Add
7, EG

6.
1. $(\forall x)(Bx \leftrightarrow Tx)$
2. $(\forall x)(Tx \leftrightarrow Zx)$
3. Ba
4. $Ba \leftrightarrow Ta$
5. $Ta \leftrightarrow Za$
6. $(Ba \to Ta) \,\&\, (Ta \to Ba)$
7. $Ba \to Ta)$
8. $(Ta \to Za) \,\&\, (Za \to Ta)$
9. $Ta \to Za$

$\lfloor Za$
1, UI
2, UI
4, Equiv
6, Simp
5, Equiv
8, Simp

          10. Ta 3, 7, MP
          11. Za 9, 10, MP

  10. 1. $(\forall x)(Mx \to Nx)$
      2. $(\forall x)(Mx \to -Nx)$
      3. Mt ⌊Za
      4. $Mt \to Nt$ 1, UI
      5. $Mt \to -Nt$ 2, UI
      6. Nt 3, 4, MP
      7. $-Nt$ 3, 5, MP
      8. Nt v Za 6, Add
      9. Za 7, 8, DS

12.10

  2. 1. $(\forall x)(Ax \to Rx)$
     2. $(\forall x)(Rx \to Ax)$ ⌊$(\forall x)(Rx \leftrightarrow Ax)$
     3. $Aa \to Ra$ 1, UI
     4. $Ra \to Aa$ 2, UI
     5. $(Ra \to Aa)$ & $(Aa \to Ra)$ 4, 3, Conj
     6. $Ra \leftrightarrow Aa$ 5, Equiv
     7. $(\forall x)(Rx \leftrightarrow Ax)$ 6, UG

  6. 1. $(\forall x)(Ax$ v $Bx)$
     2. Aa v Ba ⌊$(\forall x)(-Bx \to Ax)$
     3. Ba v Aa 1, UI
     4. $--Ba$ v Aa 2, Comm
     5. $-Ba \to Aa$ 3, DN
     6. $(\forall x)(-Bx \to Ax)$ 4, Imp
                                     5, UG

  10. 1. $(\forall x)(Ax \leftrightarrow Mx)$
      2. $(\forall x)(Mx \leftrightarrow Zx)$ ⌊$(\forall x)(Ax \to Zx)$
      3. $Aa \leftrightarrow Ma$ 1, UI
      4. $Ma \leftrightarrow Za$ 2, UI
      5. $(Aa \to Ma)$ & $(Ma \to Aa)$ 3, Equiv
      6. $Aa \to Ma$ 5, Simp
      7. $(Ma \to Za)$ & $(Za \to Ma)$ 4, Equiv
      8. $Ma \to Za$ 7, Simp
      9. $Aa \to Za$ 6, 8, HS
     10. $(\forall x)(Ax \to Zx)$ 9, UG

12.11

  2. 1. $(\exists x)(Hx$ & $Rx)$ ⌊$(\exists x)(Rx$ v $Lx)$
    → 2. Ha & Ra Assume for EI
      3. Ra & Ha 2, Comm
      4. Ra 3, Simp
      5. Ra v La 4, Add
      6. $(\exists x)(Rx$ v $Lx)$ 5, EG
      7. $(\exists x)(Rx$ v $Lx)$ 2–6, EI

Answers to Selected Exercises **543**

6. 1. $(\exists x)(Bx \mathbin{\&} Cx)$
    2. $(\forall x) -(Rx \mathbin{\&} Cx)$
    > 3. $Ba \mathbin{\&} Ca$ $\lfloor (\exists x) -Rx$ Assume for EI
    > 4. $-(Ra \mathbin{\&} Ca)$ 2, UI
    > 5. $-Ra \lor Ca$ 4, DM
    > 6. $Ca \mathbin{\&} Ba$ 3, Comm
    > 7. Ca 6, Simp
    > 8. $--Ca$ 7, DN
    > 9. $-Ca \lor -Ra$ 5, Comm
    > 10. $-Ra$ 8, 9, DS
    > 11. $(\exists x) -Rx$ 10, EG
    12. $(\exists x) -Rx$ 3–11, EI

10. 1. $(\exists x)(Ax \mathbin{\&} -Ax)$
    > 2. $Aa \mathbin{\&} -Aa$ $\lfloor (\exists y)My$ Assume for EI
    > 3. Aa 2, Simp
    > 4. $Aa \lor (\exists y)My$ 3, Add
    > 5. $-Aa \mathbin{\&} Aa$ 2, Comm
    > 6. $-Aa$ 5, Simp
    > 7. $(\exists y)My$ 4, 6, DS
    8. $(\exists y)My$ 2–7 EI

12.12A

4. 1. $(\forall x)(Ex \to -Fx)$
    2. $(\forall x)(Gx \to Fx)$
    > 3. Ga $\lfloor (\forall x)(Gx \to -Ex)$ Assume
    > 4. $Ea \to -Fa$ 1, UI
    > 5. $Ga \to Fa$ 2, UI
    > 6. Fa 3, 5, MP
    > 7. $--Fa$ 6, DN
    > 8. $-Ea$ 7, 4, MT
    9. $Ga \to -Ea$ 3–8, CP
    10. $(\forall x)(Gx \to -Ex)$ 9, UG

10. 1. $(\exists x)Mx$
    2. $(\forall x)\{Mx \lor Lx\}$
    > 3. Ma $\lfloor (\exists x)Nx$ Assume for EI
    > 4. $(Ma \lor La) \to Na$ 2, UI
    > 5. $Ma \lor La$ 3, Add
    > 6. Na 4, 5, MP
    > 7. $(\exists x)Nx$ 6, EG
    8. $(\exists x)Nx$ 3–7, EI

14. 1. $(\forall w)(-Tw \leftrightarrow -Cw)$
    2. $(\forall w)(-Rw \leftrightarrow Cw)$ $\lfloor (\forall w)(Rx \leftrightarrow -Tx)$
    3. $-Ta \leftrightarrow -Ca$ 1, UI

4. $-Ra \leftrightarrow Ca$ 2, UI
5. $(-Ta \rightarrow -Ca)$ &
   $(Ca \rightarrow -Ta)$ 3, Equiv
6. $-Ta \rightarrow -Ca$ 5, Simp
7. $(-Ca \rightarrow -Ta)$ &
   $(-Ta \rightarrow -Ca)$ 5, Comm
8. $-Ca \rightarrow -Ta$ 7, Simp
9. $(-Ra \rightarrow Ca)$ &
   $(Ca \rightarrow -Ra)$ 4, Equiv
10. $-Ra \rightarrow Ca$ 9, Simp
11. $(Ca \rightarrow -Ra)$ &
    $(Ra \rightarrow Ca)$ 9, Comm
12. $Ca \rightarrow -Ra$ 11, Simp
13. $--Ra \rightarrow -Ca$ 12, Contra
14. $Ra \rightarrow -Ca$ 13, DN
15. $Ra \rightarrow -Ta$ 14, 8, HS
16. $-Ca \rightarrow --Ra$ 10, Contra
17. $-Ca \rightarrow Ra$ 16, DN
18. $-Ta \rightarrow Ra$ 17, 6, HS
19. $(Ra \rightarrow -Ta)$ &
    $(-Ta \rightarrow Ra)$ 15, 17, Conj
20. $Ra \leftrightarrow -Ta$ 19, Equiv
21. $(\forall w)(Rx \leftrightarrow -Tx)$ 20, UG

18. 1. $(\forall x)(Bx \rightarrow -Cx)$
    2. $(\exists x)(Bx \& Dy)$ ⌊$(\exists x)(Dx \& -Bx)$
    3. $Ca \& Da$ Assume for EI
    4. $Ba \rightarrow -Ca$ 1, UI
    5. Ca 3, Simp
    6. $--Ca$ 5, DN
    7. $-Ba$ 4, 6, MT
    8. $Da \& Ca$ 3, Comm
    9. Da 8, Simp
    10. $Da \& -Ba$ 7, 9, Conj
    11. $(\exists x)(Dx \& -Bx)$ 10, EG
    12. $(\exists x)(Dx \& -Bx)$ 3–11, EI

22. 1. $-(\exists y)-Ay$
    2. $-(\exists z)-Bz$ ⌊$(\forall x)(Ax \& Bx)$
    3. $(\forall y)--Ay$ 1, QE
    4. $(\forall z)--Bz$ 2, QE
    5. $--Aa$ 3, UI
    6. Aa 5, DN
    7. $--Ba$ 4, UI
    8. Ba 7, DN
    9. $Aa \& Ba$ 6, 8, Conj
    10. $(\forall x)(Ax \& Bx)$ 9, UG

12.12B

2. 1. $(\forall x)(Cx \to -Sx)$
    2. $(\exists x)(Cx \mathbin{\&} Lx)$ ⌞ $(\exists x)(Lx \mathbin{\&} -Sx)$
    3. $Ca \mathbin{\&} La$ — Assume for EI
    4. $Ca \to -Sa$ — 1, UI
    5. Ca — 3, Simp
    6. $-Sa$ — 4, 5, MP
    7. $La \mathbin{\&} Ca$ — 3, Comm
    8. La — 7, Simp
    9. $La \mathbin{\&} -Sa$ — 6, 8, Conj
    10. $(\exists x)(Lx \mathbin{\&} -Sx)$ — 9, EG
    11. $(\exists x)(Lx \mathbin{\&} -Sx)$ — 7–10, EI

6. 1. $(\forall x)(Cx \to Wx)$
    2. $(\forall x)(Wx \to Tx)$ ⌞ $(\forall x)(-Tx \to -Cx)$
    3. $Ca \to Wa$ — 1, UI
    4. $Wa \to Ta$ — 2, UI
    5. $Ca \to Ta$ — 3, 4, HS
    6. $-Ta \to -Ca$ — 5, Contra
    7. $(\forall x)(-Tx \to -Cx)$ — 6, UG

10. 1. $(\forall x)(Sx \leftrightarrow Tx)$
    2. $(\exists x)-Tx$ ⌞ $(\exists x)-Sx$
    3. $-Ta$ — Assume for EI
    4. $Sa \leftrightarrow Ta$ — 1, UI
    5. $(Sa \to Ta) \mathbin{\&} (Ta \to Sa)$ — 4, Equiv
    6. $Sa \to Ta$ — 5, Simp
    7. $-Sa$ — 3, 6, MT
    8. $(\exists x)-Sx$ — 7, EG
    9. $(\exists x)-Sx$ — 3–8, EI

12.13

2. 1. $(-Aa \mathbin{\&} Ba) \vee (-Ab \mathbin{\&} Bb)$
    2. $Ca \vee Cb$ ⌞ $(Ca \mathbin{\&} Ba) \vee (Cb \mathbin{\&} Bb)$

Aa	Ba	Ca	Ab	Bb	Cb
T	F	T	F	T	F

6. 1. $[Ta \to (La \vee Qa)] \mathbin{\&} [Tb \to (Lb \vee Qb)]$
    2. $[(La \mathbin{\&} Qa) \to Za] \mathbin{\&} [(Lb \mathbin{\&} Qb) \to Zb]$ ⌞ $(Ta \to La) \mathbin{\&} (Tb \to Lb)$

Ta	La	Qa	Za	Tb	Lb	Qb	Zb
F	F	T	T	F	F	T	T

10. 1. $(-Ta \leftrightarrow Ra)$ & $(-Tb \leftrightarrow Rb)$
2. $--Ra \leftrightarrow La)$ &
$--(Rb \leftrightarrow Lb)$ $\quad\quad\quad |-(-Ta \leftrightarrow -(-Tb \leftrightarrow Lb))$

Ta	Ra	Tb	Rb	La	Lb
F	T	F	T	T	T

Chapter Thirteen

13.1

2. $Hfs \to Cf$
6. $Hfr \leftrightarrow Hrs$
10. $Hrf \to (Hfr \& Hsh)$

13.2A

2. UD = animals
$(\forall x)(Cx \to Hlx)$
6. UD = animals
$(\exists x)(Dx \& (Hxl))$
10. UD = animals
$(\exists x)[Dx \& (\forall y)(Py \to Ly\ x)] \to \{(\forall x)[Cx \to Dx] \to (\exists x)[Cx \& (\forall y)(Py \to Lyx)]\}$
14. UD = animals
$(\forall x)(\forall y)[(Dx \& Ry \to Hxy)] \to (\forall x)(\forall y)[(Dx \& Ry) \to Hyx]$
18. $-(\forall x)(Px \to Txx)$
22. $(\exists x)[Dx \& (\forall y)(Py \to Lyx)]$

13.2B

2. UD = views of medicine
$(\forall x)Lx$
6. UD = doctors
$(\exists x)\ Wx$
10. UD = people
$(\forall x)(\forall x)[(Hy \& -Hy) \to Lxy]$
14. UD = tests
$-(\forall x)(\forall y)[(Hy \& -Hy) \to Lxy]$

13.3

2. Two-model:
$(Maa \lor Mab) \lor (Mba \lor Mbb)$

Three-model:
(*Maa* v *Mab* v *Mac*) v (*Mba* v *Mbb* v *Mbc*) v (*Mca* v *Mcb* v *Mcc*)

6. Two-model:
(*Naa* → −*Naa*) & (*Naa* → −*Nbb*) & (*Nbb* → −*Naa*) & (*Nbb* → *Nbb*)
Three-model:
(*Naa* → −*Naa*) & (*Naa* → −*Nbb*) & (*Naa* → −*Ncc*) & (*Nbb* → *Naa*) & (*Nbb* → −*Nbb*) & (*Nbb* → −*Ncc*) & (*Ncc* → −*Naa*) & (*Ncc* → −*Nbb*) & (*Ncc* → −*Ncc*)

10. Three-model only:
{[(*Maaa* → *Maaa*) v (*Maab* → *Maba*) v (*Maac* → *Maca*)] & [(*Maba* → *Maab*) v (*Mabb* → *Mabb*) v (*Mabc* → *Macb*)] &[(*Maca* → *Maac*) v (*Macb* → *Mabc*) v (*Macc* → *Macc*)]} v {[(*Mbaa* → *Mbaa*) v (*Mbab* → *Mbba*) v (*Mbac* → *Mbca*)] & [(*Mbba* → *Mbab*) v (*Mbbb* → *Mbbb*) v (*Mbbc* → *Mbcb*)]} & [(*Mbca* → *Mbac*) v (*Mbcb* → *Mbbc*) v (*Mbcc* → *Mbcc*)]} v {[(*Mcaa* → *Mcaa*) v (*Mcab* → *Mcba*) v (*Mcac* → *Mcca*)] & [(*Mcba* → *Mcab*) v (*Mcbb* → *Mcbb*) v (*Mcbc* → *Mccb*)] & [(*Mcca* → *Mcac*) v (*Mccb* → *Mcbc*) v (*Mccc* → *Mccc*)]}

13.4A

2. 1. (∀x) (∀y) *Lxy*
    2. (∀y) *Lay*
    3. *Lab*
    4. *Lab* v *Laa*

    | *Lab* v *Laa*
    1, UI
    2, UI
    3, Add

6. 1. (∀x) (∀z) (*Rzx* ↔ *Lxz*)
    2. (∀x) (∀z) (*Mxz* ↔ *Lxz*)
    3. (∀z) (*Rza* ↔ *Maz*)
    4. *Rba* ↔ *Mab*
    5. (∀z) (*Maz* ↔ *Laz*)
    6. *Mab* ↔ *Lab*
    7. (*Rba* → *Mab*) & (*Mab* → *Rba*)
    8. *Rba* → *Mab*
    9. (*Mab* → *Lab*) & (*Lab* ↔ *Mab*)
    10. *Mab* → *Lab*
    11. *Rba* → *Lab*
    12. (*Mab* → *Rba*) & (*Rba* → *Mab*)
    13. *Mab* → *Rba*
    14. (*Lab* → *Mab*) & (*Mab* → *Lab*)
    15. *Lab* → *Mab*
    16. *Lab* → *Rba*
    17. (*Rba* → *Lab*) & (*Lab* → *Rba*)
    18. *Rba* ↔ *Lab*

    | (∀x) (∀z) (*Rzx* ↔ *Lxz*)
    1, UI
    3, UI
    2, UI
    5, UI
    4, Equiv
    7, Simp
    6, Equiv
    9, Simp
    8, 10, HS
    7, Comm
    12, Simp
    9, Comm
    14, Simp
    13, 15, HS
    11, 16, Conj
    17, Equiv

19. $(\forall z)(Rza \to Laz)$	18, UG
20. $(\forall x)(\forall z)(Rzx \leftrightarrow Lxz)$	19, UG

10.
1. $(\forall x)(\forall y)(\forall z) -Mxyz$	$\lfloor (\forall x)(\forall y)(\forall z)(Mxyz \to Rxy)$
2. $(\forall y)(\forall z) -Mayz$	1, UI
3. $(\forall z) -Mabz$	2, UI
4. $-Mabc$	3, UI
5. $-Mabc$ v Rab	4, Add
6. $Mabc \to Rab$	5, Imp
7. $(\forall z)(Mabz \to Rab)$	6, UG
8. $(\forall y)(\forall z)(Mayz \to Ray)$	7, UG
9. $(\forall x)(\forall y)(\forall z)$	
$(Mxyz \to Rxy)$	8, UG

13.4B

2.
1. $(\forall x)[Ax \to (\forall y)(My \to Ny)]$	$\lfloor (\forall x) Ax \to (\forall y)(My \to Ny)$
▶ 2. $(\forall x) Ax$	Assume
3. Aa	2, UI
4. $Aa \to (\forall y)(My \to Ny)$	1, UI
5. $(\forall y)(My \to Ny)$	3, 4, MP
6. $(\forall x) Ax \to (\forall y)$	

6.
1. $(\exists y) By \to -(\exists x) Hx$	$\lfloor (\forall y)[(\exists x) Bx \to -Hy]$
▶ 2. $(\exists x) Bx$	Assume to show $-Ha$
▶ 3. Bb	Assume
4. $(\exists y) By$	3, EG
5. $(\exists y) By$	3–4, EI
6. $-(\exists x) Hx$	1, 5, MP
7. $(\forall x) -Hx$	6, QE
8. $-Ha$	7, UI
9. $(\exists x) Bx \to -Ha$	2–8, CP
10. $(\forall y)[(\exists x) Bx \to -Hy]$	9, UG

10.
1. $(\forall x)(\exists y)(Ax$ v $By)$	$\lfloor (\forall x)Ax$ v $(\exists y) By$
▶ 2. $-(\forall x) Ax$	Assume
3. $(\exists x) -(\forall x)$	2, QN
▶ 4. $-Aa$	Assume
5. $(\exists y)(Aa$ v $By)$	1, UI
▶ 6. Aa v Bb	Assume
7. Bb	4, 6, DS
8. $(\exists y) By$	7, EG
9. $(\exists y) By$	6–8, EI
10. $(\exists y) By$	4–9, EI
11. $-(\forall x) Ax \to (\exists y) By$	2–10, CP

Answers to Selected Exercises

12. $--(\forall x)\,Ax$ v $(\exists y)\,By$ 11, Imp
13. $(\forall x)\,Ax$ v $(\exists y)\,By$ 12, DN

14.
1. $(\exists x)[Gx\ \&\ (\forall y)\,Mxy]$ $\lfloor (\exists x)(Gx\ \&\ Mxb)$
2. $Ga\ \&\ (\forall y)\,May$ Assume
3. $(\forall y)May\ \&\ Ga$ 2, Comm
4. $(\forall y)May$ 3, Simp
5. Mab 4, UI
6. Ga 2, Simp
7. $Ga\ \&\ Mab$ 5, 6, Conj
8. $(\exists x)(Gx\ \&\ Mxb)$ 7, EG
9. $(\exists x)\,(Gx\ \&\ Mxb)$ 2–8, EI

18.
1. $(\forall x)[Fx \rightarrow (\forall y)\,(Gy \rightarrow Hxy)]$
2. $(\exists x)[Fx\ \&\ (\exists y)\,-Hxy]$ $\lfloor (\exists x)\,-Gx$
3. $Fa\ \&\ (\exists y)\,-Hay$ Assume
4. $Fa \rightarrow (\forall y)\,(Gy \rightarrow Hay)$ 1, UI
5. Fa 3, Simp
6. $(\forall y)\,(Gy \rightarrow Hay)$ 4, 5, MP
7. $(\exists y)\,-Hay\ \&\ Fa$ 3, Comm
8. $(\exists y)\,-Hay$ 7, Simp
9. $-Hab$ Assume
10. $Gb \rightarrow Hab$ 6, UI
11. $-Gb$ 9, 10, MT
12. $(\exists x)\,-Gx$ 11, EG
13. $(\exists x)\,-Gx$ 9–12, EI
14. $(\exists x)\,-Gx$ 3–13, EI

13.5

2. Prove T4: $[P \rightarrow (\forall x)Fx] \Leftrightarrow (\forall x)(P \rightarrow Fx)$.
First: show $[P \rightarrow (\forall x)Fx] \Rightarrow (\forall x)(P \rightarrow Fx)$.
1. $P \rightarrow (\forall x)Fx$ $\lfloor (\forall x)(P \rightarrow Fx)$
2. P To show Fa
3. $(\forall x)\,Fx$ 1, 2, MP
4. Fa 3, UI
5. $P \rightarrow Fa$ 2–4, CP
6. $(\forall x)\,(P \rightarrow Fx)$ 5, UG

Second: show $[(\forall x)\,(P \rightarrow Fx)] \Rightarrow [P \rightarrow (\forall x)Fx]$
1. $(\forall x)(P \rightarrow Fx)$ $\lfloor P \rightarrow (\forall x)Fx$
2. P To show $(\forall x)\,Fx$
3. $P \rightarrow Fa$ 1, UI
4. Fa 2, 3, MP
5. $(\forall x)Fx$ 4, UG
6. $P \rightarrow (\forall x)\,Fx$ 2–5, CP

6. Prove T8: $[(\exists x) Fx \lor (\exists x) Gx] \Leftrightarrow (\exists x)(Fx \lor Gx)$.

1. $(\exists x)Fx \lor (\exists x)Gx$	$\lfloor (\exists x)(Fx \lor Gx)$
2. $-(\exists x)(Fx \lor Gx)$	Assume for contradiction
3. $(\forall x)-(Fx \lor Gx)$	2, QE
4. $(\forall x)(-Fx \& -Gx)$	3, DM
5. $-Fa \& -Ga$	4, UI
6. $-Fa$	5, Simp
7. $(\forall x)-Fx$	6, UG
8. $-Ga \& -Fa$	5, Comm
9. $-Ga$	8, Simp
10. $(\forall x)-Gx$	9, UG
11. $-(\exists x) Fx$	7, QE
12. $-(\exists x) Gx$	10, QE
13. $-(\exists x)Fx \& -(\exists x)Gx$	11, 12, Conj
14. $-[(\exists x)Fx \lor (\exists x)Gx]$	13, DM
15. $[(\exists x) Fx \lor (\exists x)Gx] \&$ $-[(\exists x)Fx \lor (\exists x)Gx]$	1, 14, Conj
16. $(\exists x)(Fx \lor Gx)$	2–16, Reductio

1. $(\exists x) (Fx \lor Gx)$	$\lfloor (\exists x)Fx \lor (\exists x)Gx$
2. $-(\exists x)Fx$	To show $(\exists x) Gx$
3. $Fa \lor Ga$	Assume for EI
4. $(\forall x)-Fx$	2, QE
5. $-Fa$	4, UI
6. Ga	3, 5, DS
7. $(\exists x)Gx$	6, EG
8. $(\exists x)Gx$	3–7, EI
9. $-(\exists x)Fx \to (\exists x) Gx$	2–7, CP
10. $--(\exists x)Fx \lor (\exists x)Gx$	9, IMP
11. $(\exists x)Fx \lor (\exists x)Gx$	10, DN

13.7

2. Not reflexive, not symmetric, but it is transitive.

13.8

2. $h = b$

6. $(\forall x)(\forall y)[(Qxh \& Qyh) \to (x = y)]$

10. $(\exists x)Sxa \& (\forall x)(\forall y)[(Sxa \& Sya) \to (x = y)]$

14. $Sac \to (\forall x) (Sxc \to (x = a))$

18. $a = (\daleth x) (Sxc)$

13.9

2.
1. $(\forall x)Mxx$	
2. $a = b$	$\lfloor Mab$
3. Maa	1, UI
4. Mab	3, 2, LL

Answers to Selected Exercises

6. 1. $a = b$
    2. $a = a$
    3. $b = a$

    $\lfloor b = a$
      II
    1, 2, LL

10. 1. $(\forall x)(Hxc \to Mxc)$
    2. $(\exists x) Hxc$
    3. $c = a$
    4. Hbc
    5. $Hbc \to Mbc$
    6. Mbc
    7. Mba
    8. $(\exists x) Mxa$
    9. $(\exists x) Mxa$

    $\lfloor (\exists x) Mxa$
    Assume
    1, UI
    4, 5, MP
    3, 6, LL
    7, EG
    4–8, EI

14. 1. $Pb \to Sa$
    2. $c = a$
    3. $-Sc$
    4. $b = r$
    5. $-Pb \to Sc$
    6. $-Pb$
    7. $-Pr$

    $\lfloor -Pr$
    1, 2, LL
    3, 5, MT
    6, 4, LL

Appendix

A.1

2. a = Efficient cause
   b = Final cause
   c = Formal cause

6. a = Efficient cause, proximate
   b = Efficient cause, remote

A.2

2. Method of agreement: in the cultures cited, violence is committed by men. Since the cultures differ in other respects, it is the male nature that presumably is the common factor.

6. Method of concomitant variation: level of testosterone varies with level of violence, so testosterone causes aggression/violence.

Index

abusive form, 92
accent, 169
accident, 142
algorithmic, 24
ambiguous, 167
amphiboly, 168
analogy, 135, 237
announcement, 45
the antecedent, 273
appeal to fear, 97
appeal to ignorance, 89
appeal to sex, 104
appeal to the crowd, 102
appealing to pity, 99
apples and oranges, 115
arguing against the person, 92
argument, 22, 69
argument form, 309
argumentum ad baculum, 97
argumentum ad hominem, 92
argumentum ad ignorantiam, 89
argumentum ad misericordiam, 99
argumentum ad populum, 102
argumentum ad verecundiam, 129
artificial language, 252
associated conditional, 298
association, 326

bad appeal to authority, 129
balance-indicators, 29
bandwagon argument, 102
begging the question, 117
biased description, 155
biconditional, 283
bound, 421

caricature, 95
Carroll diagram, 361
categorical propositions, 410
categorical statements, 378
categorical syllogism, 387
causal chain, 482
causal claim, 143
characterizing expressions, 402–403
circular argument, 117
cognitive science, 2
coining definition, 181
collective use, 172
commutation, 325
complement, 354
  of set A, 354
composition, 172
compounds, 19
computer programming languages, 252
conclusion-indicators, 27
conditional, 44
conditional assertion, 19
Conditional Proof, 332
conjunction, 256, 315
conjunctions, 19

the consequent, 273
contingent, 293
contradictories, 134, 383
contraposition, 341
contraries, 134, 383
converging arguments, 39
conversations
   action-oriented, 65
   aesthetics-oriented, 66
   emotion-oriented, 66
   information-oriented, 65
corrective answer, 59
correlation fallacy, 144
covert negations, 262
CP, 332

decision procedure, 252
deductive logic, 71
deductively valid, 71
definiendum, 179
definiens, 179
Defining a set by enumeration, 352
defining property, 353
definite descriptions, 402
definition, 153
DeMorgan's Laws, 326
denials, 19
deontically sufficient, 76
describe, 45
detensifier, 163
dialogue, 56, 82
direct answer, 58
direct assertion, 18
discourse, 82
disjunction, 257
disjunctions, 19
Disjunctive Syllogism, 340
distributive use, 172
diverging arguments, 38
diversion, 112
division, 174
DN, 324
domain of discourse, 409
double negation, 324
DS, 340
dyadic, 471

efficient cause, 481
elimination rule, 490
empty set, 353
enthymeme, 26
equivalence, 326
equivocation, 167
exclusive sense, 255
exclusive-or, 300
exemplify, 45
existential general statement, 408
Existential Instantiation, 428
explanation, 252

fallacies
   of language, 88
   of little evidence, 88
   of no evidence, 88
fallacy, 87
the fallacy of affirming the consequent, 318
fallacy of false dilemma, 134
false analogy, 135
false cause, 144
final cause, 482
formal cause, 482

general term, 172
genetic fallacy, 95
glittering generalities, 111
global rules, 83
guilt by association, 95

hasty generalization, 140
hedging, 165
heuristics, 7
Horn symbol, 301
HS, 340
Hypothetical Syllogism, 340

identity, 472
ignoratio elenchi, 111
ignoring the issue, 111
implication, 341
inclusive sense, 264
indexicals, 261
individual constants, 404
individual variables, 404
inductive logic, 71
inductively strong, 71
inference rule, 313
inference rules, 308
instance, 424
intersection, 354
   of two sets, 354

The joint method of agreement and difference, 485

laden, 155
Leibniz's Law, 478
linked support, 36
loaded language, 157
loaded question, 116

local rules, 83
logic, 15, 16
logic graphs, 351
logical equivalence, 295
logical equivalences, 324
logistical systems, 252

major premise, 387
major term, 387
material cause, 481
material equivalence, 283, 295
mathematics, 13
metalanguage, 253
metalogic, 13
The method of agreement, 485
The method of concomitant variation, 485
The method of difference, 485
The method of natural deduction, 308
The method of residues, 485
middle term, 387
minor premise, 387
minor term, 387
mob appeal, 104
modalities, 164
modality, 78
modus ponens, 317
Modus Tollens, 339
monadic, 471
MT, 339

natural languages, 252
negation, 255
non-truth-functional compound, 254
nonalgorithmic, 24
nonsequiters, 127
numerical constants, 404
numerical variables, 404

object language, 253

parallel arguments, 38
particular affirmative, 378
particular negative, 379
petitio principii, 117
PL, 253
poisoning the well, 94
polyadic, 471
pooh-poohing, 88
positive assertion, 44
post hoc ergo propter hoc, 144
precising definition, 181
predicate class, 378
predicate constants, 404
predicates, 402

prefix form, 304
premise cluster, 37
premise-indicators, 27
presupposition, 58
pronouns, 402
proof procedure, 252
proper names, 402
propositional function, 422
propositional logic, 253
proximate cause, 482
pseudocode, 357

QL, 351, 401
qualifier, 164
quality of a statement, 379
Quantificational Logic, 351, 401
quantifiers, 351, 407
The quantity of a categorical statement, 379
question, 57

raising a red herring issue, 112
Reductio ad Absurdum, 332, 336
referring, 402
reflexive, 471
reformative definition, 181
refutation by logical analogy, 84
remote cause, 482
repeated assertion, 44
rephrase, 45
reportive definition, 180
retroductively strong, 77
rhetoric, 11
rhetorical questions, 18

scope, 421
sentence, 17
serial argumentation, 39
set, 352
Set equality, 376
Set inclusion, 376
set membership, 375
set theory, 351
sets, 351
Sheffer stroke, 301
shifting the burden of proof, 89
simplification, 314
single support, 35
slippery slope, 114
snob appeal, 102
sophistry, 11
sound, 72
special pleading, 151
standard form, 387
statement, 17

statement functions, 404
strawman, 113
subalternation, 384
subcontraries, 383
subject class, 379
subject existence assumption, 384
subjects, 402–403
suggestive definition, 181
summary, 46
syllogism, 387
symbolic logic, 72
symmetric, 472

tautologies, 293
testify, 96
transitive, 472

truth-functional compound, 254
tu quoque, 94

UI, 427
understate a claim, 163
union, 353
    of two sets, 353
universal affirmatives, 378
universal general statement, 408
universal generalization, 431
Universal Instantiation, 427
universal negative, 379
universal set, 353

vague, 154
variables, 256